HEATING, VENTILATING and AIR CONDITIONING LIBRARY
Volume 1

by James E. Brumbaugh

THEODORE AUDEL & CO.
a division of
HOWARD W. SAMS & CO., INC.
4300 West 62nd Street
Indianapolis, Indiana 46268

FIRST EDITION

Sixth printing—1982

Copyright © 1976 by Howard W. Sams & Co., Inc., Indianapolis, Indiana 46268. Printed in the United States of America.

International Standard Book Number: 0-672-23248-0
Library of Congress Catalog Card Number: 76-29155

FOREWORD

The purpose of this series is to provide the layman with an introduction to the fundamentals of installing, servicing, and repairing the various types of equipment used in residential heating, ventilating, and air conditioning systems. Consequently, it was not written only for the engineer or skilled serviceman, but also for the average homeowner. A special effort was made to remain consistent with the terminology, definitions, and practices of the various professional and trade associations involved in the heating, ventilating, and air conditioning fields.

Volume 1 begins with a description of the principles of thermal dynamics and ventilation, and proceeds from there with a general description of the various types of heating systems used in residences and small commercial buildings. Volume 2 contains descriptions of the working principles of various types of equipment and other types of components used in these systems. Volume 3 includes detailed instructions for installing, servicing, and repairing these different types of equipment and components. Those sections in Volume 3 which deal with air conditioning follow a similar format.

The author wishes to acknowledge the cooperation of the many organizations and manufacturers for their assistance in supplying valuable data in the preparation of this book. Every effort was made to give credit and courtesy lines for materials and illustrations used in this edition.

JAMES E. BRUMBAUGH

Contents

—heat transmission loss formula—computing total heat loss—the average value method—heat loss tabulation forms—estimating fuel requirements and heating costs—the heat loss formula—the corrected heat loss formula—the degree day formula—the NEMA formula—other heating costs—determining utility rates

Chapter 5

Chapter 6

Chapter 7

Chapter 8

one-pipe circuit system with loop—two-pipe steam heating system —two-pipe divided circuit system—vapor steam heating systems —one (atmospheric) vapor systems—closed vapor systems—vacuum steam heating systems—natural vacuum systems—mechanical vacuum system—combined atmospheric pressure and vacuum systems—exhaust steam heating—proprietary systems— high-pressure steam heating systems—steam boilers—control components—Hartford return connection—pipes and piping details—steam traps—pumps—heat emitting units—air supply and venting—unit heaters—air conditioning.

Chapter 9

Central hot water systems—central forced warm-air heating system—radiant heating systems—baseboard heating systems—electric unit ventilators—electric unit heaters—electric space heaters —heat pump—electric heating and cooling controls—insulation for electrically heated and cooled structures—doors and windows —further information—advantages of electric heat and cooling— disadvantages of electric heat and cooling

Chapter 10

Classifying furnaces—gravity warm-air furnaces—selecting a furnace for a new home—selecting a furnace for an older home— furnace components and controls—pipeless, floor and wall furnaces—duct furnaces—furnace installation—furnace maintenance —general troubleshooting hints

Chapter 11

Planning suggestions — chimneys — troubleshooting chimneys — draft hood—neutral pressure point adjuster—basic components— automatic controls — heat exchanger — gas burners — crossover burners—gas pilot assembly—pilot burner adjustment—gas input adjustment—high altitude adjustment—changing burner orifices —manifold pressure adjustment—combustion air adjustment—gas supply adjustment—sizing gas piping—installing gas piping— starting up instructions—blowers and motors—air filters—air conditioning—installation and maintenance check list—troubleshooting a gas-fired furnace

Chapter 12

Chapter 13

Chapter 14

Chapter 15

CHAPTER 1

Introduction

This series is an introduction to the basic principles of heating, ventilating and air conditioning. Each of these represents a systematic attempt to control the various aspects of the environment within an enclosure whether it be a room, a group of rooms, or a building.

Among those aspects of the immediate environment that man first sought to control were heat and ventilation. Attempts at controlling heat date from prehistoric times and probably first developed in colder climates where it was found necessary to produce temperatures sufficient for both comfort and health. Over the years the technology of heating advanced from simple attempts to keep the body warm to the very sophisticated systems of maintaining stabilized environments in order to reduce heat loss from the body or the structural surfaces of the room.

Ventilation (covered in Volume 3) also dates back to very early periods in history. Certainly the use of slaves to wave large fans or fan-like devices over the heads of rulers was a crude early attempt to solve a ventilating problem. Situating a room or a building so that it took advantage of prevailing breezes and winds was a more sophisticated attempt. However, it was not until the nineteenth century that any really significant advances were made in ventilating. During that period, particularly in the early stages of the Industrial Revolution, ventilating acquired increasing importance. Work efficiency and the health of the workers necessitated the creation of ventilation systems to remove contaminants from the air. Eventually, the inter relationship of heating and ventilating became such that it is now regarded as a single subject.

Air conditioning (Volume 3) is a very recent development and encompasses all aspects of environmental control. In addition to the control of temperature, both humidity (i.e. the moisture content of the air) and air cleanliness are also regulated by air conditioning. The earliest attempts at air conditioning involved the placing of wet cloths over air passages (window openings, entrances, etc.) to cool the air. Developments in air conditioning technology did not progress much further than this until the nineteenth century. From about 1840 on, several systems were devised for both cooling and humidifying rooms. These were first developed by textile manufacturers in order to reduce the static electricity in the air. Later, adaptations were made by other industries.

Developments in air conditioning technology increased rapidly in the first four decades of the nineteenth century, but widespread use of air conditioning in buildings is a phenomenom of the post-World War II period (i.e. 1945 to the present). Today, air conditioning is found not only in commercial and industrial buildings, but in residential dwellings as well. Unlike the early forms of air conditioning which were designed to cool the air or add moisture to it, modern air conditioning systems can control temperature, air moisture content, air cleanliness, and air movement. The modern systems *condition* the air, rather than simply cool it.

HEATING AND VENTILATING SYSTEMS

Many different methods have been devised for heating buildings. Each has its own characteristics and most of them have at least one objectionable aspect (e.g. high cost of fuel, expensive equipment, or inefficient heating characteristics). Most of these heating methods can be classified according to one of the following four criteria:

1. The heat conveying medium,
2. The fuel used,
3. The nature of the heat,
4. The efficiency and desirability of the method.

The term "heat conveying medium" means the substance or combination of substances that carries the heat from its point of origin to the area being heated. There are basically four mediums for conveying heat. These four mediums are:

1. Air,
2. Water,
3. Steam,
4. Electricity.

Different types of wood, coal, oil, and gas have been used as fuels for producing heat. Electricity may be considered both as a fuel and a heat conveying medium. Each type of heating fuel has its own characteristics and the advantages of one over another depends upon such variables as availability, the efficiency of the heating equipment (which in turn is dependent upon design, maintenance, and other factors), and cost. A detailed analysis of the use and effectiveness of the various heating fuels is found in Chapter 5 (HEATING FUELS)

Heating methods can also be classified with respect to the nature of the heat applied. For example, the heat may be of the exhaust steam variety or it may consist of exhaust gases from internal combustion engines. The nature of the heat applied is inherent to the heating system itself and can be determined by reading the various chapters that deal with each type of heating system (Chapters 6 thru 9) or with heat producing equipment (e.g. Chapter 11: GAS-FIRED FURNACES).

The various heating methods will differ considerably in efficiency and desirability. This is due to a number of different but often interrelated factors such as fuel cost, conveying medium employed and the type of heating unit. The integration of these interrelated components into a single operating unit is referred to as a *heating system*.

There are a great variety of heating systems due to the different conditions met with in practice, but most fall into one of the following broad classifications:

1. Warm-air heating system (Chapter 6),
2. Hot-water heating systems (Chapter 7),
3. Steam heating systems (Chapter 8),
4. Electric heating systems (Chapter 9).

11

You will note that these classifications of heating systems are based on the heat conveying method used. This is a convenient method of classification because it includes the vast majority of heating systems used today.

As was previously mentioned, ventilating is so closely inter-related with heating in its various applications that the two are very frequently approached on the basis of being a single subject. However, specific aspects of ventilating are considered in Chapter 6 (VENTILATION PRINCIPLES); and Chapter 7 (FAN SE-LECTION AND OPERATION) in Volume 3.

The type and design of ventilating system employed will depend upon a number of different factors, including:

1. Building use or ventilating purpose,
2. Size of building,
3. Geographical location,
4. Heating system used.

A building used as a residence will have a different ventilating system from one used for commercial or industrial purposes. More-over, the requirements of a ventilating system used to provide fresh air as opposed to one that must remove noxious gases or other dangerous contaminants from the enclosure results in fundamental design differences.

The size of a building is a factor that must also be considered. A large building with internal areas far from the points where out-side air would initially gain access will produce certain ventilating problems. These can usually be solved by giving special attention to the overall design of the ventilating system.

Buildings located in the tropics or semi-tropics present different ventilating problems from those found in the temperate zones. The differences are so great that they often result in different architectural forms. At least this was the case *before* the advent of widespread use of air conditioning. The typical southern house of the nineteenth century was constructed with high ceilings (heat tends to rise), large porches which sheltered sections of the house from the hot, direct rays of the sun, and large window areas to admit the maximum amount of air. They were also usually situated so that halls, major doors and sleeping areas faced the direction of the prevailing winds. Today, with air conditioning so widely used, these considerations are not as important — at least not until the power fails or the equipment breaks down.

AIR CONDITIONING

Although the major emphasis in this series has been placed on the various aspects of heating and ventilating, some attention has also been given to air conditioning. The reason for this, of course, is the increasing use of year-round air conditioning systems that provide heating, ventilating, and cooling. These systems *condition* the air by controlling its temperature (warming or cooling it), cleanliness, moisture content, and movement. This is the true meaning of the term "air conditioning." Unfortunately, it has become almost synonymous with the idea of cooling, and this is becoming increasingly misrepresentative of the true function of an air conditioning system. Air conditioning, particularly the year-round air conditioning systems are examined in detail in Chapters 8, 9 and 10 of Volume 3.

SELECTING A SUITABLE HEATING, VENTILATING, OR AIR CONDITIONING SYSTEM

There are a number of different types of heating, ventilating, and air conditioning equipment and systems available for the installation in the home. The problem is choosing the most efficient one in terms of the installation and operating costs. These factors are, in turn, directly related to one's particular heating and cooling requirements. The system must be the correct size for the home. Any reputable building contractor or heating and air conditioning firm should be able to advise you in this matter.

If you are having a heating and ventilating or air conditioning system installed in an older house, be sure to check the construction. Weather stripping is the easiest place to start. All doors and windows should be weatherstripped to prevent heat loss. Adequate weatherstripping can cut heating costs by as much as 15 to 20%. If the windows provide suitable protection (they should be double- or triple-glazed) from the winter cold, check the calking around the edge of the glass. If it is cracking or crumbling, replace it with fresh calking. You may even want to go to the expense of insulating the ceilings and outside walls. This is where a great deal of heat loss and air leakage occur.

You have several advantages when you are building your own house. For example, you may be able to determine the location of your house on the lot. This should enable you to establish the direction in which the main rooms and largest windows face. If you position your house so that these rooms and windows face south, you will gain maximum sunlight and heat from the sun during the cold winter months. This will reduce the heat requirement and heating costs. The quality of construction depends on how much

Table 1. The relationship between career category and the type or degree of education and/or training required for it.

Career Category	Type Of Work	Education/Training
Engineer	Design and Development	4 Yrs. or More of College
Technician	Practical Application	Technical Training School and/or College
Skilled Worker	Installation, Maintenance, Repair	Apprentice Program or On-The-Job (OJT) Training
Apprentice or OJT Worker	Training for Skilled Worker Position	High School Degree or Equivalency Usually Required

you wish to spend and the reliability of the contractor. My advice would be to purchase the best insulation you can afford. Your reduced heating costs will eventually pay for the added cost of the insulation. If you suspect that your building contractor cannot be trusted, you can reduce opportunities for cheating and careless work by making frequent and unexpected visits to the construction site.

CAREER OPPORTUNITIES

There are many career opportunities available to the individual in heating, ventilating, air conditioning and related fields, and they extend over several levels of education and training. Accordingly, the career opportunities open to an individual seeking employment in these fields can roughly be divided into four categories, each dependent upon a different type or degree of education and/or training. This relationship is shown in Table 1.

Engineers are usually employed by laboratories, universities and colleges, or frequently by the manufacturers of materials and equipment used in heating, ventilating, air conditioning and related industries. They receive the highest pay, but also undergo the longest periods of education and training. Their primary responsibility is designing, developing and testing the equipment and materials used in these fields. In some cases, particularly where large buildings or district heating to several buildings is employed, they also supervise the installation of the entire system. Furthermore, industry codes and standards are usually the results of research conducted by engineers.

Technicians obtain their skills through technical training schools, some college, or both. Many assist engineers in the practical application of what the latter have designed. They are particularly necessary during the developmental stages. Other technicians are found in the field working for contractors in the larger companies. Their pay often approximates that of engineers, depending on the size of the company for which they work.

Skilled workers are involved in the installation, maintenance and repair of heating, ventilating, and air conditioning equipment. Apprentices and OJT workers are in training for these skilled positions, and are generally expected to complete at least a 2 to 5 year training program. Most skilled workers and trainees are employed by local firms that install or repair equipment in residential, commercial, and industrial buildings. Some also work on the assembly lines of factories that manufacture such equipment. Their pay varies depending on the area, seniority, and the nature of the work. Most employers require that both the skilled workers and trainees have at least a high school diploma or its equivalent (e.g. the G.E.D.). The requirement for a high school diploma may be waived if the individual has already acquired the necessary skills on a previous job. The pay is lower than that earned by engineers and technicians, but compares favorably to salaries received by skilled workers or equivalent training in other occupations.

Pipefitters, plumbers, steamfitters, and sheetmetal workers all may occasionally do some work with heating, ventilating and air conditioning equipment. Both pipefitters and plumbers (especially the former) are frequently called upon to assemble and install pipes and pipe systems that carry the heating or cooling conveying medium from the source. Both are also involved in repair work, and some pipefitters can install heating and air conditioning units. Steamfitters can assemble and install hot-water or steam heating systems. The installation of boilers, stokers, oil and gas burners, radiators, radiant heating systems, and air conditioning systems can also be done by many steamfitters.

Sheetmetal workers can also assemble and install heating, ventilating, and air conditioning systems. Their skills are particularly necessary when assembling sheet metal ducts and duct systems.

Some specialty occupations such as those performed by air conditioning and refrigeration mechanics or stationary engineers are limited to certain functions in the heating, ventilating and air conditioning fields. The former is primarily involved with assembling, installing and maintaining both air conditioning and refrigeration equipment. Stationary engineers, on the other hand, maintain and operate heating, ventilating, and air conditioning equipment in large buildings and factories. Workers in both these occupations require greater skills and longer training periods than most skilled workers.

It should be readily apparent by now that the heating, ventilating and air conditioning fields offer a variety of different career opportunities. The pay is generally good and the nature of the work provides considerable job security. Both the type of work and the level at which it is done depends solely on the amount and type of education and training acquired by the individual.

PROFESSIONAL ORGANIZATIONS

A number of professional organizations have been established for those who work in the heating, ventilating and air conditioning industries, or who handle their products. These organizations (frequently referred to as *associations, societies,* or *institutes*) provide a number of different services to both members and non-members alike, including:

1. The compilation of safety codes and other standards,

2. Educational programs and training schools,

3. Market research and various technical services,
4. The compilation of industry-wide statistics,
5. The dissemination of information in the form of journals, books, pamphlets, movies, slides, and tapes.

Some of these professional organizations have established permanent libraries as resource centers for those seeking to improve their skills or who wish to keep abreast of current developments in their fields. In many instances, research programs are conducted in cooperation with laboratories, colleges, and universities.

Anyone, member or not, can write to these professional organizations for information or assistance. Most of them seem very willing to comply with any reasonable request. The only difficulty that may be encountered is determining the current name of the particular organization and obtaining its address. Unfortunately, these professional organizations have shown a strong proclivity towards mergers over the years with resulting changes of names and addresses.

The best and most current guide to the names and addresses of professional organizations is *The Encyclopaedia of Associations* which can be found in the reference department of most public libraries. It is published in three volumes, but everything you will need can be found in the first one. At the very back of this volume is a section called the "Alphabetical & Key Word Index." By looking up the key word (e.g. "heating" or "ventilating") of the subject that interests you—you will find the page number and full name of the professional organization (or organizations) concerned with this particular area.

There are over twenty different professional organizations in the United States concerned with heating, ventilating, air conditioning or related fields (e.g. fuels, piping, etc.). Many of them address themselves to the problems and interests of specific groups. For example, there are professional organizations for manufacturers, wholesalers, jobbers, distributors, and journeymen. Some organizations represent an entire industry, while others restrict their scope to only a segment of it. Every aspect of heating, ventilating, and air conditioning is covered by one or more of these professional organizations.

Among the more than twenty professional organizations currently active are the following:

1. Air-Conditioning and Refrigeration Institute (ARI).
2. Air-Conditioning and Refrigeration Wholesalers (ARW).
3. Air Moving and Conditioning Association (AMCA).
4. Air Diffusion Council (ADC).
5. American Gas Association (AGA).
6. American Society of Heating, Refrigerating and Air Conditioning Engineers (ASHRAE).
7. American Society of Mechanical Engineers (ASME).
8. Association of Industry Manufacturers (Plumbing — Heating — Cooling — Piping).
9. Electric Energy Association (EEA).
10. Hydronics Institute.
11. International Solar Energy Society.
12. Mechanical Contractors Association of America (MCAA).
13. National Association of Oil Heating Service Managers (NAOHSM).
14. National Association of Plumbing-Heating-Cooling Contractors.
15. National Environmental Systems Contractors Association (NESCA).
16. National LP-Gas Association (NLPGA).
17. National Oil Fuel Institute (NOFI).
18. National Warm Air Heating and Air Conditioning Association (NWAH & ACA).
19. Plumbing-Heating-Cooling Information Bureau (PHCIB).
20. Refrigeration Service Engineers Society (RSES).
21. Sheet Metal and Air Conditioning Contractors' National Association (SMACNA).

Some professional organizations of long standing have been merged with others or have been disbanded. For example, the Steel Boiler Institute (formerly the Steel Heating Boiler Institute) which maintained standards in the heating industry with its SBI Rating Code is now defunct. The Institute of Boiler and Radiator Manufacturers (source of the old IBR Code) merged with the Better Heating-Cooling Council to form the Hydronics Institute. A recent attempt to contact the Steam Heating Equipment Manufacturers Association has resulted in the return of a letter marked "no forwarding address." It seems very likely that it, too, has joined the list of defunct professional organizations.

Appendix A (PROFESSIONAL AND TRADE ASSOCIATIONS) at the end of this book gives a listing of these professional organizations. It also contains their present address, the names of some of their publications, and a brief synopsis of their background and whom they represent.

CHAPTER 2

Heating Fundamentals

There is still considerable disagreement about the exact nature of heat, but most authorities agree that it is a particular form of energy. Specifically, heat is a form of energy not associated with matter and *in transit* between its source and destination point. Furthermore, heat energy only exists as such between these two points. In other words, it only exists as heat energy while flowing between the source and destination.

So far this description of heat energy has been practically identical to that of work energy, the other form of energy in transit not associated with matter. The distinguishing difference between the two is that heat energy is energy in transit as a result of temperature differences between its source and destination point, whereas work energy in transit is due to other, non-temperature factors.

BRITISH THERMAL UNIT

Heat energy is measured by the British thermal unit (Btu). Each British thermal unit is regarded as equivalent to one unit of heat (heat energy).

Since 1929 British thermal units have been defined on the basis of one Btu being equal to 251.996 IT (International Steam Table) calories, or 778.26 ft.-lb. of mechanical energy units (work). Taking into consideration that one IT calorie equals $\frac{1}{860}$ of a watt-hour, one Btu is then equivalent to about $\frac{1}{3}$ watt-hour.

Prior to its redefinition in 1929, one Btu was defined as the amount of heat necessary to raise 1 lb. of water 1°F. Because of the difficulty in determining the exact value of a Btu, it was later redefined in terms of the more fundamental electrical unit.

RELATIONSHIP BETWEEN HEAT AND WORK

Energy is the ability to do work or move against a resistance. Conversely, work is the overcoming of resistance through a certain distance by the expenditure of energy.

Work is measured by a standard unit called the foot pound. The latter may be defined as the amount of work done in raising one pound the distance of one foot, or in overcoming a pressure of one pound through a distance of one foot (Fig. 1).

Fig. 1. Man raising one pound one foot to illustrate the foot-pound standard unit.

21

The relationship between work and heat is referred to as the *mechanical equivalent of heat.* That is to say, one unit of heat is equal to 778.26 foot-pounds of work.

This relationship (i.e. the mechanical equivalent of heat) was first established by experiments conducted in the nineteenth century. In 1843 Dr. James Prescott Joule (1818-1889) of Manchester, England determined by numerous experiments that when 772 foot-pounds of energy had been expended on one pound of water, the temperature of the latter had risen one degree, and the relationship between heat and mechanical work was found (Fig. 2). The value 772 foot-pounds is known as *Joule's equivalent.* More recent experiments give higher figures, and the value 778 (1 Btu = 778.26 ft.-lbs. See BRITISH THERMAL UNITS in this chapter).

Fig. 2. The mechanical equivalent of heat.

HEAT TRANSFER

When bodies of unequal temperatures are placed near each other, heat leaves the hotter body and is absorbed by the colder one until the temperature of each is equal. The rate by which the heat is absorbed by the colder body is proportional to the difference of temperature between the two bodies—the greater the difference of temperature the greater the rate of flow of the heat.

Heat is transferred from one body to another at lower temperature by any one of the following means (Fig. 3):

1. Radiation,
2. Conduction,
3. Convection.

RADIATION   CONDUCTION   CONVECTION

CONVECTION

PUMP

Fig. 3. The transfer of heat by radiation, conduction, and convection.

Radiation, insofar as heat loss is concerned, refers to the throwing out of heat in rays. The heat rays proceed in straight lines and the intensity of the heat radiated from any one source becomes less as the distance from the source increases.

The amount of heat loss through radiation will depend upon the temperature of the floor, ceiling, and walls in the room or building. The colder these surfaces are, the faster and greater will be the heat loss from a human body standing within the enclosure. On the other hand, if the wall, ceiling, and floor surfaces are *warmer* than the human body within the enclosure they form, heat will be radiated from these surfaces to the body. In these situations a person may complain that the room is too hot.

Knowledge of the (mean radiant) temperature of the surfaces of an enclosure is important when dealing with heat loss by radiation. The *mean radiant temperature* (MRT) is the weighted average temperature of these surfaces (i.e. the floor, ceiling, and walls). The significance of the mean radiant temperature is determined when compared with the clothed body of an adult (80°F. or 26.7°C.). If the MRT is below 80°F, the human body will lose heat by radiation to the surfaces of the enclosure. If the MRT is higher than 80°F, the opposite effect will occur.

Conduction is the transfer of heat through substances, as for instance, boiler plate to another substance in contact with it (Fig. 4). Conductivity may be defined as the relative value of a mate-

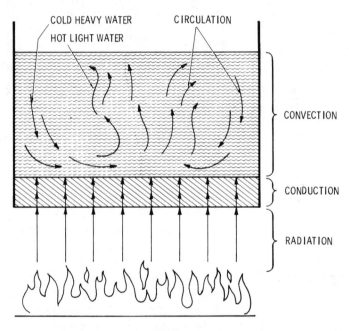

Fig. 4. Radiation, conduction, and convection in boiler operation.

rial as compared with a standard, in affording a passage through itself or over its surface for heat. A poor conductor is usually referred to as a *nonconductor* or *insulator*. Copper is an example of a good conductor. Fig. 5 illustrates the comparative heat conductivity rates of three frequently used metals. The various materials used to insulate buildings are poor conductors. It·should be pointed out that any substance, which is a good conductor of electricity, is also a good conductor of heat.

Convection is the transfer of heat by the motion of the heated matter itself. Because motion is a required aspect of the definition of convection, it can therefore take place only in liquids and gases.

Fig. 5. Conductivity of various metals.

Fig. 4 illustrates how radiation, conduction, and convection are often interrelated. Heat from the burning fuel passes to the metal of the heating surface by radiation, through the metal by conduction, and is transferred to the water by convection (i.e. circulation). Circulation is caused by a variation in the weight of the water due to temperature differences. That is, the water next to the heating surface receives heat and expands (becomes lighter) and immediately rises due to displacement by the colder and heavier water above.

Proper circulation is very important because its absence will cause a liquid, such as water, to reach the spheroidal state. This in turn causes the metal of the boiler to become dangerously overheated. A liquid that has reached the spheroidal state is easy to recognize by its appearance. When liquid is dropped upon the surface of a highly heated metal, it rolls about in spheroidal drops (Fig. 6) or masses, and without actual contact with the heated metal. This phenomenon is due to the repulsive force of heat and the intervention of a cushion of steam.

Fig. 6. Drop of water on a hot plate illustrating the spheroidal state.

SPECIFIC, SENSIBLE AND LATENT HEAT

The *specific heat* of a substance is the ratio of the quantity of heat required to raise its temperature one degree Fahrenheit to the amount required to raise the temperature of the same weight of water one degree Fahrenheit (Fig. 7). This may be expressed in the following formula:

$$\text{Specific heat} = \frac{\text{Btu to raise temp of substance } 1°F.}{\text{Btu to raise temp. of same weight water } 1°F.}$$

The standard used is water at approximately 62°F. to 63°F. which receives a rating of 1.00 on the specific heat scale. Simply stated, specific heat represents the Btu's required to raise the temperature of 1 lb. of a substance 1°F.

Sensible heat is that part of heat which provides temperature change and which can be measured by a thermometer. It is referred to as such because it can be "sensed" by instruments or the touch.

Latent heat is that quantity of heat which disappears or becomes concealed in a body while producing some change in it other than a rise of temperature. Changing a liquid to a gas, or a gas to a liquid are both activities involving latent heat. The two types of latent heat are:

1. Internal latent heat.
2. External latent heat.

These are explained in detail in a later section (see STEAM in this chapter).

1 LB. OF WATER

TEMPERATURE RISE 1^0 F

1 B. t. u. Required

A

1 LB. OF SUBSTANCE

TEMPERATURE RISE 1^0 F

B

$$\text{SPECIFIC HEAT} = \frac{B \, (B. t. u.)}{A \, (B. t. u.)}$$

Fig. 7. The principle of specific heat.

HEAT CONVEYING MEDIUMS

As was mentioned in the previous chapter, there are a variety of ways to classify heating systems. One method is based on the medium which conveys the heat from its source to the point being heated. When the majority of heating systems in use today are examined closely, it can be seen that there are only four basic heat conveying mediums involved:

1. Air,
2. Steam,

3. Water,

4. Electricity.

Air

Air is a gas consisting of a mechanical mixture of 23.2% oxygen (by weight), 75.5% nitrogen, and 1.3% argon with small amounts of other gases. It functions as the heat conveying medium for hot-air heating systems.

Atmospheric pressure may be defined as the force exerted by the weight of the atmosphere in every point with which it is in contact (Fig. 8), and is measured in inches of mercury or the corresponding pressure in lbs. per square inch.

14. 696 LBS.

ATMOSPHERIC PRESSURE
(14. 696 LBS. PER SQ. IN.)

1 SQ. IN.

29. 921 INCH
VACUUM

Fig. 8. Atmospheric pressure.

The pressure of the atmosphere is approximately 14.7 lbs. per square inch at sea level. The standard atmosphere is 29.921 inches of mercury at 14.696 lbs. per square inch. The term "inches of mercury" refers to the height to which the column of mercury in the barometer will remain suspended to balance the pressure due to the weight of the atmosphere.

28

Atmospheric pressure will vary due to elevation by decreasing approximately one half pound for every 1000 feet ascent above sea level. Atmospheric pressure in lbs. per square inch is obtained from the barometer reading by multiplying the barometer reading in inches by 0.49116. Examples are given in Table 1.

Table 1. Pressure of the atmosphere per square inch for various readings of the barometer.

Barometer (ins. of mercury)	Pressure per sq. ins., lbs.	Barometer (ins. of mercury)	Pressure per sq. ins., lbs.
28.00	13.75	29.921	14.696
28.25	13.88	30.00	14.74
28.50	14.00	30.25	14.86
28.75	14.12	30.50	14.98
29.00	14.24	30.75	15.10
29.25	14.37	31.00	15.23
29.50	14.49		
29.75	14.61		

Gauge pressure is pressure whose scale starts at atmospheric pressure. *Absolute pressure,* on the other hand, is pressure measured from true zero or the point of no pressure. When the hand of a steam gauge is at zero, the absolute pressure existing in the boiler is approximately 14.7 lbs. per square inch. Thus, by way of example, 5 lbs. pressure measured by a steam gauge (i.e. gauge pressure) is equal to 5 lbs. plus 14.7 lbs. or 19.7 lbs. per square inch of absolute pressure.

When air is compressed, both its pressure and temperature are changed in accordance with Boyle's and Charles' laws. According to Robert Boyle (1627-1691), the English philosopher and founder of modern chemistry, the absolute pressure of a gas at constant temperature varies inversely as its volume. Jacques Charles (1746-1823), on the other hand, established that the volume of a gas is proportional to its absolute temperature when the volume is kept at constant pressure.

If the cylinder in Fig. 9 is filled with air at atmospheric pressure (14.7 lbs. per square inch absolute), represented by volume A, and the piston B moved to reduce the volume to say ⅓ A, as represented by B, then according to Boyle's law, the pressure will be trebled (14.7 × 3 = 44.1 lbs absolute or 44.1 − 14.7 = 29.4 gauge pressure). According to Charles' law, a pressure gauge on the cylinder would at this point indicate a *higher* pressure than 29.4 gauge pressure because of the increase in temperature produced by compressing the air. This is called *adiabatic compression* if no heat is lost, or received externally.

Fig. 9. Elementary air compressor illustrating the phenomenon of compression as stated in Boyle's and Charles' law.

Steam

Those who design, install, or have charge of steam heating plants certainly should have some knowledge of steam, its formation and behavior under various conditions.

Steam is a colorless expansive and invisible gas resulting from the vaporization of water. The white cloud associated with steam is a fog of minute liquid particles formed by condensation. That is to say, finely divided condensation. This white cloud is caused by the exposure of the steam to a temperature lower than that corresponding to its pressure.

If the inside of a steam heating main were visible, it would be filled part way with a white cloud and in traversing the main the little particles combine, forming drops of condensation too heavy to remain in suspension and accordingly drop to the bottom of the main and drain off as condensation. This condensation flows into a drop leg of the system and finally back into the boiler, together with additional condensation draining from the radiators.

Although the word "steam" should only be applied to the saturated gas, the five following classes of steam are recognized:

1. Saturated steam,
2. Dry steam,
3. Wet steam,
4. Superheated steam,
5. Highly superheated or gaseous steam.

Three of these classes of steam (wet, saturated, and super-heated) are shown in the illustration of a safety valve blowing in Fig. 10. It should be pointed out that neither saturated steam nor superheated steam can be seen by the naked eye.

Saturated steam may be defined as steam of a temperature due to its pressure. Steam containing intermingled moisture, mist,

Fig. 10. Three types of steam.

or spray, is referred to as *wet steam. Dry steam,* on the other hand, is steam containing no moisture. It may be either saturated or superheated. Finally, superheated steam is steam having a temperature higher than that corresponding to its pressure.

The various changes which take place in the making of steam are known as *vaporization,* and are shown in Fig. 11. For the sake of illustration, only one bubble is shown in each receptacle. In actuality there is a continuous procession upward of a great multiplicity of bubbles.

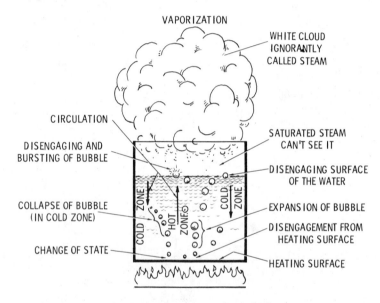

Fig. 11. The phenomenon of vaporization.

The amount of heat necessary to cause the generation of steam is the sum of the sensible heat, the internal latent heat, and the external latent heat. As mentioned elsewhere in this chapter, sensible heat is that part of the heat which produces a rise in temperature as indicated by the thermometer. The *internal latent heat* is the amount of heat that water will absorb at the boiling point without a change in temperature — that is, before vaporization begins. *External latent heat* is the amount of heat required when vaporization begins to push back the atmosphere and make room for the steam.

Another important factor to consider when dealing with steam is the boiling point of liquids. By definition, the *boiling point* is the temperature at which a liquid begins to boil (Fig. 12), and it depends upon both the pressure and nature of the liquid. For instance, water boils at 212°F. and ether at 9°F. under atmospheric pressure of 14.7 lbs. per square inch.

Fig. 12. Variation of the boiling point when pressure changes.

The relationship between boiling point and pressure is such that there is a definite temperature or boiling point corresponding to each value of pressure. When vaporization occurs in a closed vessel and there is a temperature rise, the pressure will rise until the equilibrium between temperature and pressure is re-established.

One's knowledge of the fundamentals of steam heating should also include an understanding of the role that condensation plays. By definition, *condensation* is the change of a substance from the gaseous to the liquid (or condensate) form. This change is caused by a reduction in temperature of the steam below that corresponding to its pressure.

The condensation of steam can cause certain problems for steam heating systems unless they are designed to allow for it. The water from which the steam was originally formed contained, mechanically mixed with it, $\frac{1}{20}$ or 5 percent of its volume of air (at atmospheric pressure). This air is liberated during vaporization and does not re-combine with the condensation. As a result, trouble is experienced in heating systems when attempting to get the air out and keep it out. Suitable air valves are necessary to correct the problem.

Water

Water is a chemical compound of two gases, oxygen and hydrogen in the proportion of two parts by weight of hydrogen to sixteen parts by weight of oxygen having mixed with it about 5 percent of air by volume at 14.7 lbs. absolute pressure. It may exist as ice, water, or steam due to changes in temperature (water freezes at 32°F. and boils at 212°F. when the barometer reads 29.921 inches).

One cubic foot of water weighs 62.41 lbs. at 32°F. and 59.82 lbs. at 212°F. One U.S. gallon of water (231 cubic inches) weighs 8.33111 lbs. (ordinarily expressed as 8⅓ lbs.) at a temperature of 62°F. At any other temperature the weight will, of course, be different (Table 2).

Water changes in weight with change of temperature. That is, the higher the temperature of the water, the less it weighs. It is this property of water that causes circulation in boilers and in hot water heating systems. The change in weight is due to expansion and a reduction in water volume. As the temperature rises, the water expands and results in a unit volume of water containing less water at higher temperature than lower temperature.

Fill a vessel with cold water and heat it to the boiling point. Note that boiling causes it to overflow due to expansion. Now let the water cool. You will note that when the water is cold the vessel will not be as full because the water will have contracted.

The point of maximum density of water is 39.1°F. The most remarkable characteristic of water is its expansion below *and* above its point of maximum density. Imagine one pound of water at 39.1°F. placed in a cylinder having a cross sectional area of 1 square inch (Fig. 13). The water having a volume of 27.68 cubic inches will fill the cylinder to a height of 27.68 inches. If the water is cooled it will expand, and at say 32°F. (the freezing point), will rise in the tube to a height of 27.7 inches before

Fig. 13. The point of maximum density.

Fig. 14. The principle of thermal circulation.

Table 2. Weight of water per cubic foot

Temp., deg. F.	Lb. per cu. ft.	Temp., deg. F.	Lb. per cu. ft.	Temp., deg. F.	Lb. per cu. ft.	Temp., deg. F.	Lb. per cu. ft.
32	62.41	62	62.35	91	62.10	121	61.69
33	62.41	63	62.34	92	62.08	122	61.68
34	62.42	64	62.34	93	62.07	123	61.66
35	62.42	65	62.33	94	62.06	124	61.64
36	62.42	66	62.32	95	62.05	125	61.63
37	62.42	67	62.32	96	62.04	126	61.61
38	62.42	68	62.31	97	62.02	127	61.60
39	62.42	69	62.30	98	62.01	128	61.58
40	62.42	70	62.30	99	62.00	129	61.56
41	62.42	71	62.29	100	61.99	130	61.55
42	62.42	72	62.28	101	61.98	131	61.53
43	62.42	73	62.27	102	61.96	132	61.51
44	62.42	74	62.26	103	61.95	133	61.50
45	62.42	75	62.25	104	61.94	134	61.48
46	62.41	76	62.25	105	61.93	135	61.46
47	62.41	77	62.24	106	61.91	136	61.44
48	62.41	78	62.23	107	61.90	137	61.43
49	62.41	79	62.22	108	61.89	138	61.41
50	62.40	80	62.21	109	61.87	139	61.39
51	62.40	81	62.20	110	61.86	140	61.37
52	62.40	82	62.19	111	61.84	141	61.36
53	62.39	83	62.18	112	61.83	142	61.34
54	62.39	84	62.17	113	61.81	143	61.32
55	62.38	85	62.16	114	61.80	144	61.30
56	62.38	86	62.15	115	61.78	145	61.28
57	62.38	87	62.14	116	61.77	146	61.26
58	62.37	88	61.13	117	61.75	147	61.25
59	62.37	89	62.12	118	61.74	148	61.23
60	62.36	90	62.11	119	61.72	149	61.21
61	62.35			120	61.71		

freezing. If the water is heated, it will also expand and rise in the tube, and at the boiling point (for atmospheric pressure 212°F) will occupy the tube to a height of 28.88 cubic inches.

The elementary hot water heating system in Fig. 14 illustrates the principle of thermal circulation. The weight of the hot and expanded water in the up flow column C, being less than that of the cold and contracted water in the down flow Column C' upsets the equilibrium of the system and results in a continuous circulation of water as indicated by the arrows. In other words, the

at different temperatures.

Temp., deg. F.	Lb. per cu. ft.	Temp., deg. F.	Lb. per cu. ft.	Temp., deg. F.	Lb. per cu. ft.	Temp., deg. F.	Lb. per cu. ft.
150	61.19	180	60.57	208	59.92	440	51.87
151	61.17	181	60.55	209	59.90	450	51.28
152	61.15	182	60.53	210	59.87	460	51.02
153	61.13	183	60.51	211	59.85	470	50.51
154	61.11	184	60.49	212	59.82	480	50.00
155	61.09	185	60.46	214	59.81	490	49.50
156	61.07	186	60.44	216	59.77	500	48.78
157	61.05	187	60.42	218	59.70	510	48.31
158	61.03	188	60.40	220	59.67	520	47.62
159	61.01	189	60.37	230	59.42	530	46.95
160	60.99	190	60.35	240	59.17	540	46.30
161	60.97	191	60.33	250	58.89	550	45.66
162	60.95	192	60.30	260	58.62	560	44.84
163	60.93	193	60.28	270	58.34	570	44.05
164	60.91	194	60.26	280	58.04	580	43.29
165	60.89	195	60.23	290	57.74	590	42.37
166	60.87	196	60.21	300	57.41	600	41.49
167	60.85	197	60.19	310	57.08	610	40.49
168	60.83	198	60.16	320	56.75	620	39.37
169	60.81	199	60.14	330	56.40	630	38.31
170	60.79	200	60.11	340	56.02	640	37.17
171	60.77	201	60.09	350	55.65	650	35.97
172	60.75	202	60.07	360	55.25	660	34.48
173	60.73	203	60.04	370	54.85	670	32.89
174	60.71	204	60.02	380	54.47	680	31.06
175	60.68	205	59.99	390	54.05	690	28.82
176	60.66	206	59.97	400	53.62	700	25.38
177	60.64	207	59.95	410	53.19	706.1	19.16
178	60.62	......	....	420	52.74	......	....
179	60.00			430	52.33		

heavy low temperature water sinks to the lowest point in the boiler (or system) and displaces the light high temperature water, thus causing continuous circulation as long as there is a temperature difference in different parts of the boiler (or system). This is referred to as *thermal circulation*.

Electricity

Electricity differs from air, steam, or water in that it does not actually convey heat from one point to another; therefore, including it in a list of heat conveying mediums can be misleading at first glance.

Electricity can best be defined as a quantity of electrons either in motion or in a state of rest. When these electrons are at rest, they are referred to as being *static* (hence the term "static electricity"). Electrons in *motion* move from one atom to another creating an electrical current, and thereby a medium for conveying *energy* from one point to another. Many different types of devices have been created to change the energy conveyed by an electric current into heat, light, and other forms of energy. Electric-fired furnaces and boilers are examples of these devices used to produce heat.

CHAPTER 3

Insulation Principles

This chapter is not concerned with the theoretical aspects of heat loss or gain, or with heating and cooling calculations *per se*. These problems are discussed in considerable detail in other chapters of this book, particularly in Chapter 2 (HEATING FUNDAMENTALS), Chapter 4 (HEATING CALCULATIONS, and Chapter 9, (Volume 3) (AIR CONDITIONING CALCULATIONS). The purpose of this chapter is to describe the various *types* of insulation used in construction, the methods of applying insulation, and the thermal properties of various building materials.

THE NEED FOR INSULATION

An uninsulated structure, or one that is poorly insulated, will experience a heat gain in the summer (or a heat loss in the winter) as high as 50%. The exact percentage will depend upon the types of materials used in its construction, because all construction material will have some insulating effect. In any event, it has been observed that approximately 30% of heat gain (or loss) is experienced through ceilings and about 70% occurs through walls, glass, window and door cracks, or as a result of air ventilation.

Any attempt to heat or cool an *uninsulated* structure would obviously be terribly inefficient *and expensive*. Obviously, the possibility of anyone being so shortsighted as to attempt to install a heating or cooling system in such a structure is fairly remote, but inadequate insulation in existing structures, or the failure to provide for proper insulation in proposed construction, is a common occurrence. The provision for some form of barrier to reduce the rate of heat flow to an acceptable level is therefore of prime importance. Insulation materials serve this purpose with varying degrees of effectiveness. As much as 80 to 90% of heat loss (or gain) through ceilings and 60% of heat loss (or gain) through walls can be prevented by properly insulating these areas.

The inefficiency of ordinary building materials to resist the passage of heat brought about the need for the development of materials specifically designed for insulation. A *good* insulating material should be lightweight, contain numerous air pockets, and exhibit a high degree of resistance to heat transmission. It should also be specifically treated to resist fire and the attacks of rodents and insect. Finally, a good insulating material will react well to excess moisture by drying out and retaining its resistance to heat flow rather than disintegrating.

Many inexperienced workers seem to feel that the more insulation one uses the better the insulating effect. Unfortunately, this does not hold true on a one-to-one basis. For example, twice as much insulation does *not* insulate twice as well. There is apparently a law of diminishing returns operating here. The first layer of insulation is the most effective, with successive layers decreasing in the effectiveness to impede the rate of heat flow.

The capacity of a heating or cooling system is determined by the amount of heat that must be supplied (heating applications) or removed (cooling applications) to maintain the desired temperature within the structure or space. In heating applications, the amount of heat supplied to a space at constant temperature should roughly equal the amount of heat lost. In cooling applications, the heat removed from a space should be roughly equal to the amount of heat gained. Both heat loss and heat gain will be controlled by the way in which the structure is insulated. Good insulation will reduce the rate of heat flow through construction materials to an acceptable level, and this will mean that a heating or cooling system will be able to operate much less expensively and with greater efficiency. It will also mean that the capacity of the heating or cooling unit you wish to install will be less than the one required for a poorly insulated structure. This, in turn, will reduce the initial equipment and installation costs. As you can see, the type and quality of insulation in a structure is an important factor to be considered when designing a heating or a cooling system.

The roof or attic floor of any structure can be insulated, as can also the walls of any frame building, whether of stucco, clapboard, shingles, brick or stone veneer. The form in which the insulation is applied depends on whether the structure is already built or is being constructed.

PRINCIPALS OF HEAT TRANSMISSION

Heat flows through a barrier of building materials (e.g. walls, ceilings, floors, etc.) by means of conduction and radiation. When it has passed to the other side of this barrier, it continues its movement away from the source by means of conduction, radiation, and convection. The *direction* of heat flow depends on the relationship of outdoor and indoor temperatures, because warmer air will always move toward a colder space. Thus, in the summer months, the warmer outdoor air will move toward the cooler indoor spaces. This is referred to as *heat leakage,* and is described in Chapter 9, Volume 3 (AIR CONDITIONING CALCULATIONS). In the winter months, the reverse is true. The warmer indoor air moves to the cooler outdoor spaces, and this phenomenon is referred to as *heat loss* (see Chapter 4. HEATING CALCULATIONS). The purpose of insulation is to reduce the rate of heat flow to an acceptable level.

Heat Transfer Values

The *American Society of Heating, Refrigerating and Air Conditioning Engineers* (ASHRAE) has conducted considerable research into the thermal properties of building materials and this data is used in the calculation of heat loss and heat gain. Tables of computed values for many types of construction materials and combinations of construction materials are made available through ASHRAE publications. One type of table gives the design values (i.e. conductivity values, conductance values, and resistance values) of various building and insulating materials. These values are used to calculate the coefficients of heat transmission (U-values) for different types of construction (e.g. masonry walls, frame partitions, etc.). Examples of these two types of tables have been taken from the *1960 ASHRAE Guide* (see Table 1 and Table 2).

The four heat transfer values contained in these tables are: (1) the k-value (thermal conductivity); (2) the C-value (thermal conductance); (3) the R-value (thermal resistance); and (4) the U-value (overall coefficient of heat transmission). Each of these heat transfer values will be examined before their use in heating and cooling calculations is explained.

Thermal Conductivity

The *k-value* (thermal conductivity) represents the amount of heat which flows through one square foot of a homogeneous material one inch thick in one hour for each degree of temperature difference between the inside and the outside temperatures.

To illustrate, let's use an example from the ASHRAE data shown in Tables 1 and 2. You will note that the k-value for face brick listed in Table 1 is 9.0. In other words, 9.0 Btu's of heat pass through a square foot of one-inch thick face brick each hour. The thermal resistance (R-value) of this same one-inch thick face brick is the reciprocal of its k-value.

$$R = 1/k = 1/9 = 0.11$$

Remember, this is the thermal resistance for a *one-inch* thickness of face brick. Because the masonry wall in Table 2 contains a *four-inch* thickness of face brick, you will have to multiply the R-value (0.11) of the one-inch thick face brick by four (i.e. four inches) to obtain the R-value (0.44) for the larger four-inch section.

Thermal Conductance

The *C-value* (thermal conductance) represents the amount of heat flow through a square foot of material per hour per degree of temperature difference between the inside and the outside temperatures. The C-value is based on the stated construction or thickness of the material. This distinguishes it from the k-value, which is based on a one inch thickness of a homogeneous material having no fixed thickness. For example, under "loose fill" in Table 1, you will see vermiculite listed and given a k-value of

0.48. Why a k-value? It is given a k-value (rather than a C-value) because vermiculite is a loose fill-type insulating material that can be poured into a space to form whatever thickness is desired. It is convenient, then, to give numerical values (k-values) based on one inch thicknesses of these types of materials. If, say, vermiculite was used to insulate the top floor ceiling (that is to say, the attic floor) and poured to a depth of four inches, the R-value (thermal resistance) would be calculated as follows:

(1) k-value = 0.48
(2) R-value = 1/0.48 = 2.08 (1 inch)
(3) R-value = 2.08 × 4 = 8.32 (4 inches)

Under "masonry units" in Table 1, you will find hollow clay tile listed with six different *stated* thicknesses (e.g. "1 cell deep, 3 in.; 1 cell deep, 4 in.; etc."). Each is given a C-value, rather than a k-value, because the thickness of the material listed in the table is the same thickness that will be used in the construction of a structural section.

The masonry wall in Table 2 also contains gypsum plaster (sand aggregate) one-half inch thick. The C-value is given as 11.10, and its R-value as 0.09 (Table 1). In other words, the R-value (thermal resistance) is calculated as follows:

$$R = 1/C = 1/11.10 = 0.09$$

Examine the date in Table 1 once more. You will probably note that each building material has either a k-value or a C-value, but not both. Each material is offered on the basis of either a stated thickness or an arbitrary one-inch thickness. This was explained above. Note also that each type of building material has an R-value determined by either of the following two equations:

(1) $R = 1/k$
(2) $R = 1/C$

The R-value is simply the reciprocal of either the k-value or the C-value.

Thermal Resistance

The *R-value* (thermal resistance) of a building material is its *resistance* to the flow of heat. As explained above, the R-value is

Table 1. Conductivities (K), conductances (C), and resistances (R) of various building and insulating materials.

Material	Description			Density (Lb per Cu Ft)	Conductivity or Conductance		Resistance (R)	
	Position	Heat Flow	Thickness		(k)	(C)	Per inch thickness $\frac{1}{k}$	For thickness listed $\frac{1}{C}$
AIR SPACES	Horizontal	Up (winter)	¾-4 in.	—	—	1.18	—	0.85
	Horizontal	Up (summer)	¾-4 in.	—	—	1.28	—	0.78
	Horizontal	Down (winter)	¾ in.	—	—	0.98	—	1.02
	Horizontal	Down (winter)	1½ in.	—	—	0.87	—	1.15
	Horizontal	Down (winter)	4 in.	—	—	0.81	—	1.23
	Horizontal	Down (winter)	8 in.	—	—	0.80	—	1.25
	Horizontal	Down (summer)	¾ in.	—	—	1.18	—	0.85
	Horizontal	Down (summer)	1½ in.	—	—	1.07	—	0.93
	Horizontal	Down (summer)	4 in.	—	—	1.01	—	0.99
	Sloping, 45°	Up (winter)	¾-4 in.	—	—	1.11	—	0.90
	Sloping, 45°	Down (summer)	¾-4 in.	—	—	1.12	—	0.89
	Vertical	Horiz. (winter)	¾-4 in.	—	—	1.03	—	0.97
	Vertical	Horiz. (summer)	¾-4 in.	—	—	1.16	—	0.86

Table 1. Conductivities (K), conductances (C), and resistances (R) of various building and insulating materials. (Continued.)

Position	Heat Flow					
AIR SURFACES						
Still Air						
Horizontal	Up	—	—	1.63	—	0.61
Up Sloping (45°)	Up	—	—	1.60	—	0.62
Vertical	Horizontal	—	—	1.46	—	0.68
Sloping (45°)	Down	—	—	1.32	—	0.76
Horizontal	Down	—	—	1.08	—	0.92
15 mph Wind						
Any position—any direction (for winter)		—	—	6.00	—	0.17
7½ mph Wind						
Any position—any direction (for summer)		—	—	4.00	—	0.25
BUILDING BOARD						
Boards, Panels, Sheathing, Etc.						
Asbestos-cement board		120	4.0	—	0.25	—
Asbestos-cement board ....1/8 in.		—	—	33.	—	0.03
Gypsum or plaster board ....3/8 in.		50	—	3.10	—	0.32
Gypsum or plaster board ....1/2 in.		50	—	2.25	—	0.45
Plywood		34	0.80	—	1.25	—
Plywood ....1/4 in.		34	—	3.20	—	0.31
Plywood ....3/8 in.		34	—	2.12	—	0.47
Plywood ....1/2 in.		34	—	1.60	—	0.63
Plywood or wood panels ....3/4 in.		—	—	1.07	—	0.94
Wood fiber board, laminated or homogenous		26	0.42	—	2.38	—
		31	0.50	—	2.00	—
Wood fiber—hardboard type		65	1.40	—	0.72	—
Wood fiber—hardboard type ....1/4 in.		65	—	5.60	—	0.18
Wood—fir or pine sheathing ....25/32 in.		—	—	1.02	—	0.98
Wood—fir or pine ....1 5/8 in.		—	—	0.49	—	2.03

Table 1. Conductivities (K), conductances (C), and resistances (R) of various building and insulating materials. (Continued.)

		Density	K	C	R	R
BUILDING PAPER						
Vapor—permeable felt		—	—	16.70	—	0.06
Vapor—seal, 2 layers of mopped 15 lb felt		—	—	8.35	—	0.12
Vapor—seal, plastic film		—	—	—	—	Negl
FLOORING MATERIALS						
Asphalt tile	⅛ in.	120	—	24.80	—	0.04
Carpet and fibrous pad		—	—	0.48	—	2.08
Carpet and rubber pad		—	—	0.81	—	1.23
Ceramic tile	1 in.	—	—	12.50	—	0.08
Cork tile		25	0.45	—	2.22	—
Cork tile	⅛ in.	—	—	3.60	—	0.28
Felt, flooring		—	—	16.70	—	0.06
Floor tile or linoleum—av. value	⅛ in.	—	—	20.00	—	0.05
Linoleum	⅛ in.	80	—	12.00	—	0.08
Plywood subfloor	⅝ in.	—	—	1.28	—	0.78
Rubber or plastic tile	⅛ in.	110	—	42.40	—	0.02
Terrazzo	1 in.	—	—	12.50	—	0.98
Wood subfloor	25/32 in.	—	—	1.02	—	0.98
Wood, hardwood finish	¾ in.	—	—	1.47	—	0.68
INSULATING MATERIALS						
Blanket and Batt						
Cotton fiber		0.8-2.0	0.26	—	3.85	—
Mineral wool, fibrous form, processed from rock, slag or glass		1.5-4.0	0.27	—	3.70	—
Wood fiber		3.2-3.6	0.25	—	4.00	—
Wood fiber, multilayer, stitched expanding[e]		1.5-2.0	0.27	—	3.70	—

Table 1. Conductivities (K), conductances (C), and resistances (R) of various building and insulating materials. (Continued.)

Board					
Glass fiber	9.5	0.25	—	4.00	—
Wood or cane fiber					
Acoustical tile ....½ in.	—	—	0.84	—	1.19
Acoustical tile ....¾ in.	—	—	0.56	—	1.78
Interior finish (plank, tile, lath) ....½ in.	15.0	0.35	—	2.86	—
Interior finish (plank, tile, lath) ....½ in.	15.0	—	0.70	—	1.43
Roof deck slab					
Approx. ....1½ in.	—	—	0.24	—	4.17
Approx. ....2 in.	—	—	0.18	—	5.56
Approx. ....3 in.	—	—	0.12	—	8.33
Sheathing (impreg. or coated) ....½ in.	20.0	0.38	—	2.63	—
Sheathing (impreg. or coated) ....½ in.	20.0	—	0.76	—	1.32
Sheathing (impreg. or coated) ....25/32 in.	20.0	—	0.49	—	2.06
Board and Slabs					
Cellular glass	9.0	0.40	—	2.50	—
Corkboard (without added binder)	6.5-8.0	0.27	—	3.70	—
Hog hair (with asphalt binder)	8.5	0.33	—	3.00	—
Plastic (foamed)	1.62	0.29	—	3.45	—
Wood shredded (cemented in preformed slabs)	22.0	0.55	—	1.82	—
Loose Fill					
Macerated paper or pulp products	2.5-3.5	0.28	—	3.57	—
Mineral wood (glass, slag, or rock)	2.0-5.0	0.30	—	3.33	—
Sawdust or shavings	8.0-15.0	0.45	—	2.22	—
Vermiculite (expanded)	7.0	0.48	—	2.08	—
Wood fiber: redwood, hemlock, or fir	2.0-3.5	0.30	—	3.33	—

Table 1. Conductivities (K), conductances (C), and resistances (R) of various building and insulating materials. (Continued.)

ROOF INSULATION					
All types					
Preformed, for use above deck					
Approx. ....½ in.	—	—	0.72	—	1.39
Approx. ....1 in.	—	—	0.36	—	2.78
Approx. ....1½ in.	—	—	0.24	—	4.17
Approx. ....2 in.	—	—	0.19	—	5.26
Approx. ....2½ in.	—	—	0.15	—	6.67
Approx. ....3 in.	—	—	0.12	—	8.33
MASONRY MATERIALS					
Concretes					
Cement mortar	116	5.0	—	0.20	—
Gypsum-fiber concrete 87½% gypsum, 12½% wood chips	51	1.66	—	0.60	—
Lightweight aggregates including expanded shale, clay or slate; expanded slags; cinders; pumice; perlite; vermiculite; also cellular concretes	120	5.2	—	0.19	—
	100	3.6	—	0.28	—
	80	2.5	—	0.40	—
	60	1.7	—	0.59	—
	40	1.15	—	0.86	—
	30	0.90	—	1.11	—
	20	0.70	—	1.43	—
Sand and gravel or stone aggregate (oven dried)	140	9.0	—	0.11	—
Sand and gravel or stone aggregate (not dried)	140	12.0	—	0.08	—
Stucco	116	5.0	—	0.20	—

Table 1. Conductivities (K), conductances (C), and resistances (R) of various building and insulating materials. (Continued.)

MASONRY UNITS					
Brick, common	120	5.0	—	0.20	—
Brick, face	130	9.0	—	0.11	—
Clay tile, hollow:					
1 cell deep .... 3 in.	—	—	1.25	—	0.80
1 cell deep .... 4 in.	—	—	0.90	—	1.11
2 cells deep .... 6 in.	—	—	0.66	—	1.52
2 cells deep .... 8 in.	—	—	0.54	—	1.85
2 cells deep .... 10 in.	—	—	0.45	—	2.22
3 cells deep .... 12 in.	—	—	0.40	—	2.50
Concrete blocks, three oval core:					
Sand and gravel aggregate .... 4 in.	—	—	1.40	—	0.71
.... 8 in.	—	—	0.90	—	1.11
.... 12 in.	—	—	0.78	—	1.28
Cinder aggregate .... 3 in.	—	—	1.16	—	0.86
.... 4 in.	—	—	0.90	—	1.11
.... 8 in.	—	—	0.58	—	1.72
.... 12 in.	—	—	0.53	—	1.89
Gypsum partition tile:					
3 x 12 x 30 in. solid	—	—	0.79	—	1.26
3 x 12 x 30 in. 4-cell	—	—	0.74	—	1.35
4 x 12 x 30 in. 3-cell	—	—	0.60	—	1.67
Lightweight aggregate (expanded shale, clay, 3 in.	—	—	0.79	—	1.27
slate or slag; pumice) .... 4 in.	—	—	0.67	—	1.50
.... 8 in.	—	—	0.50	—	2.00
.... 12 in.	—	—	0.44	—	2.27

Table 1. Conductivities (K), conductances (C), and resistances (R) of various building and insulating materials. (Continued.)

MASONRY UNITS (Continued)					
Stone, lime or sand	—	12.50	—	0.08	—
PLASTERING MATERIALS					
Cement plaster, sand aggregate	116	5.0	—	0.20	—
Sand aggregate ½ in.	—	—	10.00	—	0.10
Sand aggregate ¾ in.	—	—	6.66	—	0.15
Gypsum plaster:					
Lightweight aggregate ½ in.	45	—	3.12	—	0.32
Lightweight aggregate ⅝ in.	45	—	2.67	—	0.39
Lightweight agg. on metal lath ¾ in.	—	—	2.13	—	0.47
Perlite aggregate	45	1.5	—	0.67	—
Sand aggregate	105	5.6	—	0.18	—
Sand aggregate ½ in.	105	—	11.10	—	0.09
Sand aggregate ⅝ in.	105	—	9.10	—	0.11
Sand aggregate on metal lath ¾ in.	—	—	7.70	—	0.13
Sand aggregate on wood lath	—	—	2.50	—	0.40
Vermiculate aggregate	45	1.7	—	0.59	—
ROOFING					
Asbestos-cement shingles	120	—	4.76	—	0.21
Asphalt roll roofing	70	—	6.50	—	0.15
Asphalt shingles	70	—	2.27	—	0.44
Built-up roofing ⅜ in.	70	—	3.00	—	0.33
Slate ½ in.	—	—	20.00	—	0.05
Sheet metal	—	400+	—	Negl	—
Wood shingles	—	—	1.06	—	0.94

Table 1. Conductivities (K), conductances (C), and resistances (R) of various building and insulating materials. (Continued.)

Material	Density	K	C	R (per inch, 1/k)	R (for thickness, 1/C)
SIDING MATERIALS (On Flat Surface)					
Shingles					
Wood, 16-in. 7½-in. exposure	—	—	1.15	—	0.87
Wood, double, 16-in., 12-in. exposure	—	—	0.84	—	1.19
Wood, plus insul. backer board ⁵⁄₁₆ in.	—	—	0.71	—	1.40
Siding					
Asbestos-cement, ¼ in., lapped	—	—	4.76	—	0.21
Asphalt roll siding	—	—	6.50	—	0.15
Asphalt insulating siding (½ in. bd.)	—	—	0.69	—	1.45
Wood, drop, 1 x 8 in.	—	—	1.27	—	0.79
Wood, bevel, ½ x 8 in., lapped	—	—	1.23	—	0.81
Wood, bevel, ¾ x 10 in., lapped	—	—	0.95	—	1.05
Wood, plywood, ⅜ in., lapped	—	—	1.59	—	0.59
Structural glass	—	—	10.00	—	0.10
WOODS					
Maple, oak, and similar hardwoods	45	1.10	—	0.91	—
Fir, pine and similar softwoods	32	0.80	—	1.25	—

Courtesy *ASHRAE 1960 Guide*

Table 2. Coefficients of heat transmission for solid masonry walls.

Resistances used are given below in this table.

Construction	Resistance (R)
1. Outside surface (15 mph wind)	0.17
2. Face brick (4 in.)	0.44
3. Common brick (4 in.)	0.80
4. Air space	0.97
5. Gypsum lath (3/8 in.)	0.32
6. Plas. (sand agg.) (1/2 in.)	0.09
7. Inside surface (still air)	0.68
Total resistance	3.47
$U = 1/R = 1/3.47 =$	0.29

See value 0.29 in boldface type in table below.

Assume plain wall—no furring or plaster.

Total resistance	3.47
Deduct 4. Air space	0.97
5. Gypsum lath (3/8 in.)	0.32
6. Plas. (sand agg.) (1/2 in.)	0.09 1.38
Total resistance	2.09
$U = 1/R = 1/2.09 =$	0.48

Table 2. Coefficients of heat transmission for solid masonry walls. (Continued)

Exterior Construction		None	Plas. 5/8 in. on Wall		Metal Lath and 3/4 in. Plas. on Furring		Gypsum Lath (3/8 in.) and 1/2 in. Plas. on Furring			Insul. Bd. Lath (1/2 in.) and 1/2 in. Plas. on Furring		Wood Lath and 1/2 in. Plas.	No.
						Interior Finish							
Resistance		U	(Sand agg.) 0.11 U	(lt. wt. agg.) 0.39 U	(Sand agg.) 0.13 U	(lt. wt. agg.) 0.47 U	No plas. 0.32 U	(Sand agg.) 0.41 U	(lt. wt. agg.) 0.64 U	No plas. 1.43 U	(Sand agg.) 1.52 U	(Sand agg.) 0.40 U	
Material	R	A	B	C	D	E	F	G	H	I	J	K	
Brick (face and common)													
(6 in.)	0.61	0.68	0.64	0.54	0.39	0.34	0.36	0.35	0.33	0.26	0.25	0.35	1
(8 in.)	1.24	0.48	0.45	0.41	0.31	0.28	0.30	0.29	0.27	0.22	0.22	0.29	2
(12 in.)	2.04	0.35	0.33	0.30	0.25	0.23	0.24	0.23	0.22	0.19	0.19	0.23	3
(16 in.)	2.84	0.27	0.26	0.25	0.21	0.19	0.20	0.20	0.19	0.16	0.16	0.20	4
Brick (common only)													
(8 in.)	1.60	0.41	0.39	0.35	0.28	0.26	0.27	0.26	0.25	0.21	0.20	0.26	5
(12 in.)	2.40	0.31	0.30	0.27	0.23	0.21	0.22	0.22	0.21	0.18	0.17	0.22	6
(16 in.)	3.20	0.25	0.24	0.23	0.19	0.18	0.19	0.18	0.18	0.16	0.15	0.18	7
Stone (lime and sand)													
(8 in.)	0.64	0.67	0.63	0.53	0.39	0.34	0.36	0.35	0.32	0.26	0.25	0.35	8
(12 in.)	0.96	0.55	0.52	0.45	0.34	0.31	0.32	0.31	0.29	0.24	0.23	0.31	9
(16 in.)	1.28	0.47	0.45	0.40	0.31	0.28	0.29	0.28	0.27	0.22	0.22	0.29	10
(24 in.)	1.92	0.36	0.35	0.32	0.26	0.24	0.25	0.24	0.23	0.19	0.19	0.24	11

Table 2. Coefficients of heat transmission for solid masonry walls. (Continued)

Hollow clay tile													
(8 in.)	1.85	0.36	0.36	0.32	0.26	0.24	0.25	0.25	0.23	0.20	0.19	0.25	12
(10 in.)	2.22	0.33	0.31	0.29	0.24	0.22	0.23	0.22	0.21	0.18	0.18	0.23	13
(12 in.)	2.50	0.30	0.29	0.27	0.22	0.21	0.22	0.21	0.20	0.17	0.17	0.21	14
Poured concrete													
30 lb per cu ft													
(4 in.)	4.44	0.19	0.19	0.18	0.16	0.15	0.15	0.15	0.14	0.13	0.13	0.15	15
(6 in.)	6.66	0.13	0.13	0.13	0.12	0.11	0.11	0.11	0.11	0.10	0.10	0.11	16
(8 in.)	8.88	0.10	0.10	0.10	0.09	0.09	0.09	0.09	0.09	0.08	0.08	0.09	17
(10 in.)	11.10	0.08	0.08	0.08	0.08	0.07	0.08	0.08	0.07	0.07	0.07	0.08	18
80 lb per cu ft													
(6 in.)	2.40	0.31	0.30	0.27	0.23	0.21	0.22	0.22	0.21	0.18	0.17	0.22	19
(8 in.)	3.20	0.25	0.24	0.23	0.19	0.18	0.19	0.18	0.18	0.16	0.15	0.18	20
(10 in.)	4.00	0.21	0.20	0.19	0.17	0.16	0.16	0.16	0.15	0.14	0.14	0.16	21
(12 in.)	4.80	0.18	0.17	0.17	0.15	0.14	0.14	0.14	0.14	0.12	0.12	0.14	22
140 lb per cu ft													
(6 in.)	0.48	0.75	0.69	0.58	0.41	0.36	0.38	0.37	0.34	0.27	0.26	0.37	23
(8 in.)	0.64	0.67	0.68	0.53	0.39	0.34	0.36	0.35	0.32	0.26	0.25	0.35	24
(10 in.)	0.80	0.61	0.57	0.49	0.36	0.32	0.34	0.33	0.31	0.25	0.24	0.33	25
(12 in.)	0.96	0.55	0.52	0.45	0.34	0.31	0.32	0.31	0.29	0.24	0.23	0.31	26

the reciprocal of the k-value or the C-value, and represents heat flow resistance expressed as a numerical value. The higher the R-value of a material (or a combination of materials) the greater its resistance to the flow of heat. Ceilings should have an R-value of R-19 or better. Floors and walls should be insulated to at least R-11 (Fig. 1).

6" (R-19) IN CEILINGS OR ATTICS

3 1/2" (R-11) in WALLS BETWEEN HEATED AND UN-HEATED AREAS

3 1/2" (R-11) IN ALL EXTERIOR WALLS

3 1/2" (R-11) IN FLOORS OVER UNHEATED AREAS

PERIMETER INSULATION AROUND SLABS OR HEATED CRAWL SPACES

Courtesy Owens-Corning Fiberglas Corp.

Fig. 1. Minimum levels of effective insulation.

Returning to the masonry wall illustrated in Table 2, you will note that all seven R-values have been added to produce the total resistance of a solid masonry wall (i.e. 3.47). In other words, the *total* resistance to heat flow through a solid masonry wall of the *same* type of construction illustrated in Table 2 is equal to the sum of the resistances of the wall components, or:

$$R_t = R_1 + R_2 + R_3 \ldots \ldots R_n$$

The total resistance (R_t) of a particular type of construction is used to determine its U-value, which represents the total *insulating* effect of a structural section. The U-value is the reciprocal of the total resistance (R_t). Using the data supplied for the masonry wall illustrated in Table 2, its U-value is determined as follows:

$$U = 1/R = 1/3.47 = 0.29$$

Overall Coefficient of Heat Transmission

The *U-value* or *U-factor* (overall coefficient of heat transmission) represents the amount of heat (in BTU's) that will pass through one square foot of a structural section per hour per degree Fahrenheit. This is the difference between air on the inside and the air on the outside of the section. The U-value is the reciprocal of the sum of the thermal resistance values (R-values) of each element of the structural section. The total resistance (R_t) for the solid masonry wall illustrated in Table 2 is 3.47. Since the U-value is the reciprocal of R_t, the following is obtained:

$$U = 1/R_t = 1/3.47 = 0.29$$

The overall coefficient of heat transmission (U-value or U-factor) is used in the formula for calculating heat loss or heat gain. The use of this formula is described in Chapter 4 (HEATING CALCULATIONS) and Chapter 9, Volume 3 (AIR CONDITIONING CALCULATIONS).

CONDENSATION

Air will always contain a certain amount of moisture (water vapor), and the amount of moisture it contains will depend upon its temperature. Generally speaking, warm air is capable of containing more moisture than cold air.

Air will lose moisture in the form of condensation when its temperature falls below its dew point. The dew point is the temperature at which moisture begins to condense, in the form of tiny droplets or dew. Any reduction in temperature below the dew point will result in condensation of some of the water vapor present in the air.

The attic is one place where condensation often occurs. In the winter, the warm air from the occupied spaces leaks past the insulation on the top floor ceiling and comes into contact with the cooler air in the attic. This causes the water vapor in the air to condense. The moisture resulting from condensation works its way down through the ceiling and walls, and eventually causes damage to the insulation, wood, and other building materials.

The solution to condensation in the attic is twofold. Adequate ventilation should be provided in the attic to remove the humid air. This can be accomplished by installing louvers at each end of the attic or in roof overhangs to provide cross ventilation (Fig. 2). The louvers should have a total area of one square foot for each 300 square feet of attic floor space. For example, if the attic has an area of 900 square feet, the total louver area should be at least 3 square feet. In addition to suitable attic ventilation, you should also install a vapor barrier between the rafters. This is particu-largely true if the attic is occupied or used extensively.

Fig. 2. Two methods of attic ventilation.

Foundation crawl spaces are also subject to condensation. A vapor barrier between the floor joists prevents excess moisture from accumulating, and causing dampness and eventually damage to the structure. Ventilating the crawl space so that the humid air can be removed is also very effective. Finally, placing an insulat-ing paper (e.g. roofing paper) on the ground will reduce the amount of moisture that is released by the soil (Fig. 3).

Vapor Barriers

Insulation must be protected against moisture or it will lose its ability to reduce thermal transmission. An unprotected insu-lation material will absorb moisture and lose its insulating value. This can be very expensive, because the insulation must be re-placed completely. In most cases, there is no way to restore it to its original condition.

A *vapor barrier* is a nonabsorbent material (e.g. plastic, as-phalt, metal foil, roofing paper, etc.) designed to protect insula-tion from moisture or to prevent moisture from entering a space (Fig. 4). It can completely enclose the insulation or be applied to only one side. If the latter situation is the case, the vapor barrier

57

must be installed so that it faces inward toward the heated rooms and spaces.

In addition to protecting insulation from absorbing moisture, a vapor barrier is also an effective means of retaining moisture

FINISHED FLOOR

SUBFLOOR

FLOOR
JOIST

VAPOR BARRIER
FACES UP

SCREENED VENT

OUTDOOR AIR

ROLL ROOFING OR
POLYETHYLENE FILM
FOR GROUND COVER

Fig. 3. Crawl space construction showing vent location and vapor barrier installation.

COLD
DRY
AIR

WARM
MOIST
AIR

COLD
DRY
AIR

WARM
MOIST
AIR

VAPOR
BARRIER

Fig. 4. The location of a vapor barrier in respect to insulation. Note that vapor barrier faces the warm moist inside air.

in the rooms and spaces in the structure, or preventing it from entering these areas. As a result, it functions as an important factor in maintaining interior humidity conditions.

If you add insulation over an existing vapor barrier, you should make a number of cuts in the barrier to avoid moisture entrapment. Remember: the name-of-the-game is *dry insulation*. Take whatever measure necessary to protect it from moisture.

TYPES OF INSULATING MATERIALS

Insulating materials are specifically designed to reduce the rate of heat transmission through ordinary construction materials to an acceptable level.

A dry material of low density is considered a good insulator; however, in addition to this characteristic, it must also have a conductivity value of less than 0.5.

The conductivity value of a material is a purely arbitrary one determined by the amount of heat that flows through a one inch thickness of a material one square foot in area with the temperature exactly one degree Fahrenheit higher on one side of the material than on the other.

Using the arbitrary conductivity value as a guide, the following materials are regarded as providing the best insulation:

1. Expanded polyurethane (0.17),
2. Bats and blankets (0.27),
3. Mineral wool (0.27),
4. Corkboard (0.30),
5. Insulating board (0.33),
6. Expanded perlite fill (0.35),
7. Wood fibre sheathing (0.38),
8. Vermiculite (0.48).

Air spaces, or air spaces bounded by either ordinary building materials or aluminum foil, also provide some insulation, but not to the degree formerly thought. Dead air spaces in building walls were once considered capable of preventing heat transmission in a manner similar to the space in the walls of a thermos bottle. Later research proved this to be a somewhat false analogy, because the air in such spaces will often circulate and transmit heat by convection.

Air circulation can be checked by filling the hollow space with an insulating material which contains a great number of small *confined* air spaces per unit volume. This stoppage of air circulation is what produces the insulating effect, and not necessarily the existence of the air space. Under these circumstances, it is obvious that the most practical method of insulation is to fill the area in the walls with a material containing these minute air spaces.

A number of different manufacturers produce insulating materials in a variety of shapes and forms for installation in houses and other buildings. Frequently, instructions for the installation of their products will also be provided by the manufacturer. Local building supply outlets and lumber yards will often be very helpful, too, and will usually recommend the best way to install the insulation material. Some of these materials and their applications are described in the following paragraphs.

Insulating Board

Insulating board (Fig. 5) is a rigid or semi-rigid synthetic product commonly referred to as *structural insulating board, rigid board, sheathing board,* and produced under a variety of tradenames. It is available in ½ to 1 inch thicknesses, and in board sizes generally 4 ft. by 8 ft.

Some insulating board is produced specifically for use on interior walls. This type is generally made from wood fibers, and one side is prefinished. Insulating board designed for use as struc-

Fig. 5. Rigid insulation board.

tural insulation is produced from compressed fibrous materials other than wood and does not have a prefinished surface.

Slab Insulation

Rigid *slab insulation* is designed especially for application to flat roofs. The minimum thickness of slab insulation is 1 inch, with widths ranging up to 2 feet and lengths up to 4 feet. Foamed plastic or mineral wool impregnated with asphalt are used in the production of slab insulation, because they provide excellent protection against moisture.

Reflective Insulation

Reflective insulation (Fig. 6) is designed to reflect heat rather than resist its rate of flow. It is this characteristic that distinguishes it from other types of insulating materials. Aluminum foil is commonly used as the reflective surface, although coated papers have also been employed.

Air spaces must be provided on either side of a layer of reflective insulation. This type of insulating material has proven to be a good moisture barrier, and exhibits a high degree of resistance to heat transmission.

Fig. 6. Reflective insulation board.

Batt-Type Insulation

Batts (Fig. 7) are essentially smaller examples of blanket insulation, ranging in thickness from 1 to 6 inches and available in 4 ft. and 8 ft. lengths.

The batts are manufactured with a vapor seal paper adhered to one face, covering the entire surface and extending 1½ inches on the long sides of the batt. These 1½ inch laps neatly folded against the membrane backing in manufacture, are turned out and

61

SOME BATTS ARE
HELD TOGETHER
BY FRICTION

STAPLES
12" APART

OVERLAP BATTS 1"

VAPOR BARRIER
ALWAYS FACES
HEATED AREA

STUD

STUD

SOLE PLATE

START FROM
BOTTOM

Fig. 7. Batt-type insulation.

tacked or stapled to the studs, rafters or joists in application. The paper backing helps to prevent the passage of vapor and resists the penetration of moisture from excess water in fresh plaster or other sources.

Blanket-Type Insulation

Blanket insulation (Fig. 8) is produced in the form of a continuous flexible strip, and is generally available in rolls or packs in the following three thicknesses: (1) thick (3 inch); (2) medium (2 inch); and (3) thin (1 inch). Widths are commonly 15

Fig. 8. Blanket-type insulation.

inches, but wider strips up to 33 inches wide can be obtained on special order. Blanket insulation lengths range from 36 to 48 feet.

Sometimes both sides of the blanket insulation strip are covered with aluminum foil or plain paper which serves as a vapor seal. Application is made by tacking or stapling through 1½ inch laps extending from both sides of the strip.

Loose Fill Type Insulation

Loose fill insulation (Fig. 9) is available in granular, fibrous, or powdered form, and is sold in bulk. (usually in bags). It is commonly produced from vermiculite or other granular materials, vegetable fibers, or mineral wools, and is especially suitable for pouring between floor joists.

Fig. 9. Loose fill type insulation.

Blown-In Insulation

Blown-in insulation (Fig. 10) is a loose fill type insulating material that is blown under pneumatic pressure into the spaces between joists and wall studs in an existing structure. Water vapor barriers are applied first, and are available in the form of a moisture resistant spray-on paint. Blown-in insulation is most commonly used to insulate interior walls. Because special equipment must be used for its application, blown-in insulation should be applied by individuals trained in its use.

SECOND FLOOR

PLASTER

RIBBAND

Fig. 10. Blown-in insulation.

BUILDING CONSTRUCTION AND LOCATION

A new building (particularly a residence) should be located so that the large windows in the main rooms face south to receive the maximum sunlight during the winter months (Fig. 11). If

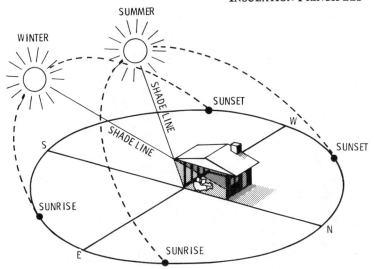

SUMMER

WINTER

SHADE LINE

SHADE LINE

SUNSET

W

SUNSET

S

SUNRISE

N

SUNRISE

E

Courtesy *U.S. Department of Agriculture.*

Fig. 11. Large glass window areas should face south to take advantage of the winter sun.

possible, the building should be built in a location that offers some natural protection from the prevailing winter winds. Tight, well insulated construction should be incorporated in the design of the building from the very beginning. Although the initial costs will be somewhat higher, they will be effectively offset by the reduction in heating and cooling costs.

If a new heating or cooling system is planned for an *older* structure, the existing insulation should be checked, and, if necessary, repaired or replaced before the new system is installed.

Installing Windows and Doors

Windows and doors account for at least 30 percent of the total heat loss (or heat gain) in a structure. Much of this results from air infiltration through cracks around the windows and doors, or in the walls of the structure. Weather stripping is used for window and door cracks, caulking for wall cracks (Fig. 12).

Cold drafts (i.e. cold air infiltration) and warm air leakage can be greatly reduced if the windows and doors of a house are properly weatherstripped and caulked. This method of reducing air leakage or infiltration is much cheaper than installing storm

Fig. 12. Caulking around windows and weatherstripping doors.

windows, and will reduce heating and cooling costs up to as high as 30 percent.

Caulk the frame around all the windows and doors and install weather strip along movable joints ,(e.g. the edges of a door). If this is done properly, a mist or fog will form on windows located on the downwind side of the structure. Not only will caulk and weatherstripping reduce heating and cooling costs, it will also provide high and healthy humidity conditions inside the structure. Typical methods of applying weatherstripping to windows are illustrated in Fig. 13.

Using storm windows or windows made from insulating glass will cut heat loss by as much as 40 to 50 percent. Although the initial cost for installing storm windows may seem high, it will be more than off-set by the lowered heating costs. The savings will be highest in those areas in which snow lies on the ground all winter.

Storm windows not only reduce heat loss in the winter, but also heat *gain* in the summer. That is to say, they are equally effective in reducing the amount of summer heat that infiltrates the interior of the structure.

Various methods for double glazing (i.e. the installing of an additional layer of glass) and adding storm sash are illustrated in Figs. 14 and 15.

FRAME

INSULATION PRINCIPLES

AIR

SASH

PUTTY

SASH

HEAD JAMB
SIMILAR

GLASS

HEAD

WEATHERSTRIP

JAMB
SIMILAR

GLASS

PUTTY

MEETING
RAIL

AIR

GLASS

MEETING
RAIL

WEATHERSTRIP

GLASS

PUTTY

WEATHER-STRIP

SILL

SASH

AIR

(A)

SILL

(B)

GROOVE

Fig. 13. Typical method of applying weatherstripping.

Overhead and Sidewall Insulation

Overhead insulation (Fig. 16) refers to the layer of insulation covering the top floor ceiling. This layer should be a *minimum* of 6 inches thick over the entire area. By carefully insulating this section of the structure, heating costs can be reduced by 10 to 15 percent. Air conditioning costs in the summer will also be reduced. The overhead insulation should be checked in an older

67

Fig. 14. Typical double glazing.

Fig. 15. Typical storm sash construction.

structure, because some building contractors will only install 3 to 4 inches of the material. Adding an additional 2 to 3 inches is not only easy to do, but well worth the effort.

Tacks or staples can be used to attach batt or blanket insulation flanges to the wood joists supporting the ceiling. Blanket, batt, *and loose* insulation require a water vapor barrier placed be-

Fig. 16. Overhead insulation.

tween the insulating material and the ceiling lath (Fig. 17). Reflective insulation is attached to the *middle* of the joists so that there is an air space both above and below it. No water vapor barrier is necessary. When using reflective insulation, additional protection is required for the insulating material, which usually takes the form of lath (below) and an attic floor (above).

Fig. 17. Three methods of insulating an attic.

Sidewall insulation (Fig. 18) refers to insulation installed in the exterior sidewalls of a house or building. In new construction, this should be incorporated as a part of the design before ground is broken. An existing house or building presents a different problem, because there is always the possibility of moisture condensation forming within the walls. If you feel you lack the expertise, then expert technical advice should be obtained before attempting to install insulation in a sidewall.

69

Fig. 18. Sidewall insulation.

INSULATING ATTICS, ATTIC CRAWL SPACES, AND FLAT ROOFS

Most houses and many other structures are constructed with an attic or attic crawl space between the roof and the ceiling of the top floor. Unless this space is properly insulated and ventilated, it will produce the following two problems:

1. Condensed moisture formation,
2. Excessive heat loss or gain.

During the winter months, the heat from the occupied lower spaces moves upward (because warm air is lighter) into the cooler attic or attic crawl space. As it moves, the heat follows a number of different pathways including:

1. Through poorly insulated ceilings,
2. Around cracks in attic stairway doors or attic pulldown stairways,
3. Through ceiling fixture holes,
4. Along plumbing vents and pipes,
5. Along air ducts,
6. Through air spaces within interior partitions.

Each of these pathways for heat transmission can be blocked by insulation. Typical attic insulation methods are illustrated in Fig. 19. Attic door cracks can be blocked with weatherstripping. Pipe, duct, vent, and fixture holes can be sealed off by stuffing them with loose insulation torn from batts or blankets.

The ceiling of the top floor should be covered with a minimum of 6 inches of suitable insulation material to be properly

Fig. 19. Attic insulation methods.

effective in reducing the heat loss from the occupied spaces during the winter months, or the heat gain during the summer. This can be installed during the initial construction stages of a new house or building, or added to the attic floor surface of an existing structure. It is not a particularly difficult thing to do (see OVERHEAD INSULATION in this chapter). Sealing or filling the air spaces within interior partitions is relatively easy to do if it is a question of new construction. An existing structure presents a more difficult problem; one that generally requires considerable experience in installing insulation materials.

Sometimes a half wall (i.e. a knee wall) is installed between the attic floor and the roof. This is most commonly insulated with blanket insulation as illustrated in Fig. 20.

The condensation of moisture in the attic or an attic crawl space is caused by the moisture in the warm air rising from the lower occupied spaces. This moisture comes in contact with the colder air of the attic, condenses and causes the insulation and other building materials to become damp. In time, this dampness will have a damaging effect. Ventilation is an effective method of reducing or eliminating condensation in attics and attic crawl spaces. The air enters and leaves through vents placed in the gables and cornices. The size of the vents will be determined by the area being ventilated.

ROOF BOARDS

RAFTER

SHINGLES

2 X 4

KNEE WALL PLATE

SEAL

BLANKET BETWEEN
KNEE WALL STUDS

SEAL

SEAL

2 X 4 SILL

JOIST

SEAL

SEAL

4 X 4 PLATE   BLANKET BETWEEN JOISTS   PLASTER

Fig. 20. Insulating a knee wall with blanket insulation.

A flat roof can be effectively insulated by installing 2-inch thick rigid insulation board before the roofing material is applied. Roofing felt should be placed between the bare wood boards of the roof deck and the insulation. Additional insulation is placed *under* the deck boards (i.e. between the deck boards and the ceiling of the room below) (Fig. 21).

RIGID INSULATION (2")   BITUMEN   ROOFING

ROOFING FELT

DECK BOARDS

LATH AND PLASTER
CEILINGS

JOIST

REFLECTIVE INSULATION

Fig. 21. Insulating a flat roof.

FOUNDATION CRAWL SPACES

Some houses and buildings are constructed so that the first floor is raised slightly above the ground, leaving a crawl space between the two. These foundation crawl spaces can be effectively insulated by attaching blanket or batt insulation between the

Fig. 22. Insulating a crawl space.

joists supporting the floor, or by nailing rigid insulation to the bottom of the joists (Fig. 22). Placing waterproof paper on the surface of the ground in the crawl space and extending it up along the inside of the foundation walls will increase the effectiveness of the insulation.

Condensation can also occur in foundation crawl spaces. The accumulation of moisture resulting from condensation can be prevented or effectively reduced by providing adequate ventilation. Foundation crawl space vents should be placed in the side walls.

SLAB CONSTRUCTION INSULATION PROBLEMS

Houses and buildings constructed on concrete slab are in direct contact with the ground. Unless they are properly insulated, the slab can become extremely cold and damp. Some form of insulation should be placed between the slab and the ground before the concrete is poured (Fig. 23). Insulation can be added to an existing structure by excavating around the edges of the

Fig. 23. Insulating concrete slab.

foundation to a level below the frost line, and adding 2 inches of waterproof insulating material to the sides of the slab (Fig. 24).

INSTALLING BATTS AND BLANKETS

Batt or blanket-type insulation is usually available faced with paper on both the front and back surfaces. Paper flanges extending along both sides of the batt or blanket are used for nailing or stapling the insulation to studs or joists.

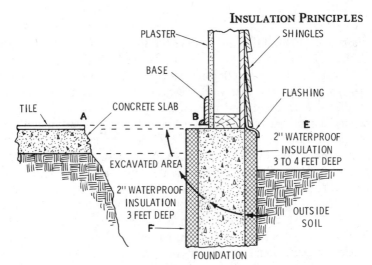

Fig. 24. Insulating the foundation on concrete slab construction.

Some batts or blankets are manufactured with a vapor barrier on one side, usually in the form of asphalt-impregnated paper. When this is the case, the surface with the vapor barrier should be installed so that it faces inward toward the heated rooms and spaces. For example, when installing insulation between top floor ceiling joists, place the insulation so that the surface with the vapor barrier faces downward toward the heated occupied rooms and spaces below.

When the batt or blanket insulation is installed directly over an insulating board panel ceiling, ordinary wood lath spaced approximately 6 inches on centers should be laid across the furring strips to support the insulation.

Insulation should be secured in place by tacking or stapling through the paper flanges extending along the sides of the batt or blanket insulation to the frame members. Space fasteners on 6 inch centers. Where lath is to be applied after insulation has been installed, fasteners sufficient only to hold the insulation in place temporarily need be used. Overlap the flanges on the face of each stud or joist to be certain there will be no break in the vapor barrier.

Insulation should be placed *under* the bridging between joists on ceilings if there is sufficient clearance. If there is insufficient clearance, cut the insulation and fit it around the bridging, but extend the vapor barrier under it. Stuff handfuls of insulation in

75

the spaces in the bridging. Remember to place the vapor barrier so that it faces down toward the occupied spaces below.

Start tacking or stapling insulation at the plate at the top of walls and work down toward the floor. Compress the insulation so that it extends around the *back* of pipes, ducts, outlet boxes, electrical receptacles, and other structural components that must remain exposed to the room (i.e. the interior) side of the walls. If it is not possible to extend the batt or blanket-type insulation behind them, then cut (or tear) the insulation and work it around the back, making certain the fit is as tight as possible. Be particularly careful to refit the vapor barrier so that there is no break.

Floors can be protected with a vapor barrier over the subfloor or between the floor joists and the subfloor. Waterproof roofing paper should be placed over the ground in foundation crawl spaces and extended up the insides of the exterior foundation walls. The exterior walls of crawl spaces should also be insulated with 2 to 3 inches of insulation. This will reduce the amount of cold air seepage entering the foundation crawl space area.

Fig. 25. Removing clapboard siding.

APPLYING BLOWN INSULATION

Blown insulation (not batt or blanket-type insulation) is recommended for insulating the exterior walls of *existing* structures. It requires the use of pneumatic equipment, and the knowledge of where to drill access holes in the exterior walls and how much insulation to use to completely fill each stud space.

The equipment and tools required for blown insulation are:

1. A blowing machine with hoses,
2. Electric drills,
3. Circular saws,
4. Screwdrivers.

The circular saws should have teeth set for wood cutting. The screwdrivers are used for removing plugs from the saws (if not self-ejecting).

The insulation when blown into the walls and attic spaces with compressed air, fills every space with heat resistant material. This method is the only practical way of insulating structures already constructed, whether of wood, brick or stone veneer, or stucco.

The insulation is usually blown into the walls at 2 lbs. per sq. in. pressure or less. This pressure gives a firm, even pack which assures maximum thermal efficiency. Furthermore, it puts the material under an initial compression so that any subsequent vibration or building movement will not cause the insulation to settle, but rather to expand and thus retain its full insulating value.

The insulation work, for the most part, is done from the outside of the building with a minimum of litter, dust or disturbance inside the building.

The openings for blowing are made near the top of wall panels, and the insulation is installed by inserting the nozzle and blowing which comes from the blowing machine, into each hole and blowing the material into the panel.

Before starting to blow the insulation into the interior spaces of the wall, soundings should be made with a plumb bob to locate cross framing. Where such framing is found, additional openings are made to assure that all spaces will be filled.

For the third story of houses having one or more rooms above the ceiling joists of the second floor, it is necessary to get between the roof and the walls of the upper story room and insulate the

77

SIDING

TO BE REMOVED

SET THIS NAIL

PRY HERE
WITH CHISEL

SHEATHING AND
BUILDING PAPER

PLASTER
AND LATH

STUD

Fig. 26. Removing shiplap siding.

side walls (knee walls) of the room from behind. This is done by boarding up the back of the studs with insulating board or by nailing diamond crosscord paper or a similar material across the back of the studs, and blowing behind it. It may be necessary to make an opening in the roof large enough to allow workmen to enter. The entire ceiling, including the portion from the wall line of the eaves to the walls of the third-story room, is insulated by blowing.

All portions around windows should be insulated thoroughly, as considerable air infiltration takes place around window framing.

In the case of open joists on unfloored attics, the man operating the nozzle stands on the joists and sprays the material to the desired thickness between the joists. Because these areas are not blown under pressure, there is no settlement problem. Therefore the one objective is to lay the material as evenly as possible in order to eliminate extra labor of leveling or screening by hand.

If the attic is floored, it is necessary to remove a few boards and insert the hose between each pair of joists, pushing the hose in as far as necessary to fill the space to the eave line. The material is then blown and as the space fills up to the required depth, the hose is slowly withdrawn.

It is recommended,as a general practice, that the size of holes made for blowing the insulation material be no greater than 3 inches, and no less than 2 inches in diameter. A 2½ inch diameter hole is sufficient to take the entire output from a 2½ inch diameter hose.

Removing building materials from the exterior walls in order to drill holes for blowing the insulation requires the skill of an experienced worker. Each type of construction (e.g. clapboard, wood siding shingles, brick veneer, etc.) presents a different problem. For example, Figs. 25, 26, 27 and 28 illustrate the problems involved in removing clapboard, siding joined with a shiplap, siding joined with a tongue-and-groove, and wood siding shingles.

A shingle should be removed by inserting a shingle ripper under the course *above* the shingles to be removed (Fig. 29). The

Fig. 27. Removing tongue-and-groove siding.

upper and lower course of nails is cut by hammering the ripper
downward and upward. Then insert the ripper under the shingle
to be removed and cut two sets of nails, upper and lower, by
pulling or hammering downward on the ripper.

Removing shingles should be done with a definite sequence
in mind. This is illustrated in Fig. 30. For example, if five shingles are to be removed from the top row, the number should start
at the left. Shingle No. 1 remains in the roof; shingles Nos. 2, 3,
4, 5, and 6 are removed; and No. 7 remains in the roof. Shingles

Fig. 28. Removing wood shingle siding.

1, 7, 8, 15, 16 and 22 remain in place. Remove 2, 3, 4, 5, 6, 9,
10, 11, 12, 13, 14, 17, 18, 19, 20 and 21 for opening (more if
necessary).

Two methods for removing brick veneer are illustrated in
Fig. 31. Care must be taken to match the color of the old mortar
when refilling the hole.

The method of blowing exterior stud walls, after the removal
of that part of the outside surface essential for the preparation

Fig. 29. Removing a shingle with a ripper.

B = NAILING STRIPS
C = ROOF RAFTERS

Fig. 30. Sequence suggested for removing shingles.

of holes for blowing, is usually a standard procedure. It is recommended that no greater vertical height than 7 feet be blown from any one wall opening. Blowing upward from any opening should not exceed a height of 2 feet.

81

Fig. 31. Two methods used for removing brick veneer.

CHAPTER 4

Heating Calculations

The size of the heating system is directly related to the amount of heat lost from the house or building. *All* structures lose heat to the outdoors or to adjacent unheated or partially heated spaces when the temperatures of the outdoor air or adjacent spaces are colder than those inside the structure. The heat within the building is normally lost by transmission through the building materials and by infiltration around doors and windows.

The loss of heat from a structure must be replaced at the same rate that it is lost. Consequently, determining the correct size of the heating system, and the rated capacity of the heating plant required by the system is very important. It should be obvious that an oversized heating plant will cost more to install than a smaller one and will provide more heat than the structure requires. On the other hand, an undersized heating plant lacks the capacity to provide sufficient heat, particularly during cold spells. The heating system and the selection of the heating plant must be carefully planned in order to adequately and efficiently replace the lost heat.

This chapter will describe several methods for calculating heat loss, ranging from rule-of-thumb methods to the rather more precise method of using overall coefficients of heat transmission (U-values) computed for the various construction materials and combinations of construction materials through which heat is commonly transmitted.

USING COEFFICIENTS OF HEAT TRANSMISSION

The more precise methods of calculating heat loss from a structure require a thorough knowledge of the thermal properties of the many materials and combinations of materials used in its construction. The term "thermal property" is used here to mean the overall coefficient of heat transmission (i.e. the rate of heat flow through a material). Each type of construction material (or combinations of materials) will have its own coefficient of heat transmission. Before continuing any further, it would be advisable to review the sections of Chapter 3 (INSULATION PRINCI-PLES) that specifically pertain to the problem of heat loss (e.g. see PRINCIPLES OF HEAT TRANSMISSION; THERMAL CONDUCTANCE; THERMAL RESISTANCE; etc.).

The ASHRAE and other authorities suggest the following basic steps (in the sequence given) for calculating heat loss:

1. Decide upon the desired inside air temperature for the structure.

2. Obtain the winter outside design temperature for the location of the structure from published lists or a local weather bureau.

3. Determine the design temperature difference (the difference between the temperatures found in Steps 1 and 2).

4. Identify on the heat loss worksheet each room or space in the structure.

5. List every structural section in each identified room or space that has an outer surface exposed to the outdoors or to an unheated or partially heated space.

6. Determine the coefficient of heat transmission (U-value) for each structural section (e.g. walls, glass, ceiling, etc.).

7. Calculate the infiltration heat loss for each identified room or space.

8. Calculate the *total* area for each exposed surface (i.e. the total outside wall area, the total floor area, etc.).

9. Calculate the area for each door and window in the walls, and add these figures together to obtain the total area for these openings.

10. Subtract the total area for doors and windows from the *gross* outside wall area (obtained in Step 8) to determine

the total *net* outside wall area. Enter the amount on the heat loss worksheet.

11. Multiply the total net wall area determined in Step 10 by the U-value (Step 2) by the design temperature difference (Step 3) to obtain the total heat loss for walls (expressed in Btuh).

12. Make the same calculations for the other surface areas (floors, ceilings, etc.) determined in Step 8.

13. Add the heat loss figures calculated for each surface category and the infiltration heat loss to obtain the total heat loss for the identified room or space.

14. Repeat the procedure outlined in Steps 1-11 for each identified room or space in the structure.

15. Add the various totals to obtain the total heat loss from the structure (expressed in Btuh).

These fifteen basic steps for calculating heat loss are provided as a useful means of reference for the more detailed description of the procedure contained in the paragraphs that follow.

OUTSIDE DESIGN TEMPERATURE

In heating calculations, the *outside design temperature* is the coldest outside temperature expected for a *normal* heating season. It is *not* the coldest temperature on record, but rather the lowest one recorded for a particular locale over a three to five year period.

Lists of outside design temperatures are published for selected localities throughout the United States (Table 1). If a locality is not included on the list of winter outside design temperatures, check with a local weather bureau. Using an outside design temperature from the nearest locality on the list can be misleading, because even closely located cities can differ widely in weather conditions as a result of different altitudes, the effects of large bodies of water, and other variables.

INSIDE DESIGN TEMPERATURE

The desired inside design temperature will depend upon the intended use of the space. In a large structure, such as a hotel or

Table 1. Winter outside design temperatures for major cities in the United States.

State	City	Outside Design Temperature Commonly Used	State	City	Outside Design Temperature Commonly Used
Alabama	Birmingham	10	Mississippi	Vicksburg	10
Arizona	Tucson	25	Missouri	Kansas City	−10
Arkansas	Little Rock	5		St. Louis	0
California	San Francisco	35	Montana	Billings	−25
	Los Angeles	35		Helena	−20
Colorado	Denver	−10		Miles City	−35
Connecticut	New Haven	0	Nebraska	Lincoln	−10
Dist. of Columbia	Washington	0		Valentine	−25
Florida	Jacksonville	25	Nevada	Reno	− 5
	Key West	45	New Hampshire	Concord	−15
Georgia	Atlanta	10	New Jersey	Atlantic City	5
	Savannah	20		Trenton	0
Idaho	Boise	−10	New Mexico	Albuquerque	0
Illinois	Cairo	0	New York	Albany	−10
	Chicago	−10		Buffalo	− 5
Indiana	Indianapolis	−10		New York City	0
Iowa	Des Moines	−15	North Carolina	Asheville	0
	Sioux City	−20		Charlotte	10
Kansas	Topeka	−10	North Dakota	Bismarck	−30
Kentucky	Louisville	0	Ohio	Akron	0
Louisiana	New Orleans	20		Cincinnati	0
Maine	Portland	− 5		Columbus	−10
Maryland	Baltimore	0	Oklahoma	Oklahoma City	0
Massachusetts	Boston	0	Oregon	Portland	10
Michigan	Detroit	−10	Pennsylvania	Erie	− 5
	Escanaba	−15		Harrisburg	0
	Sault Ste. Marie	−20		Philadelphia	0
Minnesota	Duluth	−20	Rhode Island	Providence	0
	Minneapolis	−20	South Carolina	Charleston	15
South Dakota	Huron	−20	Virginia	Lynchburg	5
Tennessee	Knoxville	0		Norfolk	15
Texas	Abilene	15	Washington	Seattle	15
	Austin	20		Spokane	−15
	Brownsville	30	West Virginia	Parkersburg	−10
	Corpus Christi	20	Wisconsin	Green Bay	−20
	Houston	20		Madison	−15
	Dallas	0	Wyoming	Cheyenne	−15
Utah	Salt Lake City	−10			
Vermont	Burlington	−10			

a hospital, there will be more than one inside design temperature, because there is more than one type of space usage. For example, hospital kitchens will generally have an inside design temperature of about 66°F. The wards, on the other hand, will range from 72°F to 74°F.

Residences will generally have a single inside design temperature for the entire structure, with 70°F or 71°F being the most commonly used temperatures.

DESIGN TEMPERATURE DIFFERENCE

The *design temperature difference* is the variation in degrees Fahrenheit between the outside and the inside design temperatures. It is used in the heat transmission loss formula (see below), and is a crucial factor in heating calculations.

Be careful that you obtain the degree *difference* between the two temperatures, and do not simply subtract the smaller figure from the larger one. For example, if the outside and inside design temperatures are −20°F and 70°F respectively, the design temperature difference will be 90 degrees (20 degrees below zero plus 70 degrees above zero).

DETERMINING COEFFICIENTS OF HEAT TRANSMISSION

The *overall coefficient of heat transmission* or U-value, as it is commonly designated, is a specific value used for determining the amount of heat lost from various types of construction. It represents the time rate of heat flow and is expressed in Btu's per hour per square foot of surface per degree Fahrenheit temperature difference between air on the inside and air on the outside of a structural section. Furthermore, the U-value is the reciprocal of the total thermal resistance values (R-values) of each element of the structural section, and may be expressed as:

$$U = 1/R_t$$

Professional organizations such as the *American Society of Heating, Refrigeration and Air-Conditioning Engineers* (ASHRAE) have already determined the U-values for a wide variety of floor, ceiling, wall, window, and door construction. Tables of these U-values are made available through ASHRAE publications (e.g.

check the *ASHRAE 1972 Handbook of Fundamentals*) found in many libraries. Many manufacturers of heating equipment also provide tables of U-values in their literature. When calculating heat loss, you have the option of selecting U-values from the tables provided by manufacturers and certain professional associations, or computing them yourself. The latter method, if done correctly, is the more precise one.

CALCULATING NET AREA

Having determined the design temperature difference and computed (or selected) an overall heat transmission coefficient for each construction material or combinations of materials, you are now ready to calculate the net area of each surface exposed to the outside or adjacent to an unheated or partially heated space.

The best procedure for calculating surface area is to work from a building plan. If one is not available, you will have to make your own measurements. Measurements for calculating net area are taken from *inside* surfaces (i.e. inside room measurements). You will not be concerned with structural surfaces (e.g. walls, ceilings, floors) between rooms and spaces heated at the same temperature, because no heat transmission occurs where temperatures are constant.

Calculating the total area for each surface should be done as follows:

1. Multiply room length by room width to determine floor and ceiling area.
2. Multiply room length (or width) by room height to determine the outside wall area for each room.
3. Multiply door width by door height to determine the surface area for each door.
4. Multiply window width by window height to determine the surface area for each window.

Whether you use room length or room width in calculating the outside wall area will depend upon which wall surface is exposed to the outside. In some cases (e.g. corner rooms), both are used and require at least two separate calculations (i.e. room length × room height and room width × room height).

Add the calculated surface area of each outside wall (see

Step 2 above) to obtain the *gross wall area* for the structure. Subtract the sum of all door and window surface areas from the gross wall area. The result will be the *net wall area* for the structure. Multiply the net wall area by the heat loss in Btu's per hour per square foot to calculate the heat loss through the walls.

HEAT TRANSMISSION LOSS FORMULA

The heat loss (expressed in Btuh) of a given space is determined by multiplying the coefficient of heat transmission by the area in square feet by the design temperature difference (i.e. the difference between the indoor and the outdoor design temperatures). Because a heating system must supply an amount of heat equal to the amount of heat lost in order to maintain a constant indoor design temperature, heat loss is approximately equal to heat required. This may be expressed by the following formula:

$$H_t = AU (t_i - t_o)$$

Where:

H_t = Heat loss transmitted through a structural section (e.g. roof, floor, ceiling, etc.) expressed in Btu's per hour. It represents both the heat lost and the heat required,

A = Area of structural components in square feet,

U = Overall coefficient of heat transmission,

T_o = Outdoor design temperature,

T_i = Indoor design temperature.

COMPUTING TOTAL HEAT LOSS

The commonly accepted procedure for calculating the total heat loss from a structure is to calculate the heat loss for each room or space *separately* and then add the totals.

The heat loss calculation worksheet you use should contain a column in which each room or space can be identified separately. Under the identification are listed those structural sections through which heat transmission losses occur. When applicable, these will include all or most of the following:

1. Walls,
2. Glass,

3. Ceiling,
4. Floor,
5. Door(s).

In addition to the five types of structural sections listed above, each identified room or space should also include a line for heat loss by air infiltration. Figure 1 illustrates how your worksheet should appear at this point.

TYPE OF ROOM OR SPACE	TYPE OF STRUCTURAL SECTION
BEDROOM NO. 1	WALLS GLASS AREA CEILING FLOOR DOOR AREA INFILTRATION

Fig. 1. Tabulation worksheet.

Now that you have identified the room or space, have listed the structural sections through which heat loss occurs, and have calculated the rate of air infiltration, you must now determine the net surface area and the U-value for each structural section. The design temperature difference must also be determined and entered for each structural section. Except where surfaces are exposed to partially heated spaces (e.g. a garage, attic, or basement), each structural section will have the same design temperature difference. An example of how your worksheet should look at this point is illustrated in Fig. 2.

Each identified room or space should also include the calculated air infiltration heat loss (see INFILTRATION HEAT

TYPE OF ROOM OR SPACE	TYPE OF STRUCTURAL SECTION	NET AREA/ AIR VOLUME	COEFFICIENT	TEMP. DIFF.	HEAT LOSS (BTU/h)
BEDROOM NO. 1	WALLS	235 SQ.FT.	0.26	90	5599
	GLASS AREA	45 SQ.FT.	0.45	90	1823
	CEILING	300 SQ.FT.	0.17	40	2040
	FLOOR	—	—		
	DOOR AREA	—	—		
	INFILTRATION	1800 CFH	0.018	90	2916

Fig. 2. Tabulation worksheet.

LOSS below). The amount of heat loss due to air infiltration will depend upon the size, type, and number of windows and doors, and other variables.

The bedroom given, as an example in Fig. 1, has two exposed walls (8 ft. X 15 ft. and 8 ft. X 20 ft.). The number of air changes suggested for a room with two exposed walls is 1½ changes per hour.

The volume of air infiltration for the bedroom can be calculated as follows:

$$
\begin{array}{ccc}
\begin{matrix}\text{Volume} \\ \text{of Air} \\ \text{Infiltration}\end{matrix} & = & \begin{matrix}\text{Number} \\ \text{of Air} \\ \text{Changes}\end{matrix} & \times & \dfrac{\text{Ceiling Area} \times \text{Ceiling Height}}{2}
\end{array}
$$

$$
1.5 \quad \times \quad \dfrac{300 \text{ sq. ft.} \times 8 \text{ ft.}}{2}
$$

Knowing that the volume of air infiltration is 1800 cfh, the heat loss in Btuh's can be calculated as follows:

$$
\begin{aligned}
\text{Heat loss} &= 0.018 \times Q(t_i - t_o) \\
&= 0.018 \times 1800 \times 90 \\
&= 2916 \text{ Btuh}
\end{aligned}
$$

The *total* heat loss for bedroom No. 1 (as it would be designated on the worksheet) is 12,378 Btuh. If there is a door in one of the outside walls, its heat loss would also have to be calculated and entered on the worksheet. Furthermore, the door area (in sq. ft.) would have to be subtracted from the surface area for the walls, because the latter is always a *net* figure.

After calculating the heat loss for each room or space in the structure, the results are added to obtain the total heat loss (in Btuh) for the structure.

Loss in Doors and Windows

It is not easy to compute the coefficient of heat transmission (U-value) for a door, because door construction varies considerably in size, thickness, and material. A fairly accurate rule-of-thumb method is to use the U-value for a comparable size *single-pane* glass window. If you wish to compute the coefficient of heat transmission for a door, the necessary data can be found in *ASHRAE* publications. The same source can be used for computing the coefficients of heat transmission of windows.

Loss in Basements

The amount of heat lost from a basement will depend upon a number of different variables. An important consideration, for example, is ground temperature, which will function as the "outdoor" design temperature in the heat loss transmission formula (see above). The ground temperature will vary with geographical location (Table 2).

Table 2. Ground temperatures below the frost line.

State	City	Ground Temperature Commonly Used	State	City	Ground Temperature Commonly Used
Alabama	Birmingham	66	Nevada	Reno	52
Arizona	Tuscon	60	New Hampshire	Concord	47
Arkansas	Little Rock	65	New Jersey	Atlantic City	57
California	San Francisco	62	New Mexico	Albuquerque	57
	Los Angeles	67	New York	Albany	48
Colorado	Denver	48		New York City	52
Connecticut	New Haven	52	North Carolina	Greensboro	62
Dist. of Columbia	Washington	57	North Dakota	Bismarck	42
Florida	Jacksonville	70	Ohio	Cleveland	52
	Key West	78		Cincinnati	57
Georgia	Atlanta	65	Oklahoma	Oklahoma City	62
Idaho	Boise	52	Oregon	Portland	52
Illinois	Cairo	60	Pennsylvania	Pittsburgh	52
	Chicago	52		Philadelphia	52
	Peoria	55	Rhode Island	Providence	52
Indiana	Indianapolis	55	South Carolina	Greenville	67
Iowa	Des Moines	52	South Dakota	Huron	47
Kentucky	Louisville	57	Tennessee	Knoxville	61
Louisiana	New Orleans	72	Texas	Abilene	62
Maine	Portland	45		Dallas	67
Maryland	Baltimore	57		Corpus Christi	72
Massachusetts	Boston	48	Utah	Salt Lake City	52
Michigan	Detroit	48	Vermont	Burlington	46
Minnesota	Duluth	41	Virginia	Richmond	57
	Minneapolis	44	Washington	Seattle	52
Mississippi	Vicksburg	67	West Virginia	Parkersburg	52
Missouri	Kansas City	57	Wisconsin	Green Bay	44
Montana	Billings	42		Madison	47
Nebraska	Lincoln	52	Wyoming	Cheyenne	42

The usual method for calculating heat loss transmission through basement walls is to view them as being divided into two sections (Fig. 3). The upper section extends from the frost line to the basement ceiling and includes those portions of the wall exposed to outdoor air temperatures. The heat loss for this section of the basement wall is calculated in the same way that other surfaces exposed to the outside air are calculated. The section of the wall extending from the frost line to the basement floor and the floor itself can be calculated on the basis of the ground temperature. The ground temperature is substituted for the outside design temperature when calculating heat loss.

Fig. 3. Basement Cross-section.

Loss in Slab Construction

The heat loss for houses and small buildings constructed on concrete slab at or near grade level is computed on the basis of heat loss per foot of exposed edge. For example, a concrete slab measuring 20 ft. × 25 ft. would have an exposed edge of 90 ft. This represents the measurement completely around the perimeter of the exposed edge of the slab. The heat loss will depend upon the thickness of insulation along the exposed edge and the outside design temperature *range*. This type of information is available from ASHRAE publications. To obtain the heat loss in Btuh, simply multiply the total length of exposed edge by the heat loss in Btuh per lineal foot. For example, a 90 ft. exposed edge with 2-inch edge insulation at an outdoor design temperature of 35°F would be calculated as follows:

$$90 \text{ lin. ft.} \times 45 \text{ Btuh/lin. ft.} = 4050 \text{ Btuh}$$

Infiltration Heat Loss

During the heating season, a portion of heat loss is due to the infiltration of cooler outside air into the interior of the structure through cracks around doors and windows and other openings that are not a part of the ventilating system. The *amount* of air entering the structure by infiltration is important in estimating the requirements of the heating system, but the *composition* of this air is equally important.

A pound of air is composed of both dry air and moisture particles which are combined (*not* mixed) so that each retains its individual characteristics. The distinction between these two basic components of air is important, because each is involved with a different type of heat: dry air with specific heat and moisture content with latent heat.

The heating system must be designed with the capability of warming the cooler infiltrated *dry air* to the temperature of the air inside the structure. The amount of heat required to do this is referred to as the sensible heat loss, and is expressed in BTU's per hour. The two methods used for calculating heat loss by air infiltration are: (1) the crack method and (2) the air change method.

The Crack Method — The crack method is the most accurate means of calculating heat loss by infiltration, because it is based on actual air leakage through cracks around windows and doors, and takes into consideration the expected wind velocities in the area in which the structure is located. The air-change method (see below) does not consider wind velocities, which makes it a less accurate means of calculation.

Calculating heat loss by air infiltration with the crack method involves the following basic steps:

1. Determine the type of window or door (see Table 3).
2. Determine the wind velocity and find the air leakage from Table 3.
3. Calculate the lineal feet of crack.
4. Determine the design temperature difference.

The data obtained in these four steps are used in the following formula:

$$H = 0.018 \times Q(t_i - t_o) \times L$$

Where:

H = Heat loss, or heat required to raise the temperature of air leaking into the structure to the level of the

Table 3. Infiltration rate through various types of windows. Expressed in cubic feet per foot of crack per hour. The infiltration rate through cracks around closed doors is generally estimated at twice that calculated for a window.

Type of Window	Remarks	Wind Velocity, Miles per Hour					
		5	10	15	20	25	30
Double-Hung Wood Sash Windows (Unlocked)	Around frame in masonry wall—not calked	3	8	14	20	27	35
	Around frame in masonry wall—calked	1	2	3	4	5	6
	Around frame in wood frame construction	2	6	11	17	23	30
	Total for average window, non-weatherstripped, $\frac{1}{16}$-in. crack and $\frac{3}{64}$-in. clearance. Includes wood frame leakage	7	21	39	59	80	104
	Ditto, weatherstripped	4	13	24	36	49	63
	Total for poorly fitted window, non-weatherstripped, $\frac{3}{32}$-in. crack and $\frac{3}{32}$-in. clearance. Includes wood frame leakage	27	69	111	154	199	249
	Ditto, weatherstripped	6	19	34	51	71	92
Double-Hung Metal Windows	Non-weatherstripped, locked	20	45	70	96	125	154
	Non-weatherstripped, unlocked	20	47	74	104	137	170
	Weatherstripped, unlocked	6	19	32	46	60	76
Rolled Section Steel Sash Windows	Industrial pivoted, $\frac{1}{16}$-in. crack	52	108	176	244	304	372
	Architectural projected, $\frac{3}{32}$-in. crack	15	36	62	86	112	139
	Architectural projected, $\frac{3}{64}$-in. crack	20	52	88	116	152	182
	Residential casement, $\frac{1}{64}$-in. crack	6	18	33	47	60	74
	Residential casement, $\frac{1}{32}$-in. crack	14	32	52	76	100	128
	Heavy casement section, projected, $\frac{1}{64}$-in. crack	3	10	18	26	36	48
	Heavy casement section, projected, $\frac{1}{32}$-in. crack	8	24	38	54	72	92
Hollow Metal, Vertically Pivoted Window		30	88	145	186	221	242

Courtesy ASHRAE 1960 Guide

indoor temperature (t_i) expressed in Btu's per hour.

Q = Volume of air entering the structure expressed in cubic feet per hour (Step 2 above).

t_i = Indoor temperature.

t_o = Outdoor temperature.

0.018 = The specific heat of air (0.240) times the density of the outdoor air (approximately 0.075).

L = Lineal feet of crack.

Determine the infiltration heat loss per hour through the crack of a 3 x 5 ft. average double-hung, non-weatherstripped, wood window based on a wind velocity of 20 mph. The indoor temperature is 70°F and the outdoor temperature is 20°F.

The air leakage for a window of this type at a wind velocity of 20 mph is 59 cubic feet per foot of crack per hour. This will be the value of Q in the air infiltration formula. The lineal feet of crack is (2 × 5) plus (3 × 3), or 19 feet (the value of L in the formula). t_i = 70°F and t_o = 20°F. Substituting this data in the air infiltration formula gives the following results:

$$H = 0.018 \times Q(t_i - t_o) \times L$$
$$= 0.018 \times 59(70 - 20) \times 19$$
$$= 1.062 \times 50 \times 19$$
$$= 1008.9 \text{ Btuh} = 1009 \text{ Btuh}$$

Table 4 represents a typical wall infiltration chart for a number of different types of wall construction. As is the case with air infiltration through windows, it is also based on wind velocity (see Table 3).

Table 4. Infiltration through various types of wall construction.

Type of Wall	Wind Velocity, Miles per Hour					
	5	10	15	20	25	30
Brick Wall 8½ in. Plain Plastered	2 0.02	4 0.04	8 0.07	12 0.11	19 0.16	23 0.24
Plain 13 in. Plastered	1 0.01	4 0.01	7 0.03	12 0.04	16 0.07	21 0.10
Frame Wall, lath and plaster	0.03	0.07	0.13	0.18	0.23	0.26

Courtesy ASHRAE 1960 Guide.

Air Change Method—In the air change method, the amount of air leakage (i.e. infiltration) is calculated on the basis of an assumed number of air changes per hour per room. The number of air changes will depend upon the *type* of room, and the number of walls exposed to the outdoors. Table 5 is an example of a typical air change chart used in the air change method.

Ventilation Heat Loss

In larger structures (e.g. commercial or industrial buildings), ventilation is necessary for both health and comfort. In houses, the amount of infiltrated air is generally sufficient to supply the ventilation requirements.

When ventilation is provided, the heating system warms the outside air used for this purpose (i.e. ventilation) to the inside design temperature. The heat required will be equivalent to the heat lost by ventilation, and can be calculated with the following formula:

$$H = 0.018 \times Q(t_i - t_o)$$

Where:

H = Heat lost by ventilation.

Q = Volume of ventilation air.

0.018 = The specific heat of air (0.240) times the density of the outdoor air (approximately 0.075).

THE AVERAGE VALUE METHOD

Heat loss from a structure can also be calculated by using average values for four basic factors. These four factors and their assigned values are:

1. Wall factor (.32 Btu),
2. Contents factor (.02 Btu),
3. Glass factor (1 Btu),
4. Radiation factor (240 Btu).

Each of these four factors is based on a 1°F. temperature difference and must be multiplied by the *design* temperature difference (see DESIGN TEMPERATURE DIFFERENCE in this chapter) obtained for the structure and its geographical location.

Table 5. Fresh air requirements.

Type of Building or Room	Minimum Air Changes Per Hour	Cubic Feet of Air Per Minute Per Occupant
Attic spaces (for cooling)	12-15	
Boiler room	15-20	
Churches, auditoriums	8	20-30
College class rooms		25-30
Dining rooms (hotel)	5	
Engine rooms	4-6	
Factory buildings (ordinary manufacturing)	2-4	
Factory buildings (extreme fumes or moisture)	10-15	
Foundries	15-20	
Galvanizing plants	20-30	
Garages (repair)	20-30	
Garages (storage)	4-6	
Homes (night cooling)	9-17	
Hospitals (general)		40-50
Hospitals (children's)		35-40
Hospitals (contagious diseases)		80-90
Kitchens (hotel)	10-20	
Kitchens (restaurant)	10-20	
Libraries (public)	4	
Laundries	10-15	
Mills (paper)	15-20	
Mills (textile-general buildings)	4	
Mills (textile-dyehouses)	15-20	
Offices (public)	3	
Offices (private)	4	
Pickling plants	10-15	
Pump rooms	5	
Schools (grade)		15-25
Schools (high)		30-35
Restaurants	8-12	
Shops (machine)	5	
Shops (paint)	15-20	
Shops (railroad)	5	
Shops (woodworking)	5	
Substations (electric)	5-10	
Theaters		10-15
Turbine rooms (electric)	5-10	
Warehouses	2	
Waiting rooms (public)	4	

The average value for each factor represents the amount of heat (in Btu's) that will pass through one square foot of the material in one hour. Thus .32 Btu will pass through *a square foot* of ordinary brick wall, or through a wall where the average frame construction is used *in one hour*. It should be stressed that these are *average* values useful in a simple and quick way of calculating heat loss. The wall factor of .32 Btu's varies for different kinds of walls, and heat loss calculations based on the use of average values will only be approximately accurate. More precise heat loss calculations can be made by using coefficients of heat transmission (U-factors).

The glass factor (1 Btu) is used to represent the amount of heat that will pass through one square foot of glass in an hour.

The radiation factor (240 Btu) was originally used to represent the amount of heat in Btu given off by an ordinary free-standing cast iron radiator on the basis of the square-foot area of heating surface per hour under average conditions. The average value of the radiation factor was based upon repeated test observations that the amount of heat given off by ordinary cast iron radiators per degree difference in temperature between the steam (or water) in the radiator and the air surrounding the radiator is about 1.6 Btu per square foot of heating surface. Taking this as a basis, a steam radiator under 2½ lbs. pressure corresponds approximately to 220°F. and when surrounded by air having a 70°F. inside air temperature gives off heat at a rate of 240 Btu ($220 - 70 \times 1.6 = 240$ Btu).

Radiators became so varied in design (with corresponding variations in the sizes of heating surface areas) that it became necessary to regard 240 Btu as a standard value.

Hot-water heating systems are planned on the basis of the Btu per hour capacity of each heat emitting unit. In steam heating systems, the radiation to be used is selected on the basis of the square foot EDR (Equivalent Direct Radiation) capacity of each unit. One square foot of EDR equals 240 Btu per hour. The EDR capacity of a space is expressed in square feet and may be determined by dividing the total heat required (expressed in Btu's per hour) by 240. For example, a room requiring 12,000 Btu per hour to heat it would have a steam radiation requirement of 50 square feet of EDR capacity.

The steps involved in determining heat loss with the average value method are:

1. Decide upon the desired inside air temperature.
2. Obtain the winter outside design temperature of the location of the structure from published lists or a local weather bureau.
3. Determine the design temperature difference (the difference between the temperatures found in Steps 1 and 2).
4. Calculate the volume of the structure or space.
5. Calculate the total area (in square feet) of window glass.
6. Calculate the total area of wall surface exposed to the outdoors.
7. Deduct the total area of window glass (Step 5) from the total area of wall surface exposed to the outdoors (Step 6).
8. Multiply the volume of the structure or space (Step 4) by the contents factor (.02) to determine the Btu required to raise the temperature of the volume an amount equal to the design temperature difference (Step 3).
9. Multiply the *net* wall surface area (Step 7) by the wall factor (.32) to determine heat loss through the walls.
10. Multiply the total window glass surface area (Step 5) by the glass factor (1) to determine heat loss through glass.
11. Add the figures obtained in Steps 8-10 to obtain the total heat loss per hour for the structure or space (this is equal to the total heat *required* per hour).

The eleven steps outlined above might be more clearly understood if shown in a sample problem.

Let's assume that you must calculate the heat loss for a space 14 ft. wide × 14 ft. long × 9 ft. high. 60 sq. ft. of window glass is equally divided between two exposed walls. The inside desired temperature is 70°F. and the winter outside design temperature is −10°F. Your procedure (corresponding step-by-step to those outlined above is as follows:

1. 70°F. inside desired temperature.
2. −10°F. winter outside design temperature.
3. 80°F. design temperature difference.

4. 1764 cu. ft. volume ($14 \times 14 \times 9$).

5. 60 sq. ft. of glass.

6. 252 sq. ft. of wall surface exposed (2 sides) to the outdoors ($(14 + 14) \times 9 = 252$).

7. 192 sq. ft. of *net* wall surface exposed ($252 - 60 = 192$).

8. 2822.4 Btu ($1764 \times .02 \times 80$).

9. 4915.2 Btu ($192 \times .32 \times 80$).

10. 4800 Btu ($60 \times 1 \times 80$).

11. 12,537.6 Btu ($2822.4 + 4915.2 + 4800$).

HEAT LOSS TABULATION FORMS

Manufacturers of heating equipment will often provide heat loss tabulation forms and instructions for their use. Lists of the most commonly used U-values (or factors) are included with the forms. Figs. 5 and 6 show examples of these forms and instructions obtained from *Amana Refrigeration, Inc.*

ESTIMATING FUEL REQUIREMENTS AND HEATING COSTS

Estimating fuel requirements and heating costs is *not* an exact science. It involves too many variables to be anything other than an estimation. Moreover there are four different formulas used for making these calculations. Depending on which formula is used, it is possible to arrive at four different estimations of fuel requirements and heating costs for a specific situation. It is no wonder then that two competent engineers can submit estimates that will differ as widely as 30 percent or more. These facts are not offered to discourage anyone, but rather to present the true picture. Any attempt to calculate fuel requirements and heating costs will not produce *precise* figures—only an estimate. The problem is to make this estimate as close an approximation to the real situation as possible.

The four formulas used in calculating fuel requirements and heating costs are:

1. The Heat Loss Formula,

2. The Degree Day Formula,

Figure 4. Residential heat loss tabulation form.

RESIDENTIAL HEAT LOSS TABULATION

Job Name: _____

Location: _____

Date: _____

Computed By: _____

Inside Design Temp. _____
Outside Design Temp. _____
Temp. Difference _____
Temp. Difference +10 _____
Bathroom _____

Room	Living Room		Dining Room		Kitchen		Bedroom 1		Bedroom 2				Bath		Rooms Btu Totals
Factor	Ar. or Ln. Ft.	BTU	Ar. or Ln. Ft.	BTU	Ar. or Ln. Ft.	BTU	Ar. or Ln. Ft.	BTU	Ar. or Ln. Ft.	BTU	Ar. or Ln. Ft.	BTU	Ar. or Ln. Ft.	BTU	
1. Room Size (LxWxH)															
2. Linear Ft. Exp. Wall															
3. Floor Area															
4. Windows															
5. Doors															
6. Exp. Wall															
7. Ceiling a															
b															
8. Infiltration c															
d															
9. SUB TOTALS															
10. Sub-Total for T.D.															
11. Floor Btu for T.D.															
12. Partition Btu for T.D.															
13. Room Total Btu Loss															

14. + ____ % Duct Loss

15. Grand Total Btu Load

Windows

	U Factors
Single Pane	1.13
Double Pane, Sealed	.57
Storm Windows, Tight	.45
Storm Windows, Loose	.75
Glass Block	.50

T.D. = TEMPERATURE DIFFERENCE

Wood Doors

	U Factors	
Nom. Thickness	No Storm	Storm Door
1"	.69	.35
1½"	.52	.30
2"	.46	.28
2½"	.38	.25

Exp. Walls

	U Factors		
Frame Construction With:	Clapboard or Shingle	Brick Venr.	Stone Venr.
No Insulation	.25	.28	.30
¾" Insl. Board	.19	.21	.22
2" Insl. Batts	.11	.12	.12
3⅝" Insl. Batts	.09	.10	.10

Ceilings
(Vent: Attic Space Above)

No Insulation	.50
2" Insulation	.10
4" Insulation	.09
6" Insulation	.05

% Duct Loss Schedule

	1 Story	1½ Story	2 Story
Ducts Insul.	10	8	6
Ducts No Insul.	30	25	20

Infiltration (factor times lineal ft. of crack)

		Weather Stripped	Not Weather Stripped
Double Hung Window	(a)	.22	.35
Pivoting Windows	(b)	.30	.50
Doors	(c)	.60	1.00
Fixed Window	(d)	.13	.13

CONCRETE FLOOR ON GROUND
Btu Per Hr. Per Lineal Ft. of Edge

Outside Design Temp.	25	20	15	10	5	0	-5	-10	-15	-20	-25	-30
1" Vert. Insul. Extending Down 18" Below Surface	55	62	68	75	80	85	90	95	100	105	110	115
1" L Type Insul. Extending 12" Down and 12" Under	50	57	63	70	75	80	85	90	95	100	105	110
2" L Type Insul. Extending 12" Down and 12" Under	40	45	50	55	60	65	70	75	80	85	90	95
No edge insulation and no heat in slab	This construction not recommended.											

Double Wood Floors or Partition

With Air Space

Temp. Diff.	10	20	30	40	50
Factor No Insul.	3.4	6.8	10.2	13.6	17.0
Factor 2" Insul.	1	2	3	4	5

Courtesy Amana Refrigeration, Inc.

Figure 5. Residential heating load estimate form.

Customer .... Address ....

Buyer .... Installation by ....

Estimate Number .... Estimate by ....; Date ....

Equipment Selected ....; Model ....; Size ....

Direction House Faces ....; Gross Floor Area .... sq. ft.; Gross Inside Volume .... cu. ft.

Design Conditions:

Outside Dry-Bulb Temperature (F) _____

ITEM	AREA (Sq. Ft.) Except Item 4; Linear Ft.	FACTOR (Circle the factors applicable.) DESIGN DRY-BULB TEMPERATURE (F)									BTU/HR (Area x Factor)
		−30	−20	−10	0	10	20	25	30	35	
1. WINDOWS (total of all windows)											
Single-glass		113	102	90	79	68	57	51	45	40	
Double glass or glass block		50	45	40	35	30	25	23	20	18	
2. WALLS											
No insulation (brick veneer, frame, stucco, masonry)		25	23	20	18	15	13	11	10	9	
1 in. insulation or ²⁵⁄₃₂ in. insulating sheathing		20	18	16	14	12	10	9	8	7	
2 in. or more insulation		10	9	8	7	6	5	5	4	4	

3. ROOFS									
Pitched or flat with vented air space and:									
No insulation	31	28	25	22	19	16	14	12	11
2 in. insulation	10	9	8	7	6	5	5	4	4
4 in. insulation	7	6	6	5	4	4	3	3	2
Flat with no air space and:									
No insulation	50	45	40	35	30	25	23	20	18
$\frac{29}{32}$ in. insulation	25	23	20	18	15	13	11	10	9
1½ in. insulation	15	14	12	11	9	8	7	6	5
3 in. insulation	10	9	8	7	6	5	5	4	4
4. FLOORS									
Over—(sq. ft. area):									
Basement	0	0	0	0	0	0	0	0	0
Enclosed crawl space	23	20	18	16	14	11	10	9	8
Open crawl space	34	31	27	24	20	17	15	14	12
On ground—(linear feet):									
No edge insulation	81	73	65	57	49	41	37	32	28
1 in. edge insulation	68	61	54	48	41	34	31	27	24
2 in. edge insulation	55	50	44	39	33	28	25	22	19
5. OUTSIDE AIR									
Total Floor Area (sq. ft.)	14	13	11	10	8	7	6	6	5
6. TOTAL (Sum of 1 through 5)									

105

HEATING AND VENTILATING

INSTRUCTIONS FOR FIGS. 4 AND 5

Step 1.   Room Size—list length, width and height.

Step 2.   Linear feet of Exposed Walls—total the length of all walls exposed to outdoors and enter the sum in the space provided.

Step 3.   Floor Area—multiply length of room by width.

Step 4.   Btu Loss—windows
Factor Column—enter proper "U" factor from "window" table.
Area Column—enter total of window area(s), width x height.
Btu Column—enter Btu loss for each degree of temperature difference obtained from factor x area.

Step 5.   Factor Column—enter proper "U" factor from "Door" table.

BTU   Area Column—enter net door area obtained from width x height minus
LOSS   area of glass in door.

DOORS Btu Column—enter Btu loss for each degree of temperature difference obtained from factor x area.

Step 6.   Btu Loss—exposed Wall
Factor Column—enter proper "U" factor from "Exposed Wall" table.
Area Column—enter net wall area obtained from linear feet exposed wall (step 2) x ceiling height minus window and door area.
Btu Column—enter Btu loss for each degree of temperature difference obtained from factor x area.

Step 7.   Btu loss—ceiling
Factor Column—enter proper "U" factor from "Ceiling" table.
Area Column—enter ceiling area, length x width. (Usually the same as floor area (step 3).
Btu Columns—enter Btu loss for each degree of temperature difference obtained from factor x area.

Step 8.   Btu Loss—infiltration
In each class of infiltration: a, b, c, d take appropriate factor from "Infiltration" table x the linear feet of crack to obtain Btu per degree of temperature difference.

Step 9.   Sub-Total Btu is the sum of Btu values. Steps 4 through 8 inclusive.

Step 10.   Multiply Btu from step 9 by temperature difference taken from upper right corner of form.

Step 11.   Btu Loss—Floor
Factor from appropriate "Floor and Partition" table x area of floor = Btu Loss. If concrete floor on ground, use linear feet, step 2, in place of area of floor. NOTE: Basement assumed 60°.

Step 12.   Partition Btu
Factor from "Floor and Partition" table x area of partition (length x

height) gives partition Btu loss for temperature difference. A partition is a wall exposed to an un-heated area or room.

Step 13. Room Total Btu loss is the sum of Btu values steps 10 through 12 inclusive.

REPEAT THE STEPS FOR EACH ROOM IN THE HOME & ADD (STEP 14) TO OBTAIN THE GRAND TOTAL.

3. The NEMA Formula,
4. The Corrected Heat Loss Formula.

If one formula is used to calculate the fuel requirements and heating costs with one type of fuel (e.g. oil) and another formula is used for a second type (e.g. natural gas), the results are practically worthless for comparison purposes. For example, the NEMA Formula (created by the National Electric Manufacturers Association) will present the use of electric energy much more favorably if the results are compared with calculations for oil or natural gas based on the Heat Loss Formula. A *true* comparison is only possible when two fuels are both calculated with the *same* formula. Each of these formulas is described in the following sections.

THE HEAT LOSS FORMULA

The Heat Loss Formula results in higher percentages of total requirements because it does not take into consideration internal heat gains obtained from appliances, sunlight, the body heat of the occupants, electric lights and other sources. The Corrected Heat Loss Formula includes these factors (see the appropriate section of this chapter).

The Heat Loss Formula will include the following data:
1. Heat loss expressed in Btu's,
2. Total hours in the heating season,
3. Average winter temperature difference,
4. Btu's per unit of fuel,
5. Efficiency of utilization,
6. Difference between inside and outside design temperatures.

The product of the first three items (1-3) is divided by the product of the last three (4-6). In other words, heat loss × total hours × average winter temperature difference ÷ by the Btu's per unit of fuel × efficiency of utilization × inside design temperature − outside design temperature.

The method for calculating heat loss was described in the last chapter (see Chapter 3.), and is expressed in Btu's per square foot per hour per degree Fahrenheit design temperature difference.

The total hours in the heating season will depend on the location of the house or building. If it is located in a southern state, the heating season will be much shorter than for one located in a colder climate. Assuming that the heating season begins October 1 and ends May 1, the total number of days for which heat may be required is 212. This figure is multiplied by 24 (hours) to obtain the total hours in the heating season (i.e. 5088).

The average winter temperature difference is found by subtracting the average low temperature from the average high temperature for the location in which the house or building is situated.

The Btu per unit of fuel is determined for each type, and this information can usually be obtained from the local distributor of the fuel.

Most heating equipment will burn oil or gas at an 80 percent combustion efficiency. Electric energy is generally considered to be used at 100 percent efficiency.

The outside design temperature is the lowest temperature experienced in a locality over a three to four year period. These outside design temperatures are available for a large number of localities throughout the United States. If the house or building is located in a small town or rural area, the nearest known outside design temperature is used. This inside design temperature is the temperature to be maintained on the interior of the house or building. The difference between the two design temperatures is used in the Heat Loss Formula.

Table 6 illustrates the use of the Heat Loss Formula for three types of fuel/energy:

1. Oil,
2. Natural gas,
3. Electricity.

Table 6. Applying the heat loss formula to three types of fuels.

(1) No. 2 oil, 140,000 Btu/gal, 80% efficiency, 35,000 Btuh loss, 73° inside temperature:

$$\frac{35,000 \times 5088}{140,000 \times .80} \times \frac{73 - 41}{70 - 0} = 727 \text{ Gallons}$$

(2) Natural gas, 100,000 Btu per therm, 80% efficiency:

$$\frac{35,000 \times 5088}{100,000 \times .80} \times \frac{73 - 41}{70 - 0} = 1018 \text{ Therms}$$

(3) Electricity, 3413 Btu per WH, 100% efficiency:

$$\frac{35,000 \times 5088}{3413} \times \frac{73 - 41}{70 - 0} = 23,852 \text{ KWH}$$

Courtesy *National Oil Fuel Institute.*

THE CORRECTED HEAT LOSS FORMULA

The Corrected Heat Loss Formula was devised in 1965 by Warren S. Harris and Calvin H. Fitch of the University of Illinois. It takes into consideration internal heat gains that have shown a marked increase over the past forty to fifty years. These internal heat gains make a significant contribution to the total required heat of a house or building, a factor which makes the original 65°F heat base established some forty odd years ago too low for making a correct estimate. Furthermore, an inside average temperature of 73°F is probably more correct than the 70°F temperature previously used.

According to Harris and Fitch, the degree day base (for all areas except along the Pacific Coast) must be corrected by the percentage given in Table 7. The calculated heat loss should then be reduced by 3.5 per cent for each 1000 feet the house or building is located above sea level. If the heat loss is below 1000 Btu's per degree temperature difference, the 65°F degree day base should be further reduced to the figures given in Table 8.

Table 9 illustrates the use of the Corrected Heat Loss Formula for determining comparative fuel requirements for a No. 2 oil and natural gas.

Table 7. Reduction in standard degree days for all areas except those along the Pacific Coast.

Degree Days At 65° Base	Reduce by This Percentage	New Total
10,000	3.15%	9,685
9,500	3.20%	9,196
9,000	3.22%	8,710
8,500	3.23%	8,226
8,000	3.25%	7,740
7,500	3.80%	7,215
7,000	3.85%	6,720
6,500	4.35%	6,218
6,000	4.55%	5,727
5,500	4.60%	5,247
5,000	4.65%	4,768
4,500	4.70%	4,289
4,000	5.30%	3,788
3,500	5.35%	3,313
3,000	6.15%	2,816
2,500	7.15%	2,321
2,000	8.00%	1,840
1,500	9.50%	1,358
1,000	11.90%	881
500	15.10%	425

Courtesy National Oil Fuel Institute.

Table 8. Reduction in degree day base when calculated heat loss (Btu's per degree temperature difference) is less than 1000.

Calculated Heat Loss	Revised DD Base
200	55° F
300	60° F
400	61° F
500	62° F
600	63° F
700	64° F
800	64° F
900	64° F
1000	65° F

Courtesy National Oil Fuel Institute.

Table 9. Using the corrected heat loss formula for determining comparative fuel requirements.

Assume 500 Btuh per degree temperature difference (35,000 Btuh at 70° difference); degree days with 65° base, 5542; average indoor temperature 73°.

Correction factor from Table 4 is 4.60%.

From Table 5, revised degree day base is 62°.

Reduction in base is (65 — 62) 3°.

Multiply .0460 × 3 = .1380 and deduct from 1.0000 = .8620.

Multiply 5542 (dd at 65° base) by .8620 = 4557.

EXAMPLE:

No. 2 Oil —
    35 × 0.00304 × 4557 = 485 gallons per year.

Natural Gas —
    35 × 0.00429 × 4557 = 684 therms per year.

Courtesy National Oil Fuel Institute.

THE DEGREE DAY FORMULA

The Degree Day Formula was devised some forty odd years ago by the American Gas Association and other groups, and has since been revised (see pages 627-8 of the 1970 *ASHRAE Guide*) to reflect internal heat gains and the levels of insulation. The correction factors involved in the revision are given in Tables 10 and 11. These correction factors should also be applied to the Corrected Heat Loss Formula (see the appropriate section in this chapter).

The Degree Day Formula is based on the assumption that heat for the interior of a house or building will be obtained from sources other than the heating system (e.g. sunlight, body heat of the occupants, etc.) until the outside temperature declines to

Table 10. Unit fuel consumption per degree day per 1000 Btu's design heat loss.

	Utilization Efficiency		
	60%	70%	80%
Gas in therms	0.00572	0.00490	0.00429
Oil in gallons (141,000 Btu)	0.00405	0.00347	0.00304

Courtesy National Oil Fuel Institute.

Table 11. Correction factors for outdoor design temperatures.

To be applied to Table 1 when design temperature
is higher or lower than 0° F.

	Outside Design Temp. F.				
	— 20	— 10	0	+ 10	+ 20
Correction factor	0.778	0.875	1.000	1.167	1.400

Courtesy National Oil Fuel Institute.

65°F. At this point the heating system begins to operate. The consumption of fuel will be directly proportionate to the difference between the 65°F base temperature and the mean outdoor temperature. In other words, three times as much fuel will be used when the mean outdoor temperature is 35°F than when it is 55°F. The mean outdoor temperature can be determined by taking the sum of the highest and lowest outside temperatures during a twenty-four hour period beginning at midnight and dividing it by two. Each degree in temperature below 65°F is regarded as one "degree day."

The Degree Day Formula is applied by dividing the heat loss figure by 1000 and multiplying the result by the figure for the unit fuel consumption per degree day per 1000 Btu's design heat loss, which, in turn, is multiplied by the total number of degree days in the heating season (calculated on a 65°F base) and then by the correction factors given in Tables 10 and 11. The application of this formula is illustrated in Table 12.

Table 12. Application of the degree day formula.

35,000 Btuh loss, outside design temperature 0, no correction factor needed, degree days, 5542.

No. 2 oil, 140,000 Btu per gallon, 80% efficiency —
    35 × 0.00304 × 5542 = 590 gallons per year.

Natural gas, 100,000 Btu per therm, 80% efficiency —
    35 × 0.00429 × 5542 = 832 therms per year.

Electricity. Required Btu/year is 832 (therms) ×
    100,000 (Btu/therm) × .80 (efficiency) = 66,560,000

$$\frac{66,560,000}{3413} = 19,502 \text{ kWh.}$$

Courtesy National Oil Fuel Institute.

THE NEMA FORMULA

The NEMA Formula was created by the National Electric Manufacturers Association. A constant of 18.5 is commonly used in the formula and reflects the percentage of heat loss that must be replaced by the heating system during a 24 hour period. In other words, the heating system will provide heat 77 percent of the time during a 24 hour period (24 × 18.5), while other heat sources will supply heat the remaining 23 percent.

The NEMA Formula consists of dividing the heat loss by 3413 (the number of Btu's per kWh), and multiplying this figure by the total annual number of degree days, which figure is then multiplied by the constant (usually 18.5). The resulting product is then divided by the difference between the indoor and outdoor temperatures. The application of the NEMA Formula is illustrated in Table 13.

A major criticism of this formula by distributors of oil and gas heating fuels is that the use of the Heat Loss Formula gives a much higher kWh consumption rate than using the NEMA Formula.

OTHER HEATING COSTS

Energy requirements for non-heating purposes are an important additional factor to be considered when estimating total fuel costs for a house or building. This non-heating energy is used to operate oil and gas burners, automatic washers and dryers, water heaters, stoves, refrigerators and the variety of other appli-

Table 13. Application of the NEMA formula.

KWH = heat loss ÷ 3413
   × annual degree days
   × constant, usually 18.5

Divided by
   difference between indoor and outdoor design temperature.

EXAMPLE: $\dfrac{35,000 \text{ Btuh}}{3413} = 1030$

$\dfrac{1030 \times 5542 \times 18.5}{70 - 0} = 15,086 \text{ kWh}$

Courtesy National Oil Fuel Institute.

ances and work-saving devices deemed so necessary for modern living.

This non-heating energy is commonly electricity or gas. Electricity is used to operate thermostats and other automatic controls found in heating and cooling equipment; burners, water heaters; home laundry equipment; refrigerators; and blowers for air conditioners to mention only a few. Much of this equipment (e.g. water heaters, home laundry equipment, etc.) is also designed to be operated in conjunction with gas. Each of these appliances consumes energy that will account for a certain percentage of the homeowner's utility bill. This must be accounted for when estimating the total fuel requirement and heating cost.

The most common method of determining which appliance is the most suitable one for a particular set of circumstances is by making a comparison of their energy use. For example, the estimated usage for home laundry equipment (either a washer or a dryer) is one hour per week per occupant. In other words, a dryer would be operated five hours per week on the average in a household containing five people. For purposes of comparison, an electric clothes dryer draws 5.0 kW. per hour and a gas clothes

Table 14.   Usage estimates for gas water heating.

GAS WATER HEATING

The estimates for gas water heating, in therms per dwelling per month are:

Cold Water Temperature:		Northern Localities 40°	North Central Localities 50°	South Central Localities 60°	Southern Localities 70°
Size of Unit	Hot Water Requirements				
0-BR	30 Gallon/Day	12.8	11.2	10.0	8.5
1-BR	40 "	16.4	14.7	12.9	10.9
2-BR	50 "	19.7	17.5	15.2	13.2
3-BR	60 "	22.9	20.5	17.7	15.3
4-BR	70 "	26.6	23.6	20.6	17.6
5-BR	80 "	30.2	27.0	23.5	20.2

(2) **Demand.** Where wholesale electricity is to be used for lighting, refrigeration, cooking, and domestic hot water, the monthly demand for all purposes may be estimated at 2.65 watts per kWh.

Courtesy National Oil Fuel Institute.

dryer uses approximately 3.5 therms of gas per month. Although these figures appear insignificant, they gain importance when the total non-heating energy use for all the appliances is added together. The cost per fuel unit will depend upon the local utility rates (see DETERMINING UTILITY RATES in this chapter). Some appliances, such as water heaters, will have to be compared on a basis which includes a number of different energy requirements met under a variety of different environmental conditions (see Tables 14 and 15).

The problem for the homeowner is to determine which type of appliance burns which type of fuel most efficiently and at the lowest cost. Probably one of the poorest sources of information is the distributor of the particular appliance or the rate correspondent of the utility. They will, of course, be quite biased in favor of their own product. This is as it should be, because this is the reason they are in business. If they made a habit of recommending the products of their competitors, they would soon be out of business.

Quite often consumer magazines will give the energy use rates of different appliances, and be quite frank in their comparison. This source is probably the most objective one to be found. However, knowing the rate at which an appliance consumes energy is only half the solution. It is also necessary to determine the utility rates.

Table 15. Usage estimates for electric water heating.

ELECTRIC WATER HEATING

The estimates for electric water heating in kWh per dwelling per month are:

Cold Water Temperature:		Northern Localities 40°	North Central Localities 50°	Central Localities 60°	Southern Localities 70°
Size of Unit	Hot Water Requirements				
0-BR	30 Gallon/Day	210	185	165	140
1-BR	40 "	280	250	220	185
2-BR	50 "	350	310	270	235
3-BR	60 "	420	375	325	280
4-BR	70 "	490	435	380	325
5-BR	80 "	560	500	435	375

Courtesy National Oil Fuel Institute.

Table 16. Determining utility rates.

DEMAND CHARGE

300 × $0.60 = $180.00
50 × $0.45 = $ 22.50
_____
$202.50

Total annual demand charges — 8 × $202.50 = $1,620

COMMODITY CHARGES

Therms		Rate		Amount
3,000	×	5.0¢	=	$150.00
2,000	×	4.5¢	=	90.00
5,000				$240.00

Total commodity charges — 8 × $240.00 = $1,920.00
Fuel adjustment (40,000 × $0.01819)   727.60
_____
$2,647.60

Average gas cost —

$$\frac{\$1,620 + \$2,647.60}{40,000} = 10.6¢/\text{Therm}$$

DETERMINING UTILITY RATES

No estimation of heating costs is complete until the utility rates for the various fuels available is determined and included in the estimate. Determining utility rates depends on so many variables that it, too, only serves as a cost estimation, not a precise figure. The problem is to arrive at as accurate an estimation as possible.

Among the many variables involved in determining utility rates are the following:

1. Differing rate structures,
2. Fuel/energy adjustment factors,
3. Variations in energy block sizes and prices,
4. Rate structure qualification procedures,
5. Community utility taxes,
6. Maximum demand charges.

Differing rate structures exist for different facilities or usage. The problem is to determine which rate structure will apply to the particular situation. This information can usually be obtained by contacting the local utility, and consulting with the utility rate correspondent.

Although this is a simplified explanation, a typical utility bill is computed by adding the demand charge, energy or commodity charge and the fuel adjustment charge, and dividing by the total amount of electricity or gas used during the billing period. This produces the average electric or gas cost. This procedure is illustrated in Table 16.

CHAPTER 5

Heating Fuels

Only the principal characteristics of those heating fuels used for domestic heating purposes will be considered in any great detail in this chapter. Descriptions of *firing methods* for the more commonly used heating fuels are found in Chapter 1 (OIL BURNERS), Chapter 2 (GAS BURNERS), Chapter 3 (COAL FIRING METHODS), Volume 3, and Chapter 3, Volume 3 (STOVES, FIREPLACES, AND CHIMNEYS).

Although electricity is used as a source of power for electric-fired furnaces, its properties are such that it cannot be regarded as a heating fuel in the sense that the term is being used here; therefore it is not included in this chapter. It is, however, described more fully in Chapter 2 (HEATING FUNDAMENTALS), Chapter 9 (ELECTRIC HEATING SYSTEMS), and Chapter 14 (ELECTRIC-FIRED FURNACES).

Heating fuels can be classified as solid, liquid, or gaseous, depending upon their physical state. Examples of solid fuels are coal and coke. Fuel oils are classified as liquid fuels, and gaseous fuels will include natural gas, manufactured gas, liquified petroleum gas, and related types.

NATURAL GAS

Natural gas is a generic term commonly applied to those gases found in or near deposits of crude petroleum. It is *not* connected with the production of oil, and should always be regarded as a separate and independent entity. Natural gas is piped under pressure from the gas fields to the consumer centers, and these pipe lines now serve a considerable portion of the United States.

Natural gas is the richest of the gases and contains from 80 to 95% methane with small percentages of the other hydrocarbons. The heating value of natural gas varies from 1000 to 1200 Btu's per cubic foot, with the majority of gases averaging about 1000 Btu's per cubic foot. The caloric value will depend upon the locality.

Natural gas can be divided into three different basic types:

1. Associated gas,
2. Nonassociated gas,
3. Dissolved gas.

Associated gas is a free (undissolved) natural gas found in close contact with crude petroleum. *Nonassociated gas* is also a free gas but is not found in contact with the crude petroleum deposit. As the name suggests, *dissolved gas* is found in solution in the crude petroleum.

MANUFACTURED GAS

A manufactured gas is any gas made by a manufacturing process. Raw materials used in the production of manufactured gas include coal, oil, coke, natural gas, or one of the other manufactured gases. For example, coal gas is made by distilling bituminous coal in either retorts or by-product coke ovens. Often several of these raw materials are used together as a base for the production of a manufactured gas.

The types of manufactured gases commercially available include:

1. Coal gas or by-product coke-oven gas,
2. Oil gas,
3. Blue water gas,
4. Carbureted water gas,
5. Producer gas,
6. Reformed natural gas,
7. Liquified petroleum gas.

Most of these gases have low calorific values (generally between 500 and 1000 Btu's per cubic foot), and are produced

119

primarily for industrial use. Producer gas and liquified petroleum gas are also used for domestic purposes. The former may be used alone or in combination with other gases. Of all the manufactured gases, liquified petroleum gas enjoys the widest application in domestic heating and cooking.

LIQUIFIED PETROLEUM GAS

Liquified petroleum gas (LP gas) is a hydrocarbon mixture extracted primarily from "wet" natural gas, and sold commercially as propane, butane, bottled gas, or under a variety of different brand names. The terms "dry" and "wet" natural gas refer to the gasoline content per 1000 cubic feet. It is regarded as "dry" if it contains less than 0.1 gallons of gasoline per 1000 cubic feet, and "wet" if it contains more than 0.1 gallons.

Propane is used quite extensively for domestic heating purposes. It contains 2516 Btu's per cubic foot (per gallon), or about two and a half times the Btu content of methane (natural gas). Table 1 compares the typical properties of propane and butane. Unlike natural gas, propane is heavier than air. An undetected leak can be quite dangerous, because the escaping gas will accumulate in layers at floor level. An explosion can be set off if the gas reaches the level of the pilot flame in the furnace.

Propane is frequently delivered in bulk and stored in large, stationary tanks holding from 100 to 1000 gallons. It is then piped into the home or building in much the same manner as natural gas. For small usage (e.g. trailers or homes that use only enough gas for one or two appliances), the LP gas is delivered in 5 to 25 gallon cylinders. The suggested clearances for LP gas cylinders and storage tanks are shown in Fig. 1. The size of the

Table 1. Typical properties of LP-gas.

	Butane	Propane
Btu per cu. ft. 60 deg. F.	3280	2516
Btu per lb.	21221	21591
Btu per gal.	102032	91547
Cu. ft. per lb.	6.506	8.58
Cu. ft. per gal.	31.26	36.39
Lb. per gal.	4.81	4.24

Courtesy of National LP-Gas Association .

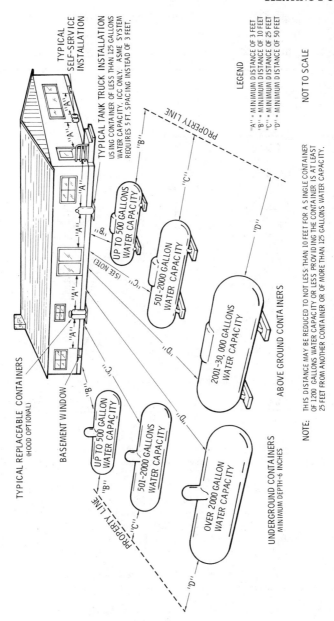

TYPICAL SELF-SERVICE INSTALLATION

TYPICAL TANK TRUCK INSTALLATION
USING CONTAINER OF LESS THAN 125 GALLONS WATER CAPACITY. ICC ONLY. ASME SYSTEM REQUIRES 5 FT. SPACING INSTEAD OF 3 FEET.

TYPICAL REPLACEABLE CONTAINERS
(HOOD OPTIONAL)

BASEMENT WINDOW

PROPERTY LINE

UP TO 500 GALLONS WATER CAPACITY

501-2000 GALLON WATER CAPACITY

2001-30,000 GALLONS WATER CAPACITY

ABOVE GROUND CONTAINERS

UP TO 500 GALLON WATER CAPACITY

501-2000 GALLONS WATER CAPACITY

OVER 2000 GALLON WATER CAPACITY

UNDERGROUND CONTAINERS
MINIMUM DEPTH-6 INCHES

LEGEND

"A" = MINIMUM DISTANCE OF 3 FEET
"B" = MINIMUM DISTANCE OF 10 FEET
"C" = MINIMUM DISTANCE OF 25 FEET
"D" = MINIMUM DISTANCE OF 50 FEET

NOT TO SCALE

NOTE: THIS DISTANCE MAY BE REDUCED TO NOT LESS THAN 10 FEET FOR A SINGLE CONTAINER OF 1200 GALLONS WATER CAPACITY OR LESS PROVIDING THE CONTAINER IS AT LEAST 25 FEET FROM ANOTHER CONTAINER OR OF MORE THAN 125 GALLONS WATER CAPACITY.

Courtesy National LP Gas Association.

Fig. 1. Suggested clearances for LP gas cylinders and storage tanks.

121

storage tank required by an installation depends upon the weather zone in which it is located (Fig. 2). Recommended tank sizes are given in Fig. 2.

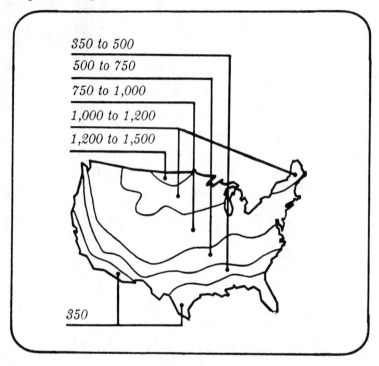

350 to 500

500 to 750

750 to 1,000

1,000 to 1,200

1,200 to 1,500

350

Courtesy National LP Gas Association.

Fig. 2. The six basic weather zones and the size of tank required to assure an adequate supply of gas.

FUEL OILS

Fuel oils are hydrocarbon mixtures obtained from crude petroleum by refining processes. They may be divided into the following six classes or grades:

1. *No. 1 fuel oil* for vaporizing pot-type and other burners designed for this type fuel.
2. *No. 2 fuel oil* for general purpose domestic heating not requiring a lighter No. 1 fuel oil.
3. *No. 3 fuel oil* (obsolete since 1948).

4. *No. 4 fuel oil* for installations not equipped for preheating.
5. *No. 5 fuel oil* for installations equipped for preheating.
6. *No. 6 fuel oil* for burner installations equipped for preheating with a high viscosity fuel.

This classification of fuel oils is based on a number of characteristics, including: (1) viscosity; (2) flash point; (3) pour point; (4) ash content; (5) carbon residue, and (6) water and sediment content.

All of these characteristics are important to consider when selecting a fuel oil. For example, the water content will determine its suitability for outdoor storage. A low viscosity indicates a fuel oil with good flow characteristics.

A No. 1 fuel oil has the lowest viscosity (1.4 - 2.2). This is a distillate oil used in atomizing and similar type burners, because it can be easily broken down into small droplets.

A No. 2 fuel oil is a distillate oil used in domestic oil burners that do not require preheating. It is slightly heavier than a No. 1 fuel oil. No. 4 fuel oil is slightly heavier than No. 2, and will work in some oil burners without preheating.

No. 2 and No. 4 fuel oils contain the highest ash content (.10 to .50% by weight respectively). The other grades contain little or no ash. Although slagging (i.e. the formation of noncombustible deposits) is commonly associated with solid fuels such as coal, this problem also occurs in oil burning equipment (see ASH, SLAG, AND CLINKER FORMATION in this chapter).

No. 5 and No. 6 (or *Bunker C*) fuel oils are heavier than the other grades and require preheating. In automatically operated oil burners, the preheating of the fuel oil must be completed before it is delivered for combustion. When the fuel oil has reached a suitable atomizing temperature, the preheating is considered completed.

PS 300 and *PS 400* are heating oils used on the West Coast, and roughly correspond to No. 5 and No. 6 fuel oils respectively. *PS 300* is the lighter of the two, and is used primarily for domestic heating purposes. *PS 400* is heavier, requires preheating, and is used for industrial purposes.

The grade of fuel oil to be used in an oil burner is specified by the manufacturer. This should also be stipulated on the label

Table 2. Btu content of the principal fuel oils.

Grade or Type	Unit	Btu
No. 1 Oil	Gallon	137,400
No. 2 Oil	Gallon	139,600
No. 3 Oil	Gallon	141,800
No. 4 Oil	Gallon	145,100
No. 5 Oil	Gallon	148,800
No. 6 Oil	Gallon	152,400
Natural Gas	Cu. ft.	950 to 1,150
Propane	Cu. ft.	2,550
Butane	Cu. ft.	3,200

Courtesy Honeywell Tradeline Controls

of the Underwriters' Laboratories, Inc., and Underwriters' Laboratories of Canada. Table 2 compares the Btu content of the principal fuel oils.

COAL

Coal is a solid fuel created from deposits of ancient vegetation that have undergone a series of metamorphic changes resulting from pressure, heat, submersion, and other natural formative processes occurring over a long period of time. Because of the different combinations of these processes acting on the vegetation and the widely differing forms of vegetation involved, coal is a complex and nonuniform substance.

The formation of peat represents a pre-coal stage of development. After increased pressure and time, lignite begins to form. The next stage is semibituminous, followed by bituminous, and then anthracite. Each of these stages is characterized by an increase in the hardness of the coal. Anthracite is the hardest of the coals. Graphite represents a post-coal developmental stage, and cannot be used for heating purposes.

Coal is often classified according to the degree of metamorphic change to which it has been subjected into the following categories:

1. Anthracite,

2. Bituminous,
3. Semibituminous,
4. Lignite.

Anthracite is a clean, hard coal that burns with little or no luminous flame or smoke. It is difficult to ignite, but burns with a uniform, low flame once the fire is started. Anthracite contains approximately 14,440 Btu's per pound, and is used for both domestic and industrial heating purposes. Its major disadvantage as a heating fuel is its cost.

Anthracite coal is divided by size into a number of different grades. Each of these grades (e.g. egg size, buckwheat size, pea size) is suitable for a specific size fire-pot. They are described in greater detail in Chapter 3, Volume 2 (COAL FIRING METHODS).

A great amount of smoke will result if it is improperly fired. The term "bituminous coal" actually covers a whole range of coals many of which have widely differing combustion characteristics. Some of the coals belonging to this classification are hard, whereas others are soft.

The available heat for bituminous coal ranges from a low of 11,000 Btu's per pound (*Indiana* bituminous) to a high of 14,100 Btu's per pound (*Pocahontas* bituminous). The heat value of the latter approximates that of anthracite (about 14,400 Btu's per pound); however, unlike anthracite coal, it is available in far greater supply, a factor that makes it a very economical solid fuel to use.

Semibituminous coal is a soft coal that ignites slowly and burns with a medium length flame. Because it provides so little smoke, it is sometimes referred to as *smokeless coal*.

Lignite (sometimes referred to as *brown coal*) ignites slowly, produces very little smoke, and contains a high degree of moisture. In structure, it is midway between peat and bituminous coal. Lignite contains approximately half the available heat of anthracite (or about 7400 Btu's per pound), and burns with a long flame. Its fire is almost smokeless, and it does not coke, a characteristic which it shares with anthracite.

Lignite is considered a low-grade fuel, and its calorific value is low when compared with the other coals. Moreover, it is difficult to handle and store.

COKE

Coke is the infusible, solid residue remaining after the distillation of certain bituminous coals, or as a by-product of petroleum distillation. Coke may also be obtained from petroleum residue, pitch. and other materials representing the residue of destructive distillation.

Coke will ignite more quickly than anthracite, but less readily than bituminous. It burns rapidly with little draft. As a result, all openings or leaks into the ash pit must be closed tightly when the coke is being burned.

Since less coke is burned per hour per square foot of grate than coal, a larger grate is required and a deep fire-pot is necessary to accomodate the thick bed of coal. Since coke contains very little hydrogen, the quick flaming combustion which characterizes coal is not produced, but the fire is nearer even and regular.

The best size of coke recommended for general use, for small fire-pots where the full depth is not over 20 inches is that which passes over a one inch screen and through a one and a half inch screen. For large fire-pots where the fuel can be fired over 20 inches deep, coke which passes over a one inch screen and through a three inch screen can be used, but a coke of uniform size is always more satisfactory.

BRIQUETTES

Briquettes are a solid fuel prepared from coal dust and fines (i.e. finely crushed or powdered coal) by adding a binder and holding it under pressure.

The binder is commonly pitch or coal tar, but other substances and materials have also been used. A briquette made with a pitch binder has the highest calorific value, exceeding that of any coal. The ash content is generally less than that of coal. Briquettes have been made from lignite coal dust and fines without a binder (only pressure being used).

COAL OIL

Coal oil is a heating fuel formerly derived from coal tar. Formerly it enjoyed widespread use, but it has now been largely replaced by petroleum based products. It is also called *paraffin oil* or *kerosene*. Today, coal oil (kerosene) is obtained from petroleum through a refining process.

WOOD AS A FUEL

A good, well-seasoned hardwood will provide approximately half as much heat value per pound as does good coal (Table 3). Wood is easy to ignite, burns with little smoke, and leaves comparatively little ash. On the other hand, it requires a larger storage space (commonly outdoors) and more labor is involved in its preparation.

Wood is commonly sold by the cord for heating purposes. A *standard cord* measures 4 feet high by 8 feet wide and contains wood cut to 4 foot lengths for a total of 128 cubic feet. Only about 70% of a standard cord actually represents the wood content, however, because of the existence of air spaces between the wood.

The heat value of wood depends upon the *type* of wood being burned. Heat values per cord for a number of different types of

Table 3. Some typical firewoods.

Name of Wood	Type of Firewood	Combustion Characteristics
Ash, white	hardwood	a good firewood
Beech	hardwood	a good firewood
Birch, yellow	hardwood	a good firewood
Chestnut	hardwood	excessive sparking. Can be dangerous.
Cottonwood	hardwood	a good firewood
Elm, white	hardwood	difficult to split, but burns well
Hickory	hardwood	slow, steady fire; best firewood
Maple, sugar	hardwood	a good firewood
Maple, red	hardwood	a good firewood
Oak, red	hardwood	slow, steady fire
Oak, white	hardwood	slow, steady fire
Pine, yellow	softwood	quick, hot fire; smokier than hardwood
Pine, white	softwood	quick, hot fire; smokier than hardwood
Walnut, black	hardwood	good firewood, but difficult to find

wood are given in Tables 4 and 5. The data used to compile the tables was obtained from "Use of Wood for Fuel" (*U.S. Department of Agriculture* Bulletin No. 753). The wood heat values are determined for cords containing approximately 90 cubic feet of solid wood (i.e. about 70% of a standard cord). Note that the heat values for both green and "dry" wood (12% moisture content) are given in Tables 4 and 5. It is important to remember that the amount of moisture in a wood will effect its burning characteristics. A green wood will burn much more slowly than a drier one, and its fire will be more difficult to start.

All woods can be divided into either hardwood or softwood types. Contrary to popular belief, the basic difference between the two groups is *not* the hardness of the wood. In other words, a wood that can be classified as a softwood is not necessarily softer than a hardwood. These terms do not refer to the physical properties of the wood but rather to its classification as a coniferous (needle or cone bearing) or deciduous (broad-leafed) tree. A softwood, for example, comes from a coniferous tree, whereas a hardwood is obtained from a deciduous tree. You will

Table 4. Heat value per cord (million Btu's) of wood with 12 percent moisture content. The equivalent coal heat values are based on one ton (2000 lbs.) of anthracite coal with a heat value of 13,000 Btu per lb.

Name of Wood	Heat Value	Equivalent Coal Heat Value
Ash, white	28.3	1.09
Beech	31.1	1.20
Birch, yellow	30.4	1.17
Chestnut	20.7	0.80
Cottonwood	19.4	0.75
Elm, white	24.2	0.93
Hickory	35.3	1.36
Maple, sugar	30.4	1.17
Maple, red	26.3	1.01
Oak, red	30.4	1.17
Oak, white	32.5	1.25
Pine, yellow	26.0	1.00
Pine, white	18.1	0.70

find that some soft wood trees produce wood that is harder than the wood obtained from some hardwood trees.

Chimney or flue fires are always a hazard when using wood as a heating fuel. Resins and soot collect in these areas over a period of time, and can be ignited if there is a flareup of the fire. The possibility of this happening can be eliminated or greatly reduced by cleaning the chimney or flue from time to time.

Softwoods (e.g. white pine, yellow pine, etc.) contain greater amounts of resin than do the hardwoods. This tends to give them a higher calorific value than the hardwoods and increased inflammability, but it also results in more smoke.

ASH, SLAG, AND CLINKER FORMATION

Ash is the noncombustible mineral residue that remains in the furnace or boiler after the fuel has been thoroughly burned. During the combustion process, the combustible portion of the fuel is consumed and the noncombustible ash remains in place. As the process continues, the ash residue is subjected to a cer-

Table 5. Heat value per cord (million Btu's) of green wood. The equivalent coal heat values are based on one ton (2000 lbs.) of anthracite coal with a heat value of 13,000 Btu per lb.

Name of Wood	Heat Value	Equivalent Coal Heat Value
Ash, white	26.0	1.00
Beech	27.1	1.04
Birch, yellow	27.2	1.05
Chestnut	19.2	0.75
Cottonwood	18.0	0.69
Elm, white	22.2	0.85
Hickory	29.0	1.12
Maple, sugar	27.4	1.05
Maple, red	23.7	0.91
Oak, red	27.5	1.06
Oak, white	28.7	1.10
Pine, yellow	23.7	0.91
Pine, white	17.3	0.67

tain degree of shrinkage. This shrinkage results in certain portions of the ash fusing together and forming more or less fluid globules or *slag*. This process is referred to as *slagging*.

Under the proper temperature conditions, the fluid slag globules can solidify and form *clinkers* in the fuel bed or deposits on the heating surfaces of the furnace or boiler. Clinkers can obstruct the necessary air flow to the fire, and result in its reduced efficiency or extinction. These deposits on the heating surfaces will reduce their heating capacity, and must be removed. Suggestions for removing clinkers are given in Chapter 3, Volume 2 (COAL FIRING METHODS).

SOOT

Soot is a fine powder consisting primarily of carbon produced by the combustion process. Because of its extreme light weight, it frequently rises with the smoke from the fire and coats the interior walls of the chimney and flue. Although the heat loss from the insulating effect of a soot layer is small (generally under 6%), it can cause a considerable rise in the stack temperature. Soot accumulation can also clog the flues, thereby reducing the draft and resulting in improper combustion. Soot may be blasted loose from the walls of the chimney or flue with a jet of compressed air, or sucked out with a vacuum cleaner. Another method is to use a brush to remove the accumulated soot layer from the walls.

CHAPTER 6

Warm-Air Heating Systems

Air is the medium used for conveying heat to the various rooms and spaces within a structure heated by a warm-air furnace. It is also the principal criterion for distinguishing warm-air heating systems from other types in use.

The warm-air furnace is a self-contained and self-enclosed heating unit which is usually (but not always) centrally located in the structure. Depending upon the design, any one of several fuels can be used to fire the furnace. Cool air enters the furnace and is heated as it comes in contact with the hot metal heating surfaces. As the air becomes warmer, it also becomes lighter, which causes it to rise. The warmer, lighter air continues to rise until it is either discharged directly into a room (as in the so-called "pipeless" gravity system), or carried through a duct system to warm-air outlets located at some distance from the furnace. After the warm air surrenders its heat, it becomes cooler and heavier. Its increased weight causes it to fall back to the furnace where it is reheated and repeats the cycle. This is a very simplified description of the operating principles involved in warm-air heating and it especially typifies those involved in gravity heating systems.

CLASSIFYING WARM-AIR HEATING SYSTEMS

A *warm-air heating system* is one in which the air is heated in a furnace and circulated through the rest of the structure

131

either by gravity or motor-driven centrifugal fans. If the former is the case, then the system is commonly referred to as a *gravity warm-air heating system.* Any system in which air circulation depends *primarily* on mechanical means for its motive force is called a *forced warm-air heating system.* The stress on the word "primarily" was intentional, because some gravity warm-air systems will use fans to supplement gravity flow, and this may prove confusing at first. In any event, one of the oldest forms of classifying warm-air heating systems has been on the basis of which method of air circulation is used: gravity or forced air.

Forced warm-air heating systems are often classified according to the type of duct arrangement used. The two basic types of duct arrangements used are:

1. Perimeter duct systems,
2. Extended plenum duct systems.

A *perimeter duct system* is one in which the supply outlets are located around the perimeter (i.e. the outer edge) of the structure close to the floor of the outside wall, or on the floor itself. The return grilles are generally placed near the ceiling on the inside wall. The two basic types of perimeter duct systems are:

(1) The perimeter-loop duct system (Fig. 1),

(2) The radial-type perimeter duct system (Fig. 2).

An *external plenum system* (Fig. 3) consists of a large rectangular duct which extends straight out from the furnace plenum in a straight line down the center of the basement, attic, or ceiling. Round supply ducts connect the plenum to the heat emitting units.

These and other modifications of duct arrangements are described in considerable detail in Chapter 7, Volume 2 (DUCTS AND DUCT SYSTEMS).

GRAVITY WARM-AIR HEATING SYSTEMS

A *gravity warm-air heating system* (Fig. 4) consists of a properly designed furnace (with casing and smoke pipe), a warm-air supply delivery system, and a cool-air return system.

In a central heating system, the warm air is delivered to the rooms and spaces being heated through a system of air ducts, the exception being a gravity floor furnace or a pipeless furnace.

REGISTER

FLOOR SURFACE
CONCRETE

1-1/2" TO 2"

2"

4"

6" MIN

8" DIAM.
ROUND
DUCT

4"

GRADE LINE

FOUNDATION

GRAVEL PASSED
OVER 3/4" SCREEN

55# ROOFING FELT
MOISTURE BARRIER

WATER PROOF INSULATION
BOARD 16" WIDE 2" THICK
RECOMMENDED

FURNACE

RADIAL
FEEDER DUCTS

PERIMETER DUCT

SLAB

Fig. 1. Perimeter-loop duct system in concrete slab construction.

Those ducts located in the basement and extending (or leading) from the furnace to the basement ceiling are referred to as *leaders*. The *stacks* are warm air ducts or pipes that connect to the leaders and extend vertically within the walls and up through the building. The warm air enters a room through registers located on the walls or floor (the former will generally connect to a stack; the latter to a leader).

133

Fig. 2. Radial-type perimeter duct system.

Fig. 3. Extended plenum system.

134

DRAFT REGULATOR
IN SMOKE PIPE

Fig. 4. Operating principle of a gravity warm-air heating system.

The cooler room air exits a room by means of return grilles and travels back through return air ducts to the furnace where it is reheated and recirculated.

A gravity warm-air heating system differs from the forced warm-air type by relying primarily on gravity to effect air circulation. The operating principle of the gravity warm-air heating system is based on the fact that the weight per unit volume of air decreases as its temperature increases, or increases as its temperature decreases. As the air is heated in the furnace, it expands and becomes lighter. Because it is lighter, it is displaced by the cooler, heavier air entering the furnace. The warmer, lighter air moves through the leaders and up through the wall stacks where

135

it enters the rooms through the registers. As the air cools, it becomes heavier and returns through the return air grilles and ducts to the furnace—hence the name *gravity* warm-air heating systems.

Air circulation in a gravity warm-air heating system depends upon the difference in temperature between the rising warm air and the cooler air that is falling and returning to the furnace for reheating. The greater the difference in temperature, the faster the movement of the air; however, natural conditions will place upper limits on the speed of the air movement, making it advisable to equip a gravity type central furnace with a fan as an integral part of its construction. Such an integral fan should be powerful enough to overcome internal duct resistance to air flow, and is *absolutely* necessary if air filters are used in the heating system. The motive force of air circulation in a standard gravity system is not strong enough to overcome the resistance of filters.

Planning a Gravity Warm-Air Heating System

A number of different trade and professional associations will provide information for planning a gravity warm-air heating system. A very useful source of information is the most recent edition of the *National Warm Air Heating and Air Conditioning Association's* Publication GRAVITY CODE AND MANUAL FOR THE DESIGN AND INSTALLATION OF GRAVITY WARM AIR HEATING SYSTEMS (Manual 5).

The following recommendations for planning such a system are offered as a basic planning guide.

1. Calculate the heat loss from each room in the structure, and add these together for the *total heat loss*.
2. Plan the leaders (i.e. the horizontal ducts or pipes leading from the furnace) so that none are longer than 10 ft. in length.
3. Keep the number of elbows in the leader to a minimum.
4. Locate the warm-air outlets (registers) and return air grilles as close to the floor in each room as possible.
5. Locate the warm air outlets (registers) on the inside walls of each room and in that part of the room closest to the furnace.

6. Select warm-air and return air ducts according to their Btu carrying capacities.

7. Select a furnace capable of delivering in Btuh a register delivery *equal* to the total heat loss calculated for the structure.

8. Locate the furnace in the *lowest* part of the structure and as near to the center as possible.

FORCED WARM-AIR HEATING SYSTEMS

A *forced warm-air heating system* (Fig. 5) consists of a furnace equipped with a blower, the necessary controls, an air duct system, and an adequate number of suitably located warm-air

FILTER

BLOWER

DRAFT REGULATOR
IN SMOKE PIPE

Fig. 5. Operating principles of a forced warm-air heating system.

registers and cold-air returns. The warm air travels from the furnace through the supply ducts to the room registers. The cold air returns through the return-air ducts to the furnace where it is reheated and recirculated. The return-air ducts are joined together before reaching the furnace, a feature distinguishing the duct arrangement in this type of heating system from that found in heating systems relying upon the gravity flow principle of air circulation (Fig. 6).

Planning a Forced Warm-Air Heating System

ASHRAE publications (see, for example, the *ASHRAE 1960 Guide*) summarize design procedures for planning several types of *forced* warm-air heating systems. The following six elements are common to each planning procedure:

1. Calculate the heat loss for each room or space, and from this data determine the total heat loss for the structure.
2. Determine the required furnace-bonnet capacity from the total heat loss of the structure.
3. Determine the location of the diffusers (warm-air outlets) on the building plan.

GRAVITY TYPE FURNACE   FORCED WARM-AIR FURNACE

Fig. 6. Duct arrangement for gravity and forced warm-air furnaces.

4. Calculate the required Btuh delivery of each diffuser.
5. Locate the position of the feeder ducts on the building plan.
6. Calculate the size of the feeder ducts on the basis of the total Btuh delivery it must supply.

Other elements in the design of a forced warm-air heating system will be specifically oriented to the characteristics of each system. For example, it is recommended that the maximum length of an extended plenum (nonexistent in a forced warm-air perimeter system) not exceed 35 ft. The locations and types of warm-air outlets and return-air inlets selected for the system will also depend upon the type of forced warm-air heating system used in the structure.

The following publications are recommended as being the most useful and authoritative sources of information for installing forced warm-air heating systems. In each case, the most current edition should be used.

1. WARM AIR PERIMETER HEATING (*National Warm Air Heating and Air Conditioning Association.* Manual 4).
2. FOUR-INCH PIPE WARM AIR PERIMETER HEAT-ING (*National Warm Air Heating and Air Conditioning Association.* Manual 10).
3. STANDARDS FOR THE INSTALLATION OF RESI-DENCE TYPE WARM AIR HEATING AND AIR CONDITIONING SYSTEMS (*National Fire Protection Association.* No. 90A).

Proprietary systems always should be installed according to the manufacturer's instructions. Any variation from these instructions increase the probability of error and future problems in the system.

PERIMETER LOOP WARM-AIR HEATING SYSTEMS

The *perimeter loop warm-air heating system* (Fig. 1) was originally designed for use in residences built on concrete slab rather than over a basement. The proven success of this type of duct arrangement in providing efficient and economical heating has resulted in its installation in all types of construction. This is

139

rapidly becoming one of the most popular forms of warm-air heating.

In the perimeter loop system, round ducts are imbedded in the concrete slab or suspended beneath the floor. The air is heated in a warm-air furnace equipped with a blower and forced through the ducts leading from the furnace to a continuous duct extending around the outer perimeter of the structure. The registers (diffusers) through which the heated air enters the various rooms and spaces are located along this outer perimeter duct. The most efficient operation can be achieved by placing these warm-air outlets on the floor next to the wall and *beneath a window* (Fig. 7). By placing the warm-air outlets here, window drafts are eliminated and the colder outside walls are kept warmed. Research shows that up to 80 percent of the heat loss from a structure can occur at these locations.

The National Warm Air Heating and Air Conditioning Association's publication WARM AIR PERIMETER HEATING (Manual 4) is probably the most authoritative and useful source of information about this subject. Ask for the most recent edition, because the manual has undergone a number of revisions.

CEILING PANEL SYSTEMS

A *ceiling panel system* is a forced warm-air heating system in which the heated air is delivered through ducts to an enclosed space above a false ceiling. There are no air supply outlets in the ceiling, and the heated air is therefore blocked from direct penetration of the occupied spaces below. Because its downward path is blocked, the heated air spreads over the entire surface area of the false ceiling. The ceiling eventually absorbs the heat and the cooler air is returned to the furnace for reheating and recirculation. The heat is transferred from the ceiling to the occupied spaces by radiation. A more appropriate name for this type of panel heating system might be "ceiling *space* panel system." Other types of ceiling panel systems are described in Chapter 1, Volume 3 (RADIANT HEATING).

A ceiling panel system of this type does not differ from the standard forced warm-air heating system except in the design of the heat emitting unit. Instead of using compact (individual heat

Fig. 7. Floor location for a warm-air outlet.

emitting units) a ceiling panel system uses a portion of the structure itself. As a result, considerable experience is necessary to install this type of heating system. It is only recommended for new construction, and never as a conversion from an existing heating system.

141

ZONING A FORCED WARM-AIR HEATING SYSTEM

The design of some residences, particularly the larger split-level homes, often causes balancing problems for the heating system. As a result, certain rooms will remain much colder than the rest of the house. This is a problem that cannot be corrected simply by turning up the heat, because the rest of the house will become too hot. The most effective solution is to divide the heating system into two separate zones, each controlled by its own thermostat. A motorized damper is installed in the existing duct system, and regulated by the thermostats to obtain the balance.

BALANCING A WARM-AIR HEATING SYSTEM

It is not always necessary to go to the expense of zoning to insure that each room receives enough heat. Sometimes a heating system can be balanced by reducing the air delivery to rooms or sections of the structure that require less heat. This results in automatically diverting more air and heat to those areas that require it.

The procedure for balancing a warm-air heating system is as follows:

1. Pick a day for balancing the system when the outdoor dry-bulb temperature is 40°F or below.

2. Open the dampers in all warm-air outlets as wide as possible (Fig. 8). The same holds true if the dampers are in the supply ducts (Fig. 9).

3. Leave the thermostat at one setting for at least three hours, and make certain the furnace blower is running.

4. Check the temperatures in all rooms. This can be done with thermometers (if you are certain they register equally), or simply by making your own judgment.

5. Leave the dampers in the coldest rooms wide open, and adjust the dampers in the warmer rooms to obtain the desired balance. After each adjustment, allow the system to stabilize for at least 30 minutes before checking the temperature or making the next adjustment.

When balancing a heating system, do not expect immediate results. A little patience makes it well worth the effort.

FLOOR DIFFUSER HAS
LEVER-CONTROLLED DAMPER

Fig. 8. Adjusting floor diffuser to full open position.

Fig. 9. Adjusting supply dampers to full operating position.

WARM-AIR FURNACES

The warm-air furnace is a self-contained heating unit designed to supply warm air to the interior of a structure. Warm-air furnaces used in central heating systems usually employ a system

143

of ducts to distribute the air to the various rooms and spaces within the structure.

These furnaces can be classified according to the method of air circulation into the following two basic categories: (1) *gravity* warm-air furnaces and (2) *forced* warm-air furnaces. A distinct advantage of the forced warm-air furnace is that its blower can move the air in any direction. As a result, this type of furnace can be located anywhere in the structure. Furthermore, the ducts do not have to be located above the heating unit as is the case with gravity warm-air furnaces. These advantages of the forced warm-air furnace have contributed to its tremendous popularity over gravity type furnaces. Both of these types are described in considerable detail in Chapter 10 (FURNACE FUNDAMENTALS).

Other criteria used for classifying warm-air furnaces include: (1) the type of fuel used (e.g. gas, oil, coal, or electricity); (2) the method of air distribution (ducts or pipeless), and (3) the method of firing (automatic or manual). Specific chapters in this book deal individually with some of these criteria (e.g. see Chapter 11. GAS-FIRED FURNACES; Chapter 12. OIL-FIRED FURNACES, etc.).

CONTROL COMPONENTS

The types of controls used in a warm-air heating system will depend upon a number of factors, including: (1) the method of air circulation used (e.g. forced or gravity); (2) the size of the structure; (3) the method used for supplying the fuel (automatic or manual) to the fire; and so on. In other words, a control system is for all intents and purposes custom designed to fullfill the requirements of a specific heating system.

All gravity and forced warm-air furnaces in central heating systems are controlled by a thermostat located in one of the rooms. If the heating system is zoned, two or more room thermostats will be used. In forced warm-air systems equipped with automatic burners or stokers, the thermostat will also control the operation of these units.

Each furnace should be equipped with a high-limit control to shut off the air or gas burner when plenum air temperatures exceed the furnace manufacturer's design limits. The high-limit control is automatic and will switch on the burner again as soon as the air temperature in the plenum has returned to normal (i.e. reached a level below the manufacturer's design limits). The high-limit control is frequently designed to operate in conjunction with the fan (blower) switch in forced warm-air furnaces.

Other types of controls used in warm-air heating systems are described in Chapter 4 (THERMOSTATS AND HUMIDI-STATS); Chapter 5 (GAS AND OIL CONTROLS); and Chapter 6 (OTHER AUTOMATIC CONTROLS) of Volume 2.

DUCTS AND DUCT SIZING

Ducts are passageways used for conveying air from a furnace or cooling unit to the rooms and spaces within a structure. The ductwork is also used to return the air to its source for recirculation.

The air is distributed to the rooms or spaces by means of grilles, registers, and diffusers. The size and location of the supply air outlets and the return air inlets are determined by such design factors as: (1) use (heating or cooling), (2) air velocity, (3) throw, (4) drop, and (5) the desired distribution pattern.

These and other aspects of air outlet and air inlet design and selection are considered more thoroughly in Chapter 7, Volume 2 (DUCTS AND DUCT SYSTEMS).

Designing and sizing a duct system is a compromise between the requirements and limitations of both the structure and the heating and cooling system. Duct sizing will be affected by a number of variables, including: (1) the architectural design of the structure, (2) space limitations, (3) the required air supply, (4) the allowable duct air velocities, (5) the desired noise level, and (6) the capacity of the blower. Adequately sized ducts represent the best possible compromise among these variables.

Accuracy in estimating the resistance to the flow of air through the duct system is important in the selection of blowers for application to such systems. Resistance should be kept as low as possible in the interest of economy. However, underestimating the resistance will result in failure of the blower to deliver the required volume of air. The various calculation methods used for sizing ducts are given in Chapter 23 (DUCTS AND DUCT SYSTEMS). ducts are given in Chapter 7, Volume 2 (DUCTS AND DUCT SYSTEMS).

A number of precautions should be taken in the design of a duct system. For example, careful study should be made of the building drawings with consideration being given to the construction of duct locations and clearances. Other recommendations that should be considered when designing a duct system can be summarized as follows:

1. Keep all duct runs as short as possible, bearing in mind that the air flow should be conducted as directly as possible between its source and delivery points, with the fewest possible changes in direction.

2. Select locations of duct outlets so as to insure proper air distribution.

3. Provide ducts with cross sectional areas which will permit air to flow at suitable velocities.

4. Design for moderate velocities in all ventilating work to avoid waste of power and to reduce noise.

COOLING WITH A WARM-AIR HEATING SYSTEM

The simplest method of cooling a house in the summer is to use the blower of a forced warm-air furnace. Most furnaces have a manual blower switch that can be used to circulate the air.

If you plan to install a central air conditioning system capable of maintaining comfortable indoor temperatures during the warmer months, you must first examine the existing furnace and duct system and determine what changes (if any) must be made to handle the cooling load.

Forced warm-air furnaces can often be converted to year-round air conditioning by installing a cooling coil in the furnace supply duct and connecting it to a compact compressor-condenser unit located outdoors. The existing duct system may be inadequate for air conditioning, but this can be modified to handle the cool air by increasing the size of certain ducts or by installing additional ducts.

Some furnaces (particularly gravity warm-air furnaces) cannot be fitted with cooling coils, *but* the existing ductwork is adequately sized for cooling. If this should be the case, the furnace can be bypassed with the cooling coils and air handling equipment installed in a short length of duct which feeds into the main

146

supply and return-air ducts at a point near the furnace. Dampers must be placed in the ducts between the furnace and the cooling coil to prevent the warm and cool air from mixing.

If both the existing furnace and ductwork cannot be used, it will be necessary to install a separate and independent air conditioning system. Adding one or more window units is the most common method of accomplishing this. Slightly more expensive is the use of a through-the-wall unit, which is installed by cutting a hole in the wall. The compressor-condenser section is located on the outside to reduce the noise level. Neither window units nor through-the-wall units use ductwork to distribute the cool air.

The most expensive separate and independent air conditioning systems are the single-package unit systems and the very popular split systems. Both use ductwork to distribute the cool air. A single-package unit system is limited to the installation of the cooling unit in the attic or basement, with the former location offering the fewest design problems. The advantage of a split system is that the cooling unit can be located anywhere inside the structure. In both systems, the preferred location of the compressor-condenser section is outdoors.

AIR CLEANING

The air cleaning equipment used in forced warm-air heating and cooling systems commonly takes the form of washable or disposable air filters, or electronic air cleaners.

Washable and disposable air filters are installed in the return air plenum of a forced warm-air furnace as shown in Fig. 10. These are dry-type filters consisting of cellulose fibers, steel wool, or some other suitable material set in a wire frame. These filters are effective only when dust concentrations are limited to 4 gr. per 1000 cubic feet of air or lower. Consequently, they are restricted in use to residences or small buildings. Air filters are not used in gravity warm-air heating systems, because they impede the rate of air flow.

Electronic air cleaners are designed for return air duct installation at the furnace, air handler (in the duct) or air conditioning unit. They generally operate on the electrostatic precipitation principle, and are capable of removing up to 95 per cent of all airborne

147

HOLDING PLATE-
SNAPS OPEN, CLOSED

RAIL-FITS INTO
FURNACE CHANNEL

MEDIA HOLDING
TABS

FRAME

FIBERGLASS MEDIA-
AIR ENTERS COLORED SIDE

Fig. 10. Examples of washable and throw-away air filters used in warm-air furnaces.

particles (e.g. dust, tobacco, smoke, etc.). The dust particles are given an electric charge when they pass through an ionizing field, and are collected when they subsequently pass between collector plates having an opposite charge. Typical installations are illustrated in Fig. 11.

The various types of air filters and electronic air cleaners used in cleaning and filtering the air are described in Chapter 14, Volume 3 (AIR CLEANERS AND FILTERS).

HUMIDIFIERS AND DEHUMIDIFIERS

Air that has a very low level of humidity is too dry for comfort, and can cause damage to walls, interior woods, and furnishings. This results from the fact that dry air will absorb moisture from other sources, including the human body. The moisture-robbing effect of dry air is particularly noticeable on the sensitive nasal and throat membranes. Moisture can be added to the air by humidification, and the device used to add moisture is called a humidifier.

A

B

HORIZONTAL FURNACE

UPFLOW LOWBOY FURNACE

A

UPFLOW HIGH
BOY FURNACE

B

COUNTERFLOW
FURNACE

B

A

Fig. 11. Typical installations of cooling coils (A) and electronic air cleaners (B) on forced warm-air furnaces.

The humidifiers used in heating and cooling systems are designed to maintain the relative humidity within the comfort zone. They are available in a number of different types, each based on a different operating principle. Pan-type humidifiers, for example, contain a reservoir of water which evaporates into the warm air flowing through the supply duct. The water level in the reservoir is controlled by a float control. The desired humidity level is determined by a humidity control setting. When the relative humidity drops below the humidity control setting, a humidifier

149

fan is actuated and air is blown over the water in the reservoir. Moisture is then picked up by the air and blown into the space to be humidified. Other types of humidifiers are: (1) the spray-type air washer and (2) the steam-type humidifiers (either air or electrically operated). Typical humidifier installations are illustrated in Fig. 12.

Sometimes air will have a humidity level that is too high. The excess moisture resulting from this condition can also cause

LEGEND: COLD AIR   HOT AIR

NOTE: THERE MAY BE SEVERAL BRANCH HOT AIR DUCTS. JUST MOUNT UNDER LARGEST CONVENIENT DUCT. RECIRCULATION OF AIR WILL UNIFORMILY HUMIDIFY HOME

STANDARD INSTALLATION

PLENUM INSTALLATION

COUNTER FLOW INSTALLATION
SLAB MOUNTED

HORIZONTAL FURNACE INSTALLATION

ALTERNATE

ALTERNATE POSITIONS

COUNTERFLOW INSTALLATION
WITH CRAWL SPACE

Fig. 12. Typical installation of power humidifiers on forced warm-air furnaces.

150

damage to walls, interior woods, and furnishings, as well as prove very uncomfortable for the occupants. Excess moisture can be removed from the air, and the device used to remove the moisture is called a dehumidifier.

Dehumidifiers operate either on the cooling or absorption method. The former method accomplishes dehumidification by an air washer with a water spray temperature lower than the dew point of the air passing through the unit. Condensation occurs and both latent and sensible heat are removed.

In the absorption method of dehumidification, sorbent materials are used for removing moisture from the air. Air to be dehumidified is drawn or blown through a screened bed of dry-solid absorbent and the water vapor in the air is caught and retained in the pores of the absorbent.

Both humidifiers and dehumidifiers are described in greater detail in Chapter 13 of Volume 3.

ADVANTAGES OF A WARM-AIR HEATING SYSTEM

By present comfort standards, gravity warm-air heating offers no special advantage and too many disadvantages to be recommended as a heating system in the types of structures popular with the public today. The architectural design of these structures *necessitates* the use of a forced warm-air heating system. Because the gravity system lacks a blower, air circulation (and, therefore, heat distribution) depends upon the temperature difference between the rising warm air and the descending cold air. Unfortunately, many of our houses and buildings are designed in such a way that a gravity system would require a considerable temperature difference to obtain the desired rate of air movement. If you design the structure *around* a gravity heating system, then quite different results are possible. For example, a residence built by a gentleman named Wendell Thomas in western North Carolina contains a gravity-type heating system using a simple wood burning stove that produces indoor temperatures ranging between 60°F and 75°F year-round. The north and west sides of the structure were set into the ground. No windows were placed in the north wall, and the windows in the other walls were two or three panes thick. As shown in Fig. 13, the warm air heated by the stove rises and gives off its heat to the room. As it loses its heat, it cools and descends along the inside surfaces of the exterior walls, passes through metal air vents in the floor, and then enters the basement where it is warmed somewhat by the slightly warmer indoor temperatures. It then rises through metal air vents placed in the floor around the stove where it is reheated and recirculated. Construction and operation details of this heating system can be obtained by sending ten cents and a self-addressed envelope to THE MOTHER EARTH NEWS, BOX 957, DES MOINES, IOWA 50304. Ask for Reprint No. 39, "THE SELF-HEATING, SELF-COOLING HOUSE."

In a structure designed to take into consideration the operating principles of a gravity-type heating system, the operating costs will be lower than any other type system.

The advantages of a *forced* warm-air heating system are as follows:

1. The installation costs are lower than those for hot-water or steam heating systems.
2. Heat delivery is generally quicker than other systems.

Fig. 13. A gravity-type system.

3. Heat delivery can be shut off immediately.

4. Air cleaning and filtering can be cheaply and easily provided.

5. The humidity level of the air can be easily controlled.

6. Air conditioning can be added without great difficulty or cost.

7. The furnace (because of its blower) can be placed in a number of different locations in the structure.

DISADVANTAGES OF A WARM-AIR HEATING SYSTEM

One of the principal disadvantages of a *gravity* warm-air heating system is that the furnace *must* be centrally located at the lowest point in the structure. This, of course, severely limits the design flexibility of the structure itself. Other disadvantages of gravity warm-air heating systems include:

1. Slow response to heat demand from the thermostat.

2. Slow air movement.

3. Air filtering is inadequate or nonexistent.

Forced warm-air heating systems depend upon blowers to circulate the air through a network of sheet metal ducts. Both the operation of the blower and the expansion and contraction of the ducts as the air temperature rises or falls can cause distracting noise.

Warm-air outlets can frequently create decorating problems. In order to operate effectively, the registers should never be blocked by furniture or carpeting. This requirement, of course, restricts furniture arrangement and the placement of carpet.

The blower in a forced warm-air heating system will create a certain amount of air turbulence. This rapid agitation of the air causes dust particles to circulate and deposit on walls, furniture, and other surfaces. Ordinary filters will remove most of the larger particles, but are not very effective against smaller ones. An electronic air filter is recommended for the latter.

Warm-air heating systems require separate hot water heaters to supply the hot water necessary for household use. This requirement for additional equipment adds to the initial installation costs for the system (in a hot-water heating system, the hot water for household use is supplied by the boiler).

Forced warm-air heating systems supply convected heat through forced air movement. The heat is supplied in bursts (rather than a continuous, sustained flow) when the room thermostat calls for it. This results in room temperatures varying up and down several degrees. The variation in temperature can be considerably reduced by setting the fan switch for continuous blower operation.

CHAPTER 7

Hot-Water Heating Systems

Hot-Water heating systems use water as the medium for conveying and transmitting the heat to the various rooms and spaces within a structure. The motive force for the water in these systems is based either on the gravity flow principle or forced circulation. The latter type (referred to as *hydronic* or *forced hot-water* heating) is the most commonly used method.

CLASSIFYING HOT-WATER HEATING SYSTEMS

Hot-water heating systems can be classified in a number of different ways depending on the criteria used. Three broad classification categories based on the following criteria are generally recognized:

1. Type of water circulation,
2. Piping arrangement,
3. Supply water temperature.

In all hot-water heating systems, the water is circulated either by forcing it through the line or by allowing it to flow naturally. The latter is referred to as a gravity hot-water heating system, because circulation results from the difference in weight (specific gravity) of the water due to temperature differences (heavy when cold; light when hot). In a *forced* hot-water heating system, the accelerated circulation of the water can result from several commonly employed methods, including: (1) using high pressures; (2) super heating the circulating water and condensing the steam; (3) introducing steam or air into the main riser pipe; (4) using a combination of pumps and local boosters; and (5) using pumps alone.

The four principal piping arrangements used in hot-water systems are as follows:

1. One-pipe system,
2 Two-pipe direct return system,
3. Two-pipe reverse return system,
4. Series-loop system.

These piping arrangements are described and illustrated in several sections of this chapter (see, for example, ONE-PIPE SYSTEM, TWO-PIPE DIRECT RETURN SYSTEM, etc.).

If a hot-water heating system uses supply water temperatures above 250°F., it is classified as a *high temperature system*. High temperature systems are used in large heating installations such as commercial or industrial buildings. A *low temperature system* is one having a supply water temperature below 250°F and is used in residences and small buildings.

ONE-PIPE SYSTEM

A *one-pipe system* (Fig. 1) is one in which a single main pipe is used to carry the hot water throughout the system. In other words, the same pipe that carries the hot water to the heat emitting units (i.e. the radiators, convectors, etc.) in the various rooms and spaces within the structure also returns the cooler water to the boiler for reheating. Each heat emitting unit is connected to the supply main by two separate branch pipes (a separate feed and return line) (Fig. 2).

The hot water flows from the boiler or heat exchanger (if a steam boiler is the heat source) to the first heat emitting unit, through it to the second unit, and so on through each of the heat emitting units in the one-pipe system until it exits the last one and returns to the boiler or heat exchanger.

One-pipe systems may be operated on either forced or gravity circulation. Special care must be taken to design the system for the temperature drop found in the heat emitting units farthest from the boiler. This is particularly true of one-pipe systems designed for gravity circulation.

A principal advantage of a one-pipe system is that one or more heat emitting units can be shut off without interfering with the flow of water to other units. This is *not* true of series-loop systems

Fig. 1 One-pipe hot-water heating system.

Fig. 2. Heat emitting unit with two separate branch pipes connected to the main supply line.

in which the units are connected in series and form a part of the supply line.

In some large one-pipe systems, zoning is possible by providing for more than one piping circuit from the boiler (Fig. 3). In such cases, each piping circuit is equipped with its own thermostat and circulating pump. Sometimes these circuits are erroneously referred to as "loops" which results in confusing them with the piping arrangement in a series-loop system.

Fig. 3. **One-pipe hot-water heating system with more than one piping circuit for zoning.**

SERIES-LOOP SYSTEM

In the *series-loop system* (Fig. 4), the heat emitting units form a part of the piping circuit (i.e. *loop*) which carries the hot water from the boiler around the rooms and spaces within the structure, and back to the boiler again for reheating. In other words, there are no branch pipes connecting the main supply pipe to the heat emitting units as in the one-pipe system. In the series-loop system, the hot water flows from the boiler through a length of the main supply pipe to the first heat emitting unit in the circuit (loop). It then flows through the unit to a length of the main supply pipe connected at the opposite end; flows through this pipe to the second heat emitting unit in the circuit, and so until the entire circuit is completed.

Fig. 4. Series-loop hot-water heating system.

The series loop system is cheaper and easier to install than other piping arrangements, because it eliminates the need for branch pipes and reduces the amount of pipe used in the main circuit to the comparatively short lengths connecting the heat emitting units. There is no need for one continuous length of main pipe.

Because the heat emitting units are connected in series and constitute a part of the main supply line, the same hot water supply passes through each unit in succession. As a result, the heat emitting unit closest to the boiler receives the hottest water, whereas those farther away receive water several degrees cooler. Furthermore, individual units cannot be shut off (unless there is a special bypass piping arrangement) without obstructing the flow of water to units further along the line.

TWO-PIPE DIRECT-RETURN SYSTEM

In a *two-pipe direct-return system* (Fig. 5) hot water re rns directly to the boiler from each heat emitting unit. In other words, the hot water supply and return mains are separate pipes. The heat emitting units are connected to the supply and return lines by separate branches (Fig. 6).

159

Fig. 5. Two-pipe direct-return hot water heating system. The water from each heat emitting unit returns directly to the boiler without passing through other units in the system.

Fig. 6. Heat emitting unit with feed line connected to the supply main and return line connected to the return main.

Each heat emitting unit represents the mid-point in a complete circuit within a two-pipe direct-return system. The boiler completes the circuit. The farther a heat emitting unit is located from the boiler, the greater the length of the piping in the circuit. Although this factor causes problems in balancing the water supply among the various circuits, balancing can be achieved in a number of ways, including: (1) balancing cocks, (2) proper pipe sizing, or (3) using a reverse-return system.

TWO-PIPE REVERSED-RETURN SYSTEM

A *two-pipe reversed-return system* (Fig. 7) achieves a balance in the water supply by creating circuits to the radiators of approximately equal length. Instead of allowing the return supply of water to proceed directly to the boiler from each radiator, the return main carries the water in the opposite direction for a predetermined distance before turning back to the boiler. The first radiator has the shortest supply main, but the longest return main. For the farthest radiator, the reverse is true. Regardless of the position of a radiator in the system, the total length of pipe within the circuit of which it forms a part will be essentially the same as that of any other circuit.

COMBINATION PIPE SYSTEMS

Sometimes two or three different types of pipe arrangements will be combined in a single heating system. For example, it is not unusual to find a two-pipe reversed-return system combined with a series-loop system. Combination pipe systems are usually

Fig. 7. Two-pipe reversed-return hot-water heating system.

161

found in commercial or industrial buildings, and are fitted to specific design needs.

ZONING A TWO-PIPE SYSTEM

Balancing problems for two-pipe hydronic heating systems in larger houses and buildings can be solved by splitting the existing system into two or more separate zones independently controlled by their own thermostats. This can be accomplished by installing a two-position valve in the hot water supply line. The valve is actuated by its own thermostat (Fig. 8).

RADIANT PANEL HEATING

Forced hot water can also be used in a radiant panel heating system (Fig. 9). In this type of heating system, no radiators or convectors are used. The hot water circulates through pipes concealed in the floor, ceiling, or walls, which function as heat emitting surfaces. Radiant baseboard systems are also used. The principles of radiant heating are described in greater detail in Chapter 11, Volume 2 (RADIANT HEATING).

Fig. 8. Zoning a two-pipe forced hot-water heating system. Thermostatically actuated zone valves are installed on the main water supply line.

Fig. 9. Radiant panel heating.

OTHER APPLICATIONS

It is possible to use the boiler of a hydronic heating system to supply heat for such purposes as snow melting, heating a swimming pool, providing domestic hot water for household use, and other applications. Separate circuits are created for each of these purposes, and controlled by their own thermostats. They are designed to tap into the main heating circuit from which they receive their supply of hot water.

Hot water for household use can be obtained by means of a heat exchanger or special coil (Fig. 10) inserted into the boiler. Cold water is run through this coil, heated as the boiler water gets hot, and is then distributed to the various hot water faucets.

In the snow melting system illustrated in Figs. 11 and 12, the circuit is filled with a water and antifreeze solution. The snow melts as soon as it touches the surface.

GRAVITY HOT-WATER HEATING SYSTEMS

All hot-water heating systems are either gravity or forced circulation types. The latter generally use pumps to force the hot water through the pipes (see FORCED HOT-WATER HEATING

HOT WATER TO TAPS

COLD WATER SUPPLY

HYDRONIC BOILER

Fig. 10. Using the heat exchanger principle to provide domestic hot water without a special tank.

SYSTEMS in this chapter). The circulation of hot water in gravity hot-water systems, on the other hand, results from the difference in weight (specific gravity) due to temperature differences. In other words, the motive force is due to the difference in density of the water at different temperatures—heavy when cold, light when hot. For this reason, gravity systems are also referred to as *thermal* or *natural* hot-water heating systems.

In order that water may transmit heat from the boiler (heating unit) to the radiators or other heat emitting devices, there must be a constant movement (circulation) of the water from the heater to the radiators and back again. As was mentioned above, this circulation in gravity systems is due to the difference in density (weight per unit volume) of water at different temperatures. For example, one cubic foot of water at say 68°F. weighs 62.31 lbs. and at 212°F. it will weigh 59.82 lbs. This difference in weight is 2.49 lbs. (62.31-59.82) which is available to cause circulation, as shown in Fig. 13.

The difference in weight is due to the expansion of water as its temperature is increased. This principle is illustrated in Fig. 14. Note that the top surface *abcd* of the cubic foot *A*, has expanded to *A'*, an increase in volume represented by the linear increase *aa'*.

In the two-pipe system illustrated in Fig. 15, the unbalanced weight between the water in the riser and downflow pipe forms a motive force which causes the water to circulate through the system as indicated by the arrows.

Fig. 11. Copper tubing imbedded in concrete driveway is connected to the hydronic system that heats the house. A small pump is used to circulate the liquid through the tubing.

The two-pipe gravity hot-water system illustrated in Fig. 16 is equipped with an expansion tank. In operation after the fire is started, the temperature of the water in the boiler rises and expands. This disturbs the equilibrium of the system, causing the colder and heavier water in the downflow pipe to flow downward,

165

pushing the warmer and lighter water in the riser upward, thus starting circulation.

A circulation or by-pass pipe is provided to form a continuous path for the flowing water in the event that all the radiators are shut off. In the absence of this provision, all the water in the upflow pipe would be pushed up into the expansion tank followed by steam.

The expansion tank provides two essential functions for this system. As the water is heated, it expands and the tank provides

Courtesy *National Better Heating-Cooling Council.*

Fig. 12. Hydronic snow melting system.

Fig. 13. Motive force in a gravity hot-water heating system.

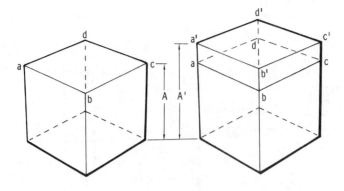

Fig. 14. The expansion of water with rise in temperature.

Fig. 15. Two-pipe gravity hot-water heating system.

167

EXPANSION TANK

BOILER

Fig. 16. Circuit two-pipe gravity hot-water heating system.

space for this increase in volume. Thus, the water as it is heated will expand from say elevation M to elevation S in the expansion tank (Fig. 15). The higher the elevation of the tank, the greater the pressure that may be brought on. the water in the boiler (due to the head). As a result, the system may be worked at higher temperature when desired.

Sometimes a special piping circuit can be built into a heating system (see TWO-PIPE DIRECT RETURN SYSTEM and TWO-PIPE REVERSE RETURN SYSTEM in this chapter). In the two-pipe gravity hot-water heating system illustrated in Fig. 16, a single main is taken from the boiler to a high point under the basement ceiling and then pitched along its run as much as possible to the return inlet of the boiler. The piping circuit formed by this main consists of a closed loop of extra large pipe in the basement having a pitch of not less than ½ inch per 10 feet of run. The main supplies all the risers to the radiators above. This arrangement is not as efficient as the two-pipe system illustrated in Fig. 15 and for this reason the circuit main should be of very liberal size to reduce friction to a minimum.

In the so-called one-pipe system shown in Fig. 17, special distribution tees are employed to deflect part of the water from the main into the radiators while letting the balance flow through the main to the next radiator.

Fig. 17. One- pipe gravity hot-water heating system.

This is called a one-pipe system because one pipe serves both inlet and outlet of each radiator. In this system, there must be an upflow side (L) and a downflow side (F). Thus, there are in reality two main pipes, but in function only *one* main pipe. The downflow may be considered as a continuation of the upflow instead of as a second pipe. The major objection to this system (as well as to the one illustrated in Fig. 16) is a lack of uniformity in heat distribution. This is primarily due to the fact that the radiators on the upflow side (L) are hotter than those on the downflow side (F).

Distribution tees (Fig. 18) are used in this one-pipe system. They are provided with a baffle tongue which deflects part of the

169

Fig. 18. Using distribution tees to connect a cast-iron radiator to the main in a one pipe hot-water heating system.

water from the main in and out of the radiator while by-passing the balance along the line as indicated by the arrows in the illustration.

Another piping system used to distribute the hot water is the overhead arrangement illustrated in Fig. 19. This is practically the same as the so-called one-pipe system (Fig. 17), except that it has a divided circuit at the high point connecting with the downflow mains. No air vents are necessary because the arrangement is such that all air works to the top and passes off into the expansion tank. One advantage to this system is a better and more uniform distribution of the heat. The hot water enters each downflow pipe at the same temperature, equalizing the heat given off by each pair of radiators.

Although the overhead piping arrangement is not suited to all classes of buildings, there are many, such as apartments, stores, office buildings, hotels, etc., where the general arrangement lends itself to this system.

Fig. 19. Overhead piping arrangement in a gravity hot-water heating system.

FORCED HOT-WATER (HYDRONIC) HEATING SYSTEMS

Water as a medium for transmitting heat to radiators and other heat emitting units gives up only sensible heat as distinguished from steam systems which heat principally by the latent heat of evaporation. The result is that the temperature of the heat emitting units of a steam system is relatively high as compared with hot water. In a hot water system, latent heat is *not* given off hence more radiating surface is needed to obtain equal heating effect.

The reasons for speeding up the flow of the water in hot-water heating systems is to increase the heating capacity of the heat emitting units; make the system more responsive to load conditions; and to allow the use of smaller pipes.

Numerous methods have been introduced to accelerate the circulation of hot water. The most commonly used methods have been based on the following techniques:

1. Introducing high pressures to gain greater temperature differences.

171

2. Super heating a part or all of the circulating water as it passes through the boiler and condensing the steam thus formed by mixing it with a portion of the cold circulating water of the return main.

3. Introducing steam or air into the main riser near the top of the system.

4. Forcing by pumps.

5. Combining pumps and local boosters.

The last two methods (i.e. pumps or a combination of pumps and local boosters) are generally the ones used in most modern forced hot-water heating systems. (See Fig. 20)

Fig. 20. Two-pipe forced hot-water heating system with separate supply and return mains.

THERMOSTAT

RADIATOR

RETURN WATER

HOT WATER SUPPLY

LOW LIMIT CONTROL

CIRCULATOR RELAY

SWITCH

HIGH LIMIT CONTROL

OIL BURNER

DOMESTIC HOT WATER

CIRCULATOR

WATER

HEAT EXCHANGER

CHANBER COMBUSTION

Fig. 21. Interior view of a typical boiler used in hot-water heating system.

HOT WATER BOILERS

The boilers used in hot water heating systems are made of cast iron or steel (Figs. 21 and 22). The cast iron boilers generally display a greater resistance to the corrosive effects of water than the steel ones do, although the degree of corrosion can be significantly reduced by chemically treating the water.

Most hot water heating boilers are designed to burn coal, gas (natural or propane) or oil. Some manufacturers will provide conversion devices for switching from one type of gas to the other (Fig. 23). Changing from coal to oil or gas is a frequent type of conversion provided for by conversion burners (see Chapter 16. BOILER AND FURNACE CONVERSIONS).

173

Never purchase an uncertified boiler. All certified boilers are approved by one of several organizations that have assumed this

Fig. 22. Examples of boilers used in hot-water heating systems.

COMBINATION
GAS VALVE

PRESS. REGULATOR
(NATURAL GAS ONLY)

BLANK PLATE
(LP GAS ONLY)

Courtesy *Honeywell Tradeline Controls*

Fig. 23. Gas conversion devices which are used for changing from one type of gas to another.

responsibility. For example, many boilers are certified by either the *Institute of Boiler and Radiator Manufacturers* (cast iron boilers) or the *Steel Boiler Institute* (steel boilers). They will be stamped "I-B-R" or "SBI" respectively. Fig. 24 illustrates some of the certifications you will encounter on boilers.

Steam boilers can also be used in hot-water heating systems, but a heat exchanger must be incorporated into the system to transfer the heat from the steam to the hot water that flows through the heat emitting units in the rooms (see HEAT EX-CHANGERS in this chapter).

HEAT EXCHANGERS

Steam can also be used as a source of heat in a hot-water heating system. If such is the case, then a heat exchanger (Fig. 25) must be used to convert the steam to hot water before it reaches the heat emitting units. The heat exchanger is commonly of the steel shell and copper tube design. The steam flows through the shell and the water through the tubes (Fig. 26). The hot water returns to the steam boiler where it is reheated and changed to steam again.

CONTROL COMPONENTS

The efficient and safe operation of a hot-water heating system requires the use of a variety of different types of controls, which

| CSA | FACTORY MUTUAL | A.G.A. | I.B.R. | ASME |

| BELGIUM | ENGLAND | CANADA | HOLLAND | GERMANY |

Fig. 24. A partial list of certification symbols for boilers.

175

can be roughly divided into either system actuating controls (e.g. the room thermostat, burner controls, circulating pump controls, etc.) or safety controls (e.g. high-limit controls, pressure-relief valve, pressure reducing valve, etc.)

The safety controls prevent damage to the system by shutting it down when pressure and temperature levels become excessive. The *high-limit control* or *aquastat* is an example of such a control. It is a device designed to operate in conjunction with the circulating pumps and is located on the hot water boiler. If the pressure or temperature of the hot water exceeds the design limits of the system, the system is shut down until conditions return to an acceptable level. Some aquastat relays provide multizone control when used with a separate circulator and a relay for each zone.

Another very important safety control for hot-water heating systems, is the *pressure relief valve*. These valves are designed to open when pressure in the boiler reaches a certain level and close when the pressure returns to a safe level again. Pressure relief valves *must* be installed in accordance with current A.S.M.E. or local codes.

Courtesy *National Better Heating-Cooling Council.*

Fig. 25. Heat exchanger used in connection with a steam boiler to provide hot water to baseboard heat emitting units.

TO STEAM BOILER

TO HOT-WATER BASEBOARD UNITS

FROM STEAM BOILER

Fig. 26. Operating principles of a heat exchanger.

These and other types of controls (e.g. main shutoff valves, pressure-reducing valves, drain cocks, etc.) are described in considerable detail in a number of different chapters in this book, including: Chapter 15 (BOILERS AND BOILER FITTINGS), Chapter 4 (THERMOSTATS AND HUMIDISTATS), and Chapter 9 of Volume 2 (VALVES AND VALVE INSTALLATION).

PIPES AND PIPE SIZING

Pipes and pipe sizing methods are described in considerable detail in Chapter 8, Volume 2 (PIPES, PIPE FITTINGS, AND PIPING DETAILS).

The selection of pipe used in hot-water heating systems depends primarily upon the following two factors: (1) the flow rate of the water and (2) the friction loss in the pipes. The *flow rate* of the water is measured in gallons per minute (gpm), and constant *friction loss* is expressed in thousandths of an inch for each foot of pipe length.

EXPANSION TANKS

Expansion tanks (Figs. 27 and 28) are installed in hot-water heating systems to provide for the expansion and contraction of the water as it changes in temperature. Water expands with the rise of temperature, and the excess volume of the water flows into the expansion tank.

Another feature of the expansion tank is that the boiling point of the water can be increased by elevating the tank. In other

Fig. 27. Open expansion tank. Because the tank is vented to the atmosphere, the surface of the water is always that of the atmosphere.

178

words, increasing the *head* (i.e. the difference in elevation between two points in a body of fluid) increases the pressure. As a result, the water can be heated to a higher temperature without generating steam which in turn causes the radiators or other heat emitting devices to give off more heat.

There are both open and closed types of expansion tanks used in hot-water heating systems. The *open* expansion tank is used on low pressure systems and the *closed* tank on high pressure systems; air in the tank above the water forms a cushion for increasing the pressure. As the temperature of the water rises, it expands and flows into the tank, thus compressing the air and increasing the pressure.

The relation between pressure and volume changes of the air should be understood. According to Boyle's Law, *at constant*

Fig. 28. Closed expansion tank. This type of expansion tank is sealed against free venting to the atmosphere.

temperature the pressure of a gas varies inversely as its volume. Thus, when the volume is reduced to half, the pressure is doubled. This is *not* gauge pressure, but absolute pressure (the pressure measured from true zero or point of no zero pressure).

In gravity hot-water heating systems, either closed or open piping arrangements can be used. In an *open* gravity system, the expansion tank is located at the *highest* point in the system (e.g. roof, attic, top floor, etc.) (Fig. 29). The expansion tank used in this piping arrangement is an open type with an overflow pipe located at the top. Provisions can be made to return the overflow water to the boiler or to discharge it through outside run-off drains.

In a *closed* gravity system, a closed, airtight expansion tank is located near the hot-water boiler. Higher pressures (and, conse-

Fig. 29. Circulating pump is located on the return water line next to the boiler.

quently, higher water temperatures) result as pressure builds up in the system. Pressure relief valves are installed on the main supply line to prevent the buildup of too much pressure.

Forced (hydronic) hot-water heating systems are closed systems with expansion tanks located near the boiler.

CIRCULATING PUMPS

In forced hot water heating systems, circulating (booster) pumps are used to force or circulate the hot water through the pipes. Fig. 29 illustrates a typical location of one of these pumps on a return line.

Circulating pumps for hot water heating systems are designed and built to handle a wide range of pumping capacities. They will vary in size from small booster-type pumps with a 5 gpm capacity to those capable of handling thousands of gallons per minute.

In order to select a suitable pump for a hot water heating system (i.e. one that will produce maximum flow without overloading the pump motor), it is necessary to correctly match the operating characteristics of the pump to the requirements of the heating system.

The correct location of a pump is crucial to its successful operation. Pump manufacturers will usually give detailed information in their literature on this point. Circulating pumps with mechanical seals are very common in hot water heating systems.

DRAINAGE

There should be a provision for adequate drainage facilities in a hot-water heating system during shut-down periods. If drainage is not provided for, there is always the danger of the unheated water freezing in the pipes during extreme cold weather. Chemical "antifreeze" solutions have also been developed for use in hot-water heating systems in an attempt to prevent the water from freezing during shut-down periods.

Pitching all pipes to a central point is an important consideration when designing the system, because it is essential to effective drainage.

HEAT EMITTING UNITS

The hot water in both gravity and forced hot-water heating systems is circulated through the pipes to the radiators or other heat emitting units from which the heat is transferred into the room.

Conventional radiators (Fig. 30) are often set on the floor or mounted on the wall. Many examples of these can still be found in older buildings. Some attempts have been made to conceal radiators by recessing them in the wall of the room or enclosing them (partially or entirely) in cabinets. Recessed radiators must have some sort of insulation between them and the wall. One-inch thick insulation board, a sheet of reflective insulation, or a combination of both is recommended for this purpose. If the radiator is partially or entirely covered with a cabinet, then openings should be provided at both the top and bottom of the cabinet for air circulation.

Convectors usually consist of finned tubes enclosed in a cabinet (Fig. 31) or baseboard unit (Fig. 32) with openings at the top and bottom. The hot water circulates through the tubes which radiate heat. Air enters the bottom of the cabinet or baseboard and exits through the openings in front.

Fig. 30. Conventional cast-iron radiator still found in many older hot-water heating systems.

FIN

Fig. 31. Finned tube connector in an upright cabinet.

Baseboard radiator units (Fig. 33) consists of long, narrow tubes directly behind a baseboard face. The hot water flows through the tubes, heats the baseboard face, and the heat is radiated from the surface into the room. Make sure that the metal cover of the baseboard unit is not too thin (an indication of a cheaply built unit). Thin covers are easily dented and water temperature changes in the heating element can cause a thin baseboard cover to be noisy.

183

Fig. 32. Baseboard radiator units.

CHANNEL FOR
HOT WATER OR STEAM

FINS

CHANNEL FOR
HOT WATER OR STEAM

TUBE

FINS

SLOTS

Fig. 33. Examples of baseboard radiator units.

AIR CONDITIONING

Air conditioning can be added to a structure heated by a hot-water heating system in a number of different ways. Central air conditioning can be added with the use of the so-called split system installation. The evaporator-blower unit is installed in the attic and the condensing unit (compressor and condenser) is located outside the house. The boiler and heat emitting units are shown in black.

Chapters 8, 9, 10 and 11, Volume 3 provide details for a variety of different types of air conditioning systems, many of which are suitable for adding to houses and buildings with existing hot-water heating systems.

MOISTURE CONTROL

If the humidity level in a structure is too low, the dry air will absorb moisture from other sources such as plaster, wood floors, wood furniture, fabrics (e.g. drapes and upholstery), and human bodies. As a result, the plaster tends to crack, the wood shrinks, the fabrics become dry and brittle, and the sensitive membranes in the nose and throat dry out faster than the human body can replace the moisture. These problems caused by dry air can be avoided by maintaining an adequate amount of moisture in the air. A hot-water heating system will generally maintain an adequate humidity level without the additional assistance of a separate humidifying unit. The humidifier illustrated in Fig. 34 is an example of the type that could be used in a hydronic heating system should one be required. It is designed with a hot water coil on the air inlet side of the unit. The heated boiler water circulates through this coil. A blower in the unit draws filtered room air over the coil and through a wetted rotating pad. The moist air is then circulated through the rooms by the blower. The humidifier is controlled by a room humidistat.

ELECTRICALLY HEATED SYSTEMS

Compact electric-fired boilers operate essentially the same way as a heating plant powered by coal, oil, or gas are also available for hot water heating systems. These boilers are characterized not only by their compactness, but also by their quiet operating characteristics.

RETURN AIR
FROM HOME

HUMIDIFIED
AIR OUT

HYDRONIC-HUMIDIFIER

HOT WATER
BOILER

Fig. 34. Typical installation of a humidifier in a hydronic heating system.

The extreme compactness of an electric-fired hot water heating system is illustrated in Fig. 35. The heat exchanger, expansion tank, and operating controls are all mounted on the wall in a relatively small space.

A central heating unit (e.g. an electric-fired boiler) can be eliminated in some systems by using thermostatically controlled electric heating components in the baseboard units. Fig. 36 illustrates a system of this type. Note that there is no boiler. The water is circulated through the system by a pump, and a uniform temperature can be maintained by the heating element in each baseboard.

Courtesy *National Better Heating-Cooling Council*

Fig. 35. Compact electric-fired boiler.

The heating system illustrated in Fig. 36 is a single-loop installation. It is also possible to divide the whole system into a series of sealed or closed units filled with a water and antifreeze solution operating on the gravity flow principle. Each unit could be controlled by a thermostat, or several units by a wall thermostat. The advantage of this type of installation is that the heating capacity can be easily increased if the house or building is enlarged.

ADVANTAGES AND DISADVANTAGES OF HOT-WATER HEATING SYSTEMS

There are several distinct advantages to using hot water heating as opposed to steam heating. For one thing, it is more flexible than low pressure (above atmospheric) steam systems because the temperature may be widely varied. Due to the low working temperature of the water, the heat from a hot water system is relatively mild and the room atmosphere is not robbed of any of its healthful qualities. Moreover, these systems function as reservoirs for storing heat because the radiators remain warm a considerable length of time after the fire in the boiler is extinguished.

Fig. 36. A single-loop installation consisting of electrically heated hydronic base-board units. Water is circulated through the entire loop by a pump.

A principal disadvantage of using hot water heating is the initial high installation cost as compared to steam and warm-air heating systems.

CHAPTER 8

Steam Heating Systems

Steam is a very effective heating medium. Until recently, this property of steam has resulted in its being the most commonly used method of heating residential, commercial, and industrial buildings. Over the past forty years or so, steam heating has been largely replaced in residences and small buildings by other types of heating systems which have often proven to be less expensive to install and operate, or which operate at similar or greater levels of efficiency in small structures.

The basic operating principles of steam heating are relatively simple. A boiler is used to heat water until it turns to steam. When the steam forms, it rises through the pipes in the heating system to the heat emitting units (radiators, convectors, etc.) located in the various rooms and spaces in the structure. The metal heat emitting units, being cooler, cause the steam to condense and return to the boiler in the form of water (condensation) for reheating.

CLASSIFYING STEAM HEATING SYSTEMS

There are a number of different methods of classifying steam heating systems, but the most commonly used ones include one or more of the following features:

1. Pressure or vacuum conditions,
2. Method of condensation flow to the boiler,
3. Piping arrangement,

4. Type of piping circuit,

5. Location of condensation returns.

Steam heating systems can be divided into low pressure and high pressure types depending on the operating pressure of the steam used in the system. A *low-pressure system* commonly operates at a pressure of 0 to 15 psig; whereas a *high-pressure system* will use operating pressures in excess of 15 psig.

Both *vapor* and *vacuum steam heating systems* operate at low pressures (i.e. 0 to 15 psig) and under vacuum conditions. The latter type system uses a vacuum pump to maintain the vacuum; the vapor system does not, relying instead on the condensation of the steam to form the vacuum.

The condensation from the heat emitting units is returned to the boiler either by gravity or some sort of mechanical means. When the former method is used, the system is referred to as a *gravity return system*. If mechanical means of returning the condensation are employed, the system is referred to as a *mechanical return system*. The three types of mechanical devices used to return the condensation in mechanical return systems are: (1) the vacuum return pump, (2) the condensation return pump, and (3) the alternating return trap. Each of these devices is described in the appropriate sections of this chapter.

Using the piping arrangement as a basis for classification, a steam heating system will be either a one-pipe or a two-pipe system. A *one-pipe system* is designed with a single main which carries the steam to the heat emitting units and the condensation back to the boiler. In other words, it functions as both a supply and a return main. In a *two-pipe system,* there is both a supply main and a return main.

The *piping circuit* may best be described as the "path" taken by the steam to the riser (or risers). In a *divided circuit* installation, two or more risers are provided for the steam supply. A *one-pipe circuit* installation, on the other hand, employs a single riser from the boiler to carry the steam supply to the heat emitting units. A *loop-circuit* installation is used when it is necessary to operate heat emitting units at locations *below* the water level of the boiler.

A steam heating system may also be classified according to the direction of steam flow in the risers (i.e. supply mains). An up-feed system is designed so that the risers are *below* the heat emitting units. In other words, the steam supply moves from the boiler up to the heat emitting units in the rooms and spaces within the structure. An up-feed system is also sometimes referred to as an *up-flow* or *under-feed system*. A *downfeed system* is one in which the steam supply flows down to the heat emitting units. In this type system, the supply main is located *above* the heat emitting units.

Sometimes the location of the condensation return is used as a basis for classifying a steam heating system. If the condensation return is located below the water level in the boiler, it is referred to as a *wet return*. A *dry return* is a condensation return located above the water level line.

GRAVITY STEAM HEATING SYSTEMS

A gravity steam heating system is characterized by the fact that the condensation is returned to the boiler from the heat emitting units by means of gravity rather than mechanical means. Both one- and two-pipe installations are used with the former being the oldest and most commonly used for a number of years. (Figs. 1 and 2). Gravity systems are generally limited to residences and small buildings where the heat emitting units can be located at least 24 inches above the water level line of the steam boiler.

Gravity steam heating systems are the cheapest and easiest to install because they are adaptable to most types of structures, but they do have a number of disadvantages inherent to the system. The principal disadvantages of gravity steam heating systems are:

1. The return lines in *one-pipe* gravity systems must be large enough to overcome the resistance offered by the steam flowing up from the boiler in the opposite direction.

2. There is the possibility of water hammer developing in one-pipe systems, because the steam and condensation must flow in opposite directions in *the same pipe*.

3. Air valves (which are required) sometimes malfunction by either spurting water or failing to open. If the latter situation is the case, excess heat will build up in the system.

191

4. Automatic control of the steam flow from the boiler results in room temperature fluctuations.

5. Comfortable room temperatures are possible by manually regulating the valves on individual heat emitting units, but this results in some inconvenience.

AIR VENT

RADIATOR VALVE

AIR VENT

STEAM MAIN

BOILER WATER LINE

WET DRIP MAIN

BOILER RETURN HEADER

Courtesy Dunham-Bush, Inc.

Fig. 1. One-pipe gravity steam heating system.

AIR
VENT

RADIATOR
VALVE

BOILER
WATER
LINE

AIR VENT

STEAM MAIN   RETURN MAIN

AIR
VENT

SWING
CHECK
VALVE

WET DRIP MAIN

RETURN HEADER

Fig. 2. A two-pipe gravity steam heating system.

6. In *two-pipe gravity systems,* the condensation return from each heat emitting unit must be separately connected to a wet return or water sealed. This is expensive.

The principal reason for the development of the two-pipe gravity system is to create a means of overcoming the resistance offered by the steam flow to the condensation returning to the boiler.

ONE-PIPE REVERSE FLOW SYSTEM

The *one-pipe reverse flow system* is the simplest and cheapest of this type to install. This system is easily distinguished by the absence of any wet or dry returns to the boiler. Supply mains from the boiler are inclined upwards and connect with the room heat emitting units, there being no other piping. This system is called "reverse flow" because the condensation flows back through the mains in a reverse direction or opposite to that of the steam flow.

The operation of a typical one-pipe reverse flow system is shown in Fig. 3. Steam from the boiler flows into the main or mains (inclined upward) through the risers to the heat emitting

Fig. 3. One pipe reverse flow heating system.

units at the bottom. The steam pushes the air out of the mains, risers, and heat emitting units, and escapes through air valves placed at the other end of the heat emitting units as shown. The condensation forming in the units flows back to the boiler through the same piping, but in the opposite or reverse direction to the steam flow.

For satisfactory operation, every precaution should be used to install the *correct* pipe sizes, especially for the mains. If the piping is too small, it will be necessary to carry excess pressure in the boiler to insure proper operation. Without the excess pressure, operation of the remote heat emitting units will be unsatisfactory. Satisfactory operation of these remote units can be achieved without excess boiler pressure if the pipes used in the system are the correct size.

The smaller the main the greater the speed of the steam flow and the greater the resistance to the flow of condensation. It is very difficult for the condensation to flow back to the boiler against the onrushing steam in a long main that is in an almost horizontal position. The main should be inclined as much as conditions in the basement will permit.

Leaky radiator valves can be a problem. When the valve does not close tightly, steam will work its way into the radiator and stop the condensation from coming out. The result is that the radiator soon fills with water and when turned on again there is difficulty getting the condensation out. This produces gurgling, hissing and the more violent effect known as "water hammer".

UPFEED ONE-PIPE SYSTEM

One-pipe steam heating systems can also be of the *upfeed* type. In a standard upfeed one-pipe steam heating system both the steam and condensation travel through the same pipes, and the heat emitting units are located *above* the supply mains (hence the name "upfeed" system).

An attempt to reduce the amount of condensation return in the pipes carrying the steam has resulted in a modification of the upfeed one-pipe system. In the modified system, the condensation is dripped at each radiator (and therefore from the main itself) into a wet return.

Fig. 4. Upfeed gravity one-pipe air-vent steam heating system.

The upfeed one-pipe system shown in Figs. 4 and 5 consists of a main or mains branching off from the boiler steam outlet and inclined downward, instead of upward in the previously described one-pipe system. This arrangement causes the condensation in the mains to flow in the same direction as the steam.

The mains connect with return pipes which carry all the condensation back to the boiler, and it is only in the risers that reverse flow of the condensation takes place, and they accordingly should be large enough to take care of this reverse flow without undue turbulence.

Steam from the boilers flow through the main or mains and accruing condensation with it, the condensation returns (hence the name) to the boiler at a much lower level as shown in Figs 4 and 5. After traversing the mains, steam flows through the risers and into the radiators. Condensation takes place in warming the radiators and drains through the risers and return pipes to boiler.

A distinction is made between a *wet* (or *sealed*) return and a *dry* return. A wet return is below the water level in the boiler whereas a dry return is above the water level. The advantage of a wet return is that it seals and prevents steam at a slightly higher

Fig. 5. One pipe under or up-feed nonrelief system.

pressure entering the return. Under most circumstances, a wet return is preferable to a dry return. The latter is necessary to clear doorways or other openings.

There are no disadvantages to a wet return when the system is properly installed and the valves maintained in tight condition. In most installations, these requirements are often lacking, resulting many non-inherent disadvantages. Among the installation problems that should be avoided are: (1) pipes too small for the job; (2) sharp turns; (3) air pockets; (4) not enough pitch to the mains; (5) not enough air valves; and (6) air valves that are too small or cheap.

UPFEED ONE-PIPE RELIEF SYSTEM

Fig. 6 shows an *upfeed one-pipe relief system* applied to an eight radiator installation. Connected to the main outlet A are two or more branch mains AB and AC, which supply the various risers. Steam is supplied to each riser, the condensation draining in the riser in reverse direction to the steam flow.

The condensation returns from the risers to the boiler by gravity through *drip pipes* which are virtually continuations of the return pipe so that the condensation will flow back into the boiler.

Steam (usually at from 1 to 5 lbs. gauge pressure) passes from the boiler to mains AB and AC. These branches being slightly inclined, any condensation will drain into the drip pipes. The steam passes through the risers to the radiators, where its heat is radiated in warming the rooms, thus causing condensation. The risers being of liberal size the condensation is carried by gravity in a direction reverse (that is, opposite) to the direction of the steam flow, and drains down the drip pipes, and returns to the boiler through the low level return pipe.

The condensation is forced back into the boiler by the pressure resulting from the greater head of the column of condensation in the drip pipes, as compared with the head of water in the boiler. Moreover, the water in the drip pipes being at a lower temperature than the water in the boiler is heavier which upsets the equilibrium of the two columns.

This system is characterized by a slight difference in pressure (pressure differential) in the different parts of the system due to frictional resistance offered by the pipe to the flow of the steam. The steam flow is variable in different parts of the system due to variable condensation and badly proportioned pipe sizes. This can be explained by Fig. 6. If the water level in the boiler is at D, then in operation with wet returns, the pressure difference will be balanced by the condensation standing at different levels in the different drip pipes, as at E and F, these levels being such that the difference in head and density will restore equilibrium. The effect of a wet return can be obtained with the dry return shown at the left, in Fig. 6, by attaching a siphon to the bottom of the drip pipe. Water from the drip falls into the loop formed by the siphon and after it is filled, overflows into the dry return.

The water will rise to different heights, G and H, in the legs of siphon to balance the difference in pressure at points P and P'. If the siphon were omitted and the drip pipe connected direct to the dry return, there would be a tendency for the condensation in the dry return to back up instead of draining into the boiler,

Fig. 6. One pipe under or up-feed relief system showing locations of dry and wet returns.

because the pressure in the drip pipe at P, is greater than the pressure in the dry return. In general, the pressure varies because there is a gradual reduction in pressure as the steam flows from the boiler to the remote parts of the system. This is caused by frictional resistance offered by the pipe and fittings (Fig. 7).

Fig. 7. Fractional resistance offered by the pipe and fittings.

The steam flows into the pipes only when condensation is taking place. The variation in pressure exists, on the other hand, only when the steam is flowing in the pipes. The effects of pressure variation can be explained with the aid of Fig. 8. With the plant in operation and condensation is taking place in the radiators and draining into the drip pipes—suppose the pressure in the boiler is 5 lbs. gauge; 4 lbs. in drip 1; and 3 lbs. in drip 2, then to balance these pressure differences the water will rise in drip 1 to (L) 2.3 ft. above the water level in the boiler, because there is a pressure difference of 1 lb. (5 − 4) and the weight of a column of water 2.3 ft. is 1 lb. for each square inch of cross section. Similarly, for drip 2, the pressure difference is 2 lbs. hence, the water will rise in this column to an elevation above the water level in the boiler equal to 2.3 × 2 or 4.6 ft. to balance the 2 lbs. pressure difference. Strictly speaking, these figures are correct only when the temperature of the water in the two drip pipes is the same as the temperature of the water in the boiler. For simplicity, the difference in weight or density of the "cold" water in the drips and hot water in the boiler were not considered. Under these circumstances, the heavy cold water in the drips would rise to lower elevations than those shown at (L) and (F) in Fig. 8.

Sometimes there are problems encountered on long lines with radiators located at the end of the line and near the level of the

Fig. 8. Effects of pressure variation in the different parts of a system.

water in the boiler. On long lines where there is considerable reduction of pressure, the water sometimes backs up into the radiator, as in Fig. 9, interfering with its operation.

Fig. 9. Effects of water backing up in radiator on end of long line.

Radiators located at elevations below the water level in the boiler may be operated by means of a steam loop (Fig. 10). In the steam loop, the condenser element may consist of a pipe radiator placed on the floor above the boiler. The liberal condensing surface thus provided will render the loop very active in removing the condensation and at the same time the heat radiated from the condenser is utilized in heating. The drop leg is provided with a drain cock (D) and the connection to the boiler with a check valve. To start the system, turn on the steam at the boiler and open (D) until the steam appears. The condensation of steam in the condenser (upper radiator) will cause a rapid circulation in the riser, carrying with it the condensation from the radiator, which, in passing over the goose neck, cannot return, but must gravitate through the upper radiator and drop leg past the check valve and into the boiler. The pipe at the bottom of the main riser which acts as a receiver for the condensation from the lower radiator should be one or two sizes larger than the pipe in the main riser.

Fig. 10. Steam loop method of operating radiator placed below water level in boiler.

DOWNFEED ONE-PIPE SYSTEM

The *downfeed one-pipe system* (also referred to as the *one-pipe over system* or simply the *downfeed system*) is characterized by having the heat emitting units located *below* the supply mains. The condensation drips through the risers, thereby keeping the supply main relatively free of the condensation.

The downfeed one-pipe system is well suited for buildings two to five stories high because if the heat emitting units were fed from below, the risers would have to be excessively large. Instead of a steam main encircling the basement, it is carried to the attic, forming a central riser for all the heat emitting units. The branches in the attic correspond to the mains in the systems already described, but do not carry any condensation from the heat emitting units. These branches connect with the drops or down flow pipes which feed the heat emitting units and drain the condensation.

As shown in Fig. 11, steam flows from the boiler up the central upflow pipe through the branches and down through the drops or downflow pipes. The branches being inclined, any water or condensation in the steam drains into the upflow riser. The steam passes to the heat emitting units through the connecting pipes. Condensation forming in the units drains through these connecting pipes into the downflow pipes. Air valves or vents are

Fig. 11. One pipe overhead feed system installed in a tall building.

provided to rid the system of air. It should be noted that there is no *reverse* flow of condensation in the drop pipes. As a result, pipe sizes can be made smaller because, with parallel flow, steam velocities can be higher than in the one-pipe system.

ONE-PIPE CIRCUIT SYSTEM

In the *one-pipe circuit system* the steam main is carried entirely around the basement, taken off from the boiler by an elbow at the high point as in Fig. 12.

Note that the main *must* incline all the way from the high point to the low point. To allow for this inclination requires twice as much rise between high and low points as with the divided circuit system; that is, where there are two mains taken off from a T connection.

The one-pipe system is adapted for a rectangular building of low or moderate size. The size of the main (since all the steam flows through it) must be larger in size than the divided circuit system. However, especially in large installations, a saving in piping may be made by installing a "tapered main." A tapered

Fig. 12. One pipe under or upfeed nonrelief circuit system.

main is one which is reduced in size along its length by connecting lengths of different size pipes with reducers. Eccentric reducers should be used to avoid water pockets which would interfere with the proper drainage of the condensation. The risers are connected by being tapped from the main at various points to serve the heat emitting units.

In a one-pipe system, the condensation drains into the main and flows in the same direction as the steam flow, being carried to the drip pipe, thence into the boiler. Since there is no return pipe as with the relief system, the circuit arrangement is less expensive to install.

Proportioning a tapered main is very important. The amount of condensation increases from the beginning to the end of the main and near the end is considerable, depending upon the number of heat emitting units. Due allowance should be made for this and too much tapering should be avoided.

ONE-PIPE DIVIDED CIRCUIT NONRELIEF SYSTEM

The *one-pipe divided circuit nonrelief system* differs from the one just described in that there are two mains at the high point taken off by a tee as in Fig. 13. These mains terminate in a U-shaped drip connection (LF) at the low point. Each should be vented with a quick vent as shown. Evidently each main takes care of only half the total condensation.

Fig. 13. One pipe under or upfeed nonrelief divided circuit system.

This steam heating system is suited to long buildings with a boiler located near the center. The end of each main is connected to a separate drip pipe connected with a common return, giving separate seals for each end.

For proper operation, these ends should not be at a lower elevation than 14 inches above the boiler water line. The individual seals make the two halves of the divided circuit independent, which is desirable for unequal loads. Thus, there may be considerable difference between the pressure at (L) and (F), each being what is necessary to balance the load.

ONE-PIPE CIRCUIT SYSTEM WITH LOOP

The *one-pipe circuit system with loop* is adapted to L-shaped buildings, a circuit being used for the main building and a loop (tapped from the circuit) servicing the wing (Fig. 14). The mains are installed for the proper drainage of the condensation by providing two high points: one at the beginning of the circuit and the other at the beginning of the loop, thus giving ample margin above the boiler water line for liberal pitch in both the circuit and the loop. This is clearly shown in the illustration.

The low point of the loop is higher than the low point of the circuit, because the pressure at the end of the loop is less than the

Fig. 14. One pipe under or upfeed nonrelief circuit system with loop for L-shaped building.

pressure at the end of the circuit. This requires a longer vertical drip pipe since the liquid column rises higher to balance the lower pressure.

TWO-PIPE STEAM HEATING SYSTEMS

In two-pipe systems, separate pipes are provide for the steam and the condensation, hence they may be of smaller size than in one pipe systems where a single pipe must take care of both steam and condensation. Various piping arrangements are used in two-pipe systems (e.g. circuit, divided circuit, loop, etc.) as will best meet the requirements of the building. Steam is supplied to the heat emitting units through risers and the condensation returned through downflow or drip pipes.

TWO-PIPE DIVIDED CIRCUIT SYSTEM

A *two-pipe divided circuit system* is shown in Fig. 15. In this type of two-pipe system, steam passes from the boiler at the high point to the mains along which risers bring the steam to the heat emitting units. From the opposite side of each unit is connected a drop or drip pipe. The illustration shows a wet return on the left side and a dry return on the right side.

In the wet return, the condensation is returned to the boiler by means of individual drips provided at each connection. There is

Fig. 15. Two pipe under or upfeed relief divided circuit system.

a different arrangement for the dry return. The drip pipe from each heat emitting unit terminates in a loop or siphon which is tapped to the dry return. In operation, the condensation gradually fills the siphons and flows over into the dry return passing into the boiler drip pipe.

A two-pipe system should also be provided with a check valve so placed as to allow water to pass into the boiler but prevent undue outflow. Under certain conditions, this prevents the water in the boiler from being driven out into the return system by the boiler pressure. The disadvantage of a check valve is that it sometimes gets stuck; a problem which can interfere with the operation of the system. An equalizing pipe with a Hartford connection loop may be used in place of a check valve to avoid this situation (see HARTFORD RETURN CONNECTION on page 253 in this chapter).

VAPOR STEAM HEATING SYSTEMS

A *vapor steam heating system* is one which commonly uses steam at approximately atmospheric pressure or slightly more, and which operates under a vacuum condition without the aid of a vacuum pump (see VACUUM STEAM HEATING SYSTEMS). (Figs. 16, 17 and 18).

The steam pressure at the boiler necessary to operate a vapor system is generally very low (often less than one pound), being no more than is required to overcome the frictional resistance of the piping system. Under most operating conditions, the pressure at the vent will be zero or atmospheric.

Vapor steam heating systems may consist of various combinations of closed or open, upfeed or downfeed, and one-pipe or two-pipe arrangements, depending on the requirements of the installation. Any of these combinations will have certain advantages and disadvantages.

OPEN (ATMOSPHERIC) VAPOR SYSTEMS

A vapor system with a return line open to the atmosphere *without* a check, trap, or other device to prevent the return of air is sometimes referred to as an *open* or *atmospheric system*.

RADIATOR
VALVE

RADIATOR
TRAP

EQUALIZER   CHECK VALVE

TRAP

AIR ELIMINATOR
AIR CHECK   STOP COCK
STEAM HEADER   STEAM MAIN
BLEEDER

BOILER WATER LINE   24" MINIMUM
                                                    RETURN MAIN

2"   6"

SWING
CHECK VALVE

RETURN   SCALE POCKET
HEADER

SCALE POCKET   WET DRIP MAIN
DRAW OFF

Courtesy *Dunham-Bush, Inc.*

Fig. 16. A vapor type steam heating system.

An open vapor system is frequently used when the steam is
delivered from its source under high pressure. When this is the
case, pressure reducing valves should be installed in the system to
reduce the pressure of the steam to a suitable operating level. An
open vapor system is also used when there is no need to return the

Courtesy *Dunham-Bush, Inc.*

Fig. 17. Piping of a vapor system to a boiler with only one supply tapping.

condensation to the boiler (i.e. when it is wasted within the system). A condensation-return pump should be used when the system design requires the return of the condensation to the boiler.

In an open vapor system, the pressure at the boiler is one to five ounces or enough to overcome the frictional resistance of the piping system. The pressure at vent is zero gauge or atmospheric. In operation, steam is maintained at about five ounces pressure in the boiler by the action of the automatic damper regulator. The amount of heat desired at the radiators is regulated by the degree of opening of the supply valve. Steam enters at the top of the radiator and pushes out the air through the outlet connection which is open to the atmosphere. The condensation returns to the boiler by gravity. This system has the advantage of heat adjustment at the radiator, but the divitalizing effect in the air is somewhat greater than in the vacuum systems because the steam entering the radiators is at a higher temperature than the steam of lower pressure in the vacuum system. It is, however, simple.

Fig. 18. Piping of a vapor system to a boiler with two or more supply tappings.

Fig. 19 illustrates a very simplified mechanically controlled vapor steam heating system. Though it does not present latest practice, it does present control principles very plainly.

211

Fig. 19. Open atmospheric vapor system.

The success of a vapor steam heating system depends upon the proper working of the automatic damper regulator in keeping the boiler pressure within proper limits. To accomplish pressure regulation, the dampers are controlled by a float working in a float chamber in communication with the water space in the boiler, as shown in Fig. 19.

When the pressure in the boiler is the same as that of the atmosphere (i.e. zero gauge pressure), the water level in the float chamber (Fig. 19) is the same as that in the boiler and the index hand points to zero.

As steam generates, the steam pressure increases and the water level in the boiler is forced downward. The latter action causes the level in the float chamber to rise until the pressure due to the difference (AB) (Fig. 19) of water levels balances that in the boiler.

The float in rising, connected as it is by pulleys and chains to the dampers, closes the ash pit damper, thus checking the draft and preventing a further increase of steam pressure.

In this system, the steam feed is connected to the top of the radiators and the air and condensation is taken from the bottom, because steam is lighter than either air or condensation. Accordingly, when steam is admitted it floats on top of the air, thus, driving the air out through the lower connection.

The chief feature of a vapor system is that the amount of heat given off by each radiator may be regulated by the steam valve. Thus, in Fig. 19, the valve of radiator (C) is opened just a little, which will admit only just enough steam to heat a larger portion of the radiator; with valve wide open on (E), the entire radiator is heated.

The kind of radiator used is the downflow type in which steam enters at one end at the top and the air and condensation passes out at the other end at the bottom.

As steam enters a cold radiator it forces the cool air in the radiator out through the trap into the return piping. The operation of a typical downflow radiator is shown in Figs. 20, 21 and 22. Fig. 20 shows the steam entering, air passing out through thermostatic retainer valve. Fig. 21 shows more steam entering and condensation and the balance of the air passing out through the trap, the action progressing until (as in Fig. 22) the radiator is full of steam.

As the radiator warms up, the steam gives off heat and in so doing condensation takes place. The condensation being heavier than steam falls to the bottom of the radiator and flows to the trap through which it passes into the return piping. After the air is forced out, the steam fills the radiator and follows the condensa-

Fig. 20. Steam entering top of radiator and pushing air out the bottom.

Fig. 21. Condensation and air leaving the radiator.

Fig. 22. Radiator filled with steam.

tion to the trap. The trap closes when the steam enters it because the steam is hotter than the water. This excess heat expands the valve control element, closing and holding the valve against its seat with a positive pressure, thus preventing the steam from flowing into the return piping.

The trap closes once the radiator is completely filled with steam, and heat is given off as the steam condenses. The condensation thus formed, which is cooler than the steam, flows in a steady stream to the trap which it slightly chills, causing it to open and allowing the condensation to pass out into the return piping (Figs. 23 and 24).

Fig. 23. Progressive action of thermostatic check or trap from beginning of air entrance to closing of valve by expansion of actuating element.

When properly working, the trap adjusts itself to a position corresponding to the temperature of the condensation just as a thermometer does to the room temperature, and permits a continuous flow of condensation from the heat emitting units (Fig. 25).

CLOSED VAPOR SYSTEMS

A vapor system with a return line closed to the atmosphere is sometimes referred to as a *closed vapor* system. The condensation returns by gravity flow to a receiving device (an alternating receiver or boiler-return trap), where it is discharged into the boiler. The condensation from the alternating receiver is discharged against the boiler pressure.

Because air cannot enter a closed vapor system, a moderate vacuum is created by the condensing steam. As a result, steam is produced at lower temperatures and the system will continue to provide heat after the boiler fire has died down.

STEAM IN TRAP

VALVE CLOSED

CONDENSATE COLLECTING

FLOWING INTO TRAP

CONDENSATE FLOODING TRAP

VALVE OPEN

TRAP DISCHARGING

Fig. 24. Progressive action of thermo-static check or trap closed position to open position due to contraction of actuation element when chilled by relatively cool condensation.

Fig. 26 shows the arrangement of a typical upfeed two-pipe vapor system with an automatic return trap. The heat emitting units discharge their condensation through thermostatic traps to the dry return pipe. These systems operate at a few ounces pressure and above, but those with mechanical condensation return devices may operate at pressures upward of 10 lbs. per square inch. The simplest method of venting the system consists of a ¾ inch pipe with a check valve opening outward. Most systems employ various forms of vent valves, which allow air to pass and prevent its return. A dry return is provided so that the air will easily go out the vent pipe.

VACUUM STEAM HEATING SYSTEMS

A *vacuum steam heating system* is one which operates with steam at pressures less than that of the atmosphere. The object of

CONDENSATE

WIDE OPEN

PARTIALLY OPEN

CONTINUOUS FLOW OF CONDENSATE

Fig. 25. Detail of trap showing the valve in intermediate position for ideal continuation flow of condensation.

TRAP

TRAP

GIVE GOOD PITCH

TRAP

TRAP

DRY RETURN

SUPPLY MAIN

AIR ELIMINATING AND PRESSURE EQUALIZING DEVICE

BOILER WATER LINE

RISER DRIPPED

WET RETURN

HARTFORD RETURN CONNECTION

Fig. 26. Two pipe upfeed vapor system with automatic return trap.

217

such systems is to take advantage of low working temperatures of the steam at these low pressures, giving a mild form of heat such as is obtained with hot water heating systems. Vacuum systems such as these which operate at all times at pressures less than atmospheric should not be confused with the combined atmospheric and vacuum systems described in the next section. There are distinct design differences between the two.

A distinction should be made between a vacuum system and a subatmospheric system. The latter differs from an ordinary vacuum system in that they maintain a controlled partial vacuum on both the supply and return sides of the system, instead of only on the return side. In the vacuum system, steam pressure above that of the atmosphere exists in the supply mains and heat emitting units practically at all times. The subatmospheric system is characterized by atmospheric pressure or higher existing in the steam supply piping and heat emitting units only during severe weather (Figs. 27 and 28).

There are a number of different methods of classifying vacuum systems (e.g. one-pipe or two-pipe, vacuum pressure or subatmospheric, etc.). For the purposes of this chapter they will be classified according to the type of vacuum: natural vacuum systems and mechanical vacuum systems.

NATURAL VACUUM SYSTEMS

Any standard one- or two-pipe steam system may be converted into a *natural* vacuum system by replacing the ordinary air valve with a mercury seal, or connecting thermostatic valves to the radiator return outlet on radiators and providing a damper regulator on coal burning boilers adapted to vacuum working. The mercury seal system is shown in Fig. 29.

A mercury seal is virtually a barometer consisting (as shown in Fig. 30) of a tube (A) which dips just below the surface of the mercury in a cup (B). When the steam is raised in the boiler to pressures above atmospheric, it drives all air out of the system, the air leaving by bubbling through the mercury in cup (B).

If the fire is allowed to go out, the steam will condense and produce a vacuum, provided all pipe fitting has been carefully done and the valve stuffing boxes are tightly packed.

218

Fig. 27. A typical variable vacuum steam heating system installation.

219

TRAP

DRIP TRAP

BOILER WATER LEVEL

RETURN

ROOM THERMOSTATS

PUMP CONTROL

VACUUM HEATING PUMP

AIR DISCHARGE

VALVE

STEAM CONTROL VALVE

STEAM MAIN

OUTDOOR THERMOSTAT

Fig. 28. A subatmospheric steam heating system.

Fig. 29. Natural vacuum mercury seal system.

Fig. 30. Operation of mercury seal.

In Fig. 29 the loop at (C) prevents water being carried over into the seal pipe when purging the system of air. If air should again enter the system, it can be expelled by raising the steam pressure above atmospheric. In very cold weather, the system can be operated at pressures above atmospheric by closing valve (D). When fires are banked for the night, valve (D) may be opened and the system worked as a vacuum system. The flexibility of vacuum systems is in sharp contrast with low pressure systems where steam disappears from the radiators as the temperature

221

drops below 212°F. According to weather demands, the radiators may be kept at any temperature from say to 150°F. to 220°F.

Another method for maintaining a natural vacuum is by using thermostatic valves instead of a mercury seal. A thermostatic valve is one having an expansion element which operates to close the valve when heated by hot steam and to open the valve when chilled by the relatively cold condensation.

The two kinds of thermostatic valves used are the single unit (or retainer) valves and two unit or combined retainer and air check valves sometimes called master thermostatic valves.

Fig. 31 shows the details of a master thermostatic valve, consisting of a thermostatic unit and an air check. The thermostatic unit has an expanding element, the air check consisting of a group seat poppet check valve that is practically air tight and will retain the vacuum within the system for a considerable length of time.

The air check operates when excess pressure is generated in the boiler to purge the system of air, the check at other times remaining closed.

Fig. 31. A natural vacuum system with retainer and master thermostatic valve showing sectional views of the valves.

222

The thermostatic valve remains open while the system is being purged of air and condensation but closes when steam enters the valve chamber—it retains vacuum in the air line.

Fig. 32 shows the operation of the natural vacuum system with retainer and master thermostatic valves. Individual thermostatic retainer valves (A, B and C) are placed in the outlet of each radiator which pass air or water, but close to steam. At the end of the air line is a master thermostatic valve (D), which operates when the system is purged of air by excess pressure.

The drip should be proportioned to prevent water entering the air line in case of high vacuum in such a manner that the vertical distance (M), between water level in the boiler and the lowest point of air line is not less than two feet for each inch of vacuum to be carried in the system.

Fig. 32. Natural vacuum system with retainer and master thermostatic valves.

The successful operation of natural vacuum systems depends largely on efficient damper regulators on coal burning boilers (i.e. efficient draft control), for unless the fire is held in proper check, the pressure will rise and break the vacuum. This can waste fuel, for there may be sufficient heat in the boiler to supply steam to the system with a five, or even ten inch vacuum and hold that heat in the system for hours.

Automatic damper regulators are designed to act by pressure, temperature or a combination of these two. Fig. 33 shows a regulator which acts on the pressure principle. It consists of a diaphragm connected at (B) to a lever fulcrumed at (A) and having a weight (W) free to slide along a slot between the stops.

In starting, the weight is placed on the left side of the lever as shown (Fig. 33). This tilts the lever (position LF) and opens the damper. The weight is adjusted by the stop so that sufficient pressure is produced to clear the system of air before the regulator trips to position (L'F') (shown in dotted lines) and closes the damper.

The regulator is gradually closing as the pressure comes on. When entirely closed, the weight slides to the right and remains in this position until the vacuum in the system becomes strong enough to gradually open the damper—just enough to maintain a vacuum.

Fig. 33. Damper regulator for natural vacuum system operating on the pressure principle.

In the morning, the regulator may be set to open position from the floor above by the pull chain (M). This generates pressure and purges the system of any air which might have accumulated, and then the regulator weight automatically goes to the vacuum side of the regulator and maintains the vacuum heat until more fuel is required or further regulation necessary. Temperature controls or damper regulators which depend on temperature changes for their operation may also be used (Fig. 34). Since the temperature of steam increases with the pressure, evidently the expansion and contraction of a rod exposed to the steam can be made to operate the damper.

Fig. 34 shows the construction of a typical thermostatic regulator. The "expansion element" or rod is fastened at (A) in a closed cylindrical chamber through which steam from the boiler passes to the main. The end (B) is free to move, passing out of the chamber through a stuffing box. The motion of the rod is considerably magnified by the bell crank lever, which is connected to the damper by a chain attached at (C).

In operation, as the pressure of the steam rises, so does its temperature and the rod (which is made of a metal having a higher coefficient of expansion than that of the cylindrical cham-

Fig. 34. Damper regulator for natural vacuum system operating on the temperature principle.

ber) end (B), will move to the right, thus causing end (C) of the lever to descend, closing the damper.

When the pressure falls, the rod contracts and the spring which keeps the bell end in contact with the rod causes end (C) of the lever to rise and open the damper.

The level will assume some intermediate position in actual operation, thus holding the steam at some pre-determined pressure. which may be varied by means of the screw adjustment(D).

The major objection to regulation by temperature is that there is no provision for securing excess pressure to purge the system of air in starting. This must be done by hand control of the damper. This objection can be overcome by the method of combined pressure and temperature regulation, which employs pressure for starting and temperature for running. In starting, the thermostatic portion of the regulator is closed off from the system during which pressure is generated sufficient (about one pound) to purge the air from the system. After purging, the regulator automatically opens a valve to the thermostatic position, which then maintains the temperature desired, its range embracing both vacuum and low pressure operation.

MECHANICAL VACUUM SYSTEM

A *mechanical vacuum system* is one in which an ejector or pump is used to maintain the vacuum. The hook up of the ejector system is shown in Fig. 35. The ejector, which may be operated either by steam or water is started before steam is turned on in the system. Thus, after the air is removed, steam will quickly fill the heat emitting units full of steam since the air is automatically removed as fast as it accumulates.

The system commonly used in exhaust heating employs an alleged or so called vacuum pump which ejects the air and condensation from the system. In operation, this device pumps out most of the air (or other gas) from the condenser maintaining a partial vacuum. The pump cannot obtain a perfect vacuum because each stroke of the pump piston or plunger removes only a certain fraction of the air, depending on the percentage of clearance in the pump cylinder, resistance of valves, etc., hence. theoretically an infinite number of strokes would be necessary to obtain a perfect vacuum (not considering resistance of the valves, clearance, etc.).

Fig. 35. Mechanical vacuum ejector system.

Condensation of steam creates the vacuum, and the pump which removes the air maintains the vacuum. A "wet" pump (that is, one which removes both air and condensation) is the type generally used. A "dry" pump removes only the air.

The essential features of a mechanical vacuum pump system are shown in Fig. 36. This system is of the fractional valve distribution type. In operation, air, being heavier than steam, passes off through thermostatic retainer valves to the pump. When the steam reaches the retainer valves they close automatically to prevent the steam passing into the dry return line to the pump and thereby breaking the vacuum. The condensation is pumped from the receiver back into the boiler by a feed pump, passing on its way through a feed water heater, where it is heated by the exhaust steam from the air and feed pumps.

Fig. 36. Mechanical vacuum air pump system as applied to fractional valve distribution.

COMBINED ATMOSPHERIC PRESSURE AND VACUUM SYSTEMS

A combined atmospheric and vacuum system works at pressures in the boiler at from one to five ounce gauge pressure (that is, above atmospheric). This pressure is needed on coal burning installations to operate the damper regulator.

The desired vacuum in the heat emitting units is obtained by "throttling" the supply steam with the unit feed valves. The working principle of this system is shown in the elementary sketch found in Fig. 37.

In operation, when steam is raised in the boiler it passes through the steam main risers and supply valves to the heat emitting units.

The proper working of this system is obtained by an automatic device or trap which closes against the pressures of either steam or condensation and allows air to pass out, but not the steam. The trap (Fig. 37) consists of three elements: (1) a diaphragm valve (point L); (2) a float valve (point A); and (3) a thermostatic

228

valve (point F). Connection (R), in Fig. 37, connects the steam outlet of the boiler to the diaphragm. When there is no pressure in the boiler, diaphragm valve (L) is held closed by the spring.

Fig. 37. Combined atmospheric pressure and vacuum system.

When the fire in the boiler is started and the air in the boiler expands, the diaphragm is inflated and moves the valve spring to the right (against the action of the spring) which opens valve (L). This makes a direct opening through float valve (A) and thermostatic valve (F), thus opening the system to the atmosphere. Valve (L) remains open as long as there is a fraction of an ounce pressure on the boiler. When steam forms and passes through the system, it drives all the air out of the system through the three open valves (L, A, and F).

The heat of the steam causes the expansion element of valve (F) to expand and close the valve, thus the system is filled only with steam.

The vacuum is obtained on the principle that the steam admitted to the radiators condenses while giving off heat through the radiator walls. This causes a tremendous reduction in volume of the steam remaining in the radiators, resulting in a pressure reduction in radiators, less than atmospheric—that is, a vacuum is formed. The steam condenses because of a reduction in temperature below that corresponding to the pressure of the steam.

229

When the radiator gives off heat in heating the room, the temperature of the steam in the radiator is lowered. This reduction in temperature causes some of the steam to condense in a sufficient amount to restore equilibrium between temperature and pressure of the steam.

The pressure falls because the temperature falls. If a closed flask containing steam and water is allowed to stand for a length of time, the atmosphere being at a lower temperature than that inside, the flask will abstract heat from the steam and water, but the heat will leave the steam quicker than the water. The result is a continuous condensation of the steam and reevaporation of the water, during which process the temperature of the whole mass and the boiling point is gradually lowered until the temperature inside of the flask is the same as that outside. This process is accomplished by a gradual decrease in pressure. Figs. 38, 39 and 40 illustrate the effect of pressure on the boiling point.

A one-pipe system may be converted into a combined system by replacing the air vent on the heat emitting units and making the piping absolutely tight. In this conversion, a "compound" gauge recording both pounds of steam and inches of vacuum is required (Fig. 41).

Fig. 38. Equilibrium between the steam and the water. Equalized temperature.

COOLING WATER

REDUCED PRESSURE

RAPID BOILING TO
REESTABLISH EQUILIBRIUM

Fig. 39. Reducing the pressure by applying cooling water to the closed vessel. Reduction of pressure causes the water to boil.

PRESSURE CORRESPONDES
TO TEMPERATURE

P

T

Fig. 40. Equilibrium reestablished between pressure and temperature.

EXHAUST-STEAM HEATING

The term *exhaust-steam heating* relates to the source of the steam rather than to its distribution. In fact, after the exhaust steam enters the heating system, its action is no different from live

231

steam taken from a heating boiler, it being adapted to both low pressure and vacuum systems.

Fig. 41. Compound gauge used on combined atmospheric-vacuum systems.

The chief difference between exhaust systems and those already described are the provisions for delivering steam from the engine to the heating system free from oil and at a constant pressure, and for returning the condensation to the high pressure boiler (Fig. 42).

Fig. 43 shows the essential features of an exhaust-steam heating system having fractional control vacuum distribution.

The necessary devices between the engine and the inlet to the heating system are: (1) the oil separator; (2) the trap; (3) the back pressure valve; and (4) the pressure regulating valve. In addition, for mechanically producing the vacuum and returning the condensation to the high pressure boiler—an air pump, a receiver with vent, and a feed pump are required. A feed water heater should also be provided both for economy and to permit returning the condensation and make up feed water to boiler at the proper temperature.

In operation, exhaust steam from the engine first passes through the heater, then through the oil separator, which frees it from the lubrication oil, the latter passing off into the oil trap. The steam now enters the heating system at (A), its pressure being

Fig. 42. Typical exhaust-steam heating system with vacuum distribution showing application to a power plant containing an open feed water heater.

prevented from rising above a predetermined limit by the back pressure valve (regulated by weight (B)) and maintained at a predetermined constant pressure by the pressure regulating valve (adjusted by weight (C)).

Fig. 43. Exhaust steam-heating system with fractional valve control and automatic make up.

The pressure regulating valve is, in fact, an automatic steam "make up" valve which admits live steam to the heating system when the exhaust is not adequate to supply the demand, thus, "making up" for this deficiency and maintaining the pressure constant.

The condensation and air removed from the system at (D) by a wet pump (as distinguished from a dry pump which removes only the air). The condensation and air are discharged into a receiver where the air passes off through a vent, the condensation being pumped by a feed pump back into the boiler after passing through a feed water heater

There is a continued loss of water through various leaks, the feed pump inlet (alleged suction) is connected at (E) (Fig. 43), with the supply from the street main or other source the amount entering the system being controlled by the "make up" valve.

The construction of a typical regulating valve and its connections are shown in Fig. 44. The valve is controlled by means of a governing pipe (A) (Fig. 44) connecting the diaphragm chamber to the accumulator, the latter being connected to the heating main at the point from which the pressure regulator is to be governed.

The accumulator is always half full of water and its elevation must be such that the water line in the accumulator is level with the diaphragm, so that there will not be an unbalanced column of water to exert pressure on the diaphragm.

The water is provided to protect the diaphragm from the steam, the pressure of the water being transmitted from the surface of the water in the accumulator to the diaphragm.

The working principle of the regulating valve and its connected devices is relatively simple. In operation, when the exhaust side (F) is at the predetermined pressure this brings sufficient force against the under or water side of the diaphragm to overcome the downward thrust due to the adjustable weighted level and close the valve.

If the engine slows down or there is heavy demand for heat so that the exhaust steam is not adequate to supply the demand, the pressure in the exhaust side (F) (Fig. 44) will fall and the downward thrust of the adjustment weight will overcome the opposing pressure of the water on the diaphragm and open the valve, admitting live steam from the boiler side (L) into the exhaust side (F) in sufficient quantity to restore the pressure.

The inertia of the water in the accumulator acts as a damper to prevent over sensitiveness of the valve or hunting (i.e. the behavior of any mechanism which runs unsteadily oscillating either too far or too little in an attempt to adjust itself to momentary fluctuations of pressure or other conditions which cause this action).

The spring under the diaphragm acts to balance the downward thrust of the lever and hold the valve in closed position when the pressure is the same on both sides of the diaphragm.

Fig. 44. Pressure regulating valve and accumulator for maintaining a constant pressure in the steam heating main of an exhaust-steam heating system.

The back pressure valve is used to prevent exhaust pressure exceeding a predetermined limit. This is virtually a lever safety valve designed to work at very low pressure. Some back pressure valves are so light that they will open or close with a variation of only two ounces. The position of the weight on the lever which is fulcrummed at (F) (Fig. 45) determines the exhaust pressure at which the valve will open.

Fig. 45. Details of exhaust-steam heating system back valve.

PROPRIETARY SYSTEMS

Over the years, a number of automatic heating systems have been designed and patented by manufacturers of steam heating equipment. Because these heating systems are protected by patent, they are referred to by the name of the manufacturer. Three of the most popular of these proprietary steam heating systems have been:

1. The *Trane* Vapor System,
2. The *Dunham* Differential System,
3. *Webster* Moderator Systems.

The *Trane Vapor System* illustrated in Fig. 46 is a combined atmospheric and natural vacuum system installed for residential heating. In this system, the air and water return runs in the same direction and is practically the same length as the supply main. When properly worked out, this feature gives the same effect as

237

Fig. 46. *Trane* vapor heating system. A combined low pressure and natural vacuum steam heating system.

though each convector radiator were placed at an equal distance from the boiler. This tends to synchronize the heating effect of all the heat emitting units in the system.

When the fire is started in the boiler of the *Trane* system, the water becomes heated and steam is formed which flows through the supply main and enters the radiators, displacing the air which is heavier than the steam. The air and condensation drain from the radiators through radiator traps and return piping to a point near the boiler where the air is exhausted through the quick vents (4) and float vents (1) at the end of the steam and return mains. The water is returned to the boiler by the direct return trap (2).

As the rooms become warm, less steam is condensed and the pressure in the boiler begins to rise. The rising pressure causes the damper regulator (3) to operate, closing and opening the drafts and maintaining just the amount of pressure necessary for proper heating.

When the fire in the boiler becomes lower, condensation forms and air is prevented from entering the system and a vacuum is created. As a result, the operation changes from atmospheric to natural vacuum; hence the name "combined atmospheric and natural vacuum system." The reduced pressure of the vacuum allows the water to boil and furnish steam to the radiators at a lower temperature; a decided economy when the fire is low.

The boiler connections for *Trane* system is shown in Fig. 47. It is very important that the steam connection to the return trap be taken from the steam space of the boiler and *not* from the supply piping of the header. The top of the direct return trap (2) must be placed at least 22 inches above the water line of the boiler. In *no* case should the top of the trap be less than 4 inches below the air and water return main. The dimensions from the water line to the end of the mains and the top of the return trap are the minimum allowable. Greater clearance above the water line should be employed where possible. Where the ends of the steam and return mains occur in remote parts of the building away from the boiler, use the connections shown in Fig. 48. The ends of such mains must always be vented before dropping to the wet return. The vent pipe must be installed from the piping below the trap up to the return main when a wet return is used. If a dry return is used, the vent pipe may be omitted.

Fig. 47. Recommended boiler connections for *Trane* vapor system.

If desired, a *Hartford* connection may be used between the inlet to the boiler and the return connections from the steam and return mains (see HARTFORD RETURN CONNECTION in this chapter).

The recommended size for vertical piping immediately below the float vents (1) and the quick vents (4) is 1¼ inch pipe at least 6 inches long. This affords a separate chamber for air and water.

The connections for a dripping steam main that rises to a higher level are illustrated in Fig. 49. The trap and strainer may be omitted provided the lower main is at least 18 inches above the water line and the return is connected directly into the return header of the boiler.

Fig. 48. Pipe connections for use where steam and return mains drop to wet return in remote part of building.

STEAM MAIN

FLOAT TRAP

STRAINER

DIRT POCKET

TO RETURN MAIN

Fig. 49. Connections for dripping steam main where it rises to a higher level.

When the motor operated steam valves are used on the mains, the boiler connections should be arranged as shown in Fig. 50. An equalizer line is required to equalize between the steam main and the return main when a vacuum forms in the former after the motor operated valve closes. A swing check valve prevents the flow of steam into the return main (Fig. 51). Fig. 52 shows the arrangement of boiler connections when air is eliminated from the return main through the No. 9 vent trap. Sometimes reversed circulation resulting from rapid condensation of steam will tend to create greater vacuum in the steam main than in the return main. This can be prevented by using the boiler connections illustrated in Fig. 49.

A typical convector radiator used in the *Trane* system is shown in Fig. 53. It is equipped with an angle valve and an angle trap having horizontal laterals below the floor. Connections for *Trane* convector radiators are illustrated in Fig. 54. The upper unit shows a vertical trap and a straightway valve concealed within the convector radiator. The lower unit is equipped with an angle valve and trap with a downfeed riser dripped through the angle trap.

The *Dunham Differential Vacuum Heating System* is a simple two-pipe power vacuum (air pump) return system working normally at pressures below atmospheric (subatmospheric) in the system and employing orifice supply valves on the radiators. The principal advantage of the *Dunham* system is that it continuously

Fig. 50. Pipe connections when motor operated steam valve is used on mains.

Fig. 51. Equalizer connection.

distributes heat at a variable rate equal to heat losses from the structure. This is accomplished by reducing the capacity of the system by reducing the volume temperature of the steam.

243

Fig. 52. Connections when air is eliminated from the return main through the vent trap No. 9.

Fig. 53. *Trane* convector radiator with angle valve and angle trap having horizontal laterals below floor.

STRAIGHTWAY VALVE

VERTICAL TRAP

RETURN MAIN

STEAM MAIN

DOWNFEED RISER

ANGLE VALVE

ANGLE TRAP

ANGLE TRAP

STRAINER

DRY RETURN MAIN

Fig. 54. Typical connections for *Trane* convector radiators.

Standard radiators, pipe fittings, and boiler connections are used in the *Dunham* system. The principal features of the *Dunham* system that distinguish it from other heating systems are:

1. The traps and valves,
2. The condensation pump,
3. The controller.

Each of these components is differentially controlled; that is, each is actuated by a pressure differential (the difference in pressure in the supply piping and the pressure return piping). The

245

Dunham system distributes the steam proportionately to all radiators, the pressure range being from 2 lbs. gauge to pressures considerably lower than atmospheric. Because of the relatively constant differential in pressures between the steam and return lines, the radiators are filled with steam.

A condensation pump is connected to the differential controller and to the supply and return piping. The controller starts the pump when the pressure difference between the supply and return piping tends to fall or disappear, and stops it when the pressure differential is restored. The condensation pump is a wet air pump which handles both air and condensation.

The thermostatic radiator traps are actuated by temperature changes within the radiator. Drip traps are installed at drip points to which large volumes of condensation flow. These are combined thermostatic and float traps. A control valve regulates the admission of a *continuous* flow of steam into the heating main.

Webster Moderator Systems are special control systems used primarily with two-pipe vapor, vacuum, and vented return systems, or in modified form with some one-pipe steam heating systems.

These *Webster* moderator systems are controlled by the weather with an automatic outdoor thermostat. One or more hand operated *variators* can be used to adjust the outdoor thermostat. The two types of moderator control systems produced by *Webster* are based on the following operating principles:

1. Continuous steam flow.
2. Pulsating steam flow.

The continuous steam flow arrangement is called the *Webster* Electronic Moderator System (Fig. 55), and is suitable for medium to large buildings requiring one or more control valves. This system consists of the following basic components:

1. Outdoor thermostat,
2. Main steam control valve,
3. Variator.

The steam supply may be taken from a high or low pressure boiler or any other source. The initial pressure should not be over 15 lbs., using a reducing valve if necessary. The return piping may be either open or closed.

OUTDOOR THERMOSTAT

METERING ORIFICES
IN SUPPLY VALVES

TRAP

ELECTRIC CABLE

OPERATING
SCHEDULE

SWITCH FOR FULL HEAT,
AUTOMATIC, AND NO HEAT
POSITIONS OF MAIN
STEAM CONTROL VALVE

STEAM SUPPLY

MAIN STEAM
CONTROL VALVE

VARIATOR

ELECTRIC
CABLE

TO RADIATORS

PIPING FROM
STEAM MAIN

ELECTRIC
CABLE

CONTROL CABINET

110 V 60 Hz
SERVICE

TRANSFORMER

PIPING TO RETURN

OPEN OR
CLOSED RETURN

Fig. 55. Typical arrangement of *Webster* E-5 electronic moderator system.

The moderator control regulates the pressure difference and will function equally well regardless of whether the pressure in the return piping is at atmospheric pressure or below. The main steam

247

control valve is adjusted automatically by the moderator control which acts to reverse the direction of the motor, causing it to move the valve in the closing direction when *less* steam is required and in the closing direction when *less* steam is required and in the opening direction when *more* steam is required. The outdoor thermostat automatically varies the steam flow in accordance with changes in outdoor temperature. Depending upon the outdoor temperature, the thermostat automatically selects the position of the main steam control valve. Its position may be advanced or reduced by the variator to give more or less steam than is called for by the outdoor temperature.

Changes in pressure difference in the heating system are automatically compensated for by a pressure actuated mercury tube in the control cabinet. One end of the tube is connected to the steam supply main and the other end to the return main. If the supply pressure is unduly increased, mercury rises in the tube to unbalance resistances contained therein and the main steam valve begins to close in amount sufficient to balance resistances. A reverse action takes place when the pressure difference falls below that called for by the control equipment.

The pulsating steam flow arrangement is called the *Webster Moderator System*. This system comprises a central heat control of the pulsating flow type for new or existing steam or hot water systems. It is designed for small and medium size buildings and for zoning of large buildings. It directly controls the operation of burner, stoker, blower or draft damper motors. The basic components of this system are:

1. Outdoor thermostat,
2. Pressure difference controller,
3. Control cabinet,
4. Capillary tubing.

These four components work together to open and close a valve in the steam main or to start and stop the automatic firing device at the boiler generally through a relay. Fig. 56 shows the general arrangement of a *Webster* EH-10 moderator system for a building served by a central station or street steam. A *Webster* EH-10 moderator system used for controlling a burner on a steam boiler is illustrated in Fig. 57.

Fig. 56. Typical arrangement of *Webster* EH-10 moderator system in a building served by a central station or street steam.

The pressure difference controller maintains the correct pressure difference between supply and return piping. In combination with metering orifices, this device gives an even distribution of the steam to all radiators in the system thereby preventing over or under heating.

Control is accomplished by varying the length of intervals during which steam is delivered to the radiators. These intervals are longest in cold weather and shortest in mild weather. The timing is such that the longest *off* interval is comparatively short so that heat output from radiators is practically continuous.

249

Fig. 57. General arrangement of the *Webster* EH-10 moderator system controlling a burner of a steam boiler.

The timing mechanism inside the control cabinet is powered by a synchronous motor which turns a cam. Timing gears between motor and cam set the length of the operating cycle. Rising on the cam is a roller connected to the arm of a switch. When the roller is on the high part of the cam, the switch is in the position for opening the control valve or starting the firing equipment. When

250

the roller is on the low part of the cam, the switch is on the position for closing the control valve or stopping the firing equipment. The length of the ON interval is changed automatically by the outdoor thermostat or by adjusting the variator by hand. The average length of the cycle is 30 minutes. Other gears can be furnished for cycle lengths from 12 to 60 minutes.

A variator is included in the control cabinet for manual adjustment of the rate of heat delivery to the building. The variator changes the relationship of the switch lever and roller to the cam by moving the cam itself. This is accomplished by mounting the motor, cam and gear train on a movable carriage.

HIGH PRESSURE STEAM HEATING SYSTEMS

High pressure steam heating systems operate at pressures above 15 psig (generally in the 25 psig to 150 psig range), and are usually found in large industrial buildings in which space heating or steam process equipment (e.g. water heaters, dryers, etc.) is used. An example of a typical two-pipe high pressure steam heating system is shown in Fig. 58. This type of system is also referred to as a *medium* pressure steam heating system when the steam pressures are in the lower ranges.

One advantage of this type of heating system is that the high pressure of the steam permits the use of smaller pipe sizes. The high steam pressure also makes possible the elevation of the condensation return lines above the heating units, because the return water can be lifted into the return mains. High pressure condensation pumps and thermostatic traps are commonly used in these systems to handle the condensation and return it to the boiler.

High pressure steam heating systems are more expensive to operate and maintain than low pressure systems. In most cases, a licensed stationary engineer must be hired for the installation; a factor which tends to increase the operating cost.

STEAM BOILERS

The boiler is the source of heat for a steam heating system, and it will operate on a number of different types of fuels; however, regardless of the fuel used, the operating principle will be the same. Water is heated until it boils and changes to steam. The steam is then distributed to the heat emitting units throughout the structure either by natural or mechanical means.

Low pressure steam heating boilers are used in residences and small buildings. The design and construction of these boilers is very similar to the boilers used in hot-water heating systems in the same size structure. Both types of boilers are described in considerable detail in Chapter 15 (BOILERS AND BOILER FITTINGS).

CONTROL COMPONENTS

The controls used to regulate steam boilers and to insure their safe and efficient operation, are similar in most respects to those used with hot water boilers. A list of the principal controls used with steam boilers would include:

Fig. 58. A typical two-pipe high-pressure steam heating system.

1. ASME safety valve,
2. Steam pressure gauge,
3. Low-water cutoff,
4. Boiler water feeder,
5. High-limit pressure control,
6. Water gauge glass,
7. Water cocks,
8. Primary control (burner mounted),
9. Operating control (with tankless heater).

These and other types of boiler controls are described in considerable detail in Chapter 15 (BOILERS AND BOILER FITTINGS); Chapter 4, Volume 2 (THERMOSTATS AND HUMIDISTATS); and Chapter 9, Volume 2 (VALVES AND VALVE INTALLATION).

HARTFORD RETURN CONNECTION

A well designed steam heating system will have a *Hartford return connection* (Fig. 59) in the condensation return line. The purpose of the *Hartford* return connection (or *Hartford loop,* as

Fig. 59. The *Hartford* return connection.

it is sometimes called) is to prevent excessive loss of water from the boiler when a break-down (such as a water leak) occurs in the return line.

As shown in Fig. 59, an equalizer connects the lower outlet to the steam outlet. The *Hartford* connection is taken off from the equalizer an inch or two below the normal water level. Evidently low pressure can only draw water out of the boiler connection. This gives a low water level and it cannot recede further because steam will flow into the connection leg as the water recedes in the leg due to pressure difference.

More details about different types of boiler connections are found in Chapter 15 (BOILERS AND BOILER FITTINGS) and in Chapter 9, Volume 2.

PIPES AND PIPING DETAILS

A steam heating system requires careful planning of the piping to insure both an efficient and a safe operation. For example, the type of pipe material used (e.g. wrought iron, Schedule-40 black steel, etc.) is important, because the *capacity* of a particular weight pipe will depend on both its size and the material from which it is constructed.

The expansion of pipes when they become heated is another factor that must be considered when designing a steam heating system. Sufficient flexibility in the piping can be provided for by correctly designed offsets, slip joints or bellows, radiators and riser runouts, U-bends, or other types of expansion loops. The pitch of connections from risers must be sufficient to prevent the formation of water pockets when pipe expansion occurs.

Pipe materials, pipe sizes (and pipe sizing methods), pipe expansion rates, pipe fittings, and piping details such as wet and dry returns, drips, and connections to heat emitting units are covered in Chapter 8, Volume 2 (PIPES, PIPE FITTINGS AND PIPING DETAILS). Although valves and steam traps are often included under piping details by heating authorities, the maingly, valves are covered in Chapter 9, Volume 2 (VALVES AND VALVE INSTALLATION) and steam traps, along with pumps, siphons, strainers and similar devices.

STEAM TRAPS

A *steam trap* is an automatic device installed in a steam line to control the flow of steam, air, and condensation. In operation, it opens to expel air and condensation, and closes to prevent the escape of steam. All steam traps operate on the principle that the pressure within the trap at the time of discharge will be slightly in excess of the pressure against which the trap must discharge.

The principal functions of steam traps in steam heating systems include: (1) *draining condensation* from the piping system, radiators, and steam processing equipment; (2) *returning condensation* to the boiler; (3) *lifting condensation* to a higher elevation in the heating system; and (4) *handling condensation* from one pressure to another.

Steam traps may be classified on the basis of their operating principles as follows:

1. Float traps,
2. Bucket traps,
3. Thermostatic traps,
4. Float and thermostatic traps,
5. Flash traps,
6. Impulse traps,
7. Lifting traps,
8. Boiler return traps.

PUMPS

Pumps are used in steam heating systems to dispose of the condensation or return it to the boiler, and to discharge excess air and noncondensable gases to the atmosphere. The specific function (or functions) of a pump will depend on the type of steam heating system in which it is used. The two basic types of steam heating pumps are:

1. Condensation return pumps.
2. Vacuum heating pumps.

HEAT EMITTING UNITS

A *heat emitting unit* is a device which transmits heat to the interior of a room or space. The two types of heat emitting units used in steam heating systems are radiators and convectors. Simply defined, a *radiator* is a heat emitting unit which transmits heat from a direct heating surface principally by means of radiation. A *convector,* on the other hand, may be defined as a heat emitting unit which transmits heat from a heating surface principally by means of convection. The heating surface of a convector is usually of the extended finned tube construction.

A detailed description of the heat emitting units used in steam heating systems is contained in Chapter 2, Volume 3 (RADIATORS, CONVECTORS AND UNIT HEATERS).

AIR SUPPLY AND VENTING

Two types of venting occur in heating systems. One type deals with the passing to the outdoors of smoke and gases resulting from the burning of combustible fuels (e.g. coal, oil, and gas), and is described in the various chapters on furnaces and boilers. Another type of venting deals with the relief of pressure in steam heating systems by allowing a certain amount of air to escape (i.e. be vented) from the radiators. Venting radiators is described in Chapter 2, Volume 3 (RADIATORS, CONVECTORS AND UNIT HEATERS).

UNIT HEATERS

A *unit heater* is essentially a forced draft convector. A centrifugal fan or propeller is used to force the air over the heating surface and into the room or space through deflector vanes. The principal operating components of a steam unit heater are shown in Fig. 60. Unit heaters find their widest application in industrial and commercial buildings, gymnasiums, field houses, auditoriums, and other types of large buildings.

All types of unit heaters are described in considerable detail in Chapter 2, Volume 3 (RADIATORS, CONVECTORS AND UNIT HEATERS).

Fig. 60. Principle components of a steam unit heater.

AIR CONDITIONING

Central air conditioning can be added to a steam heated structure by installing a water chiller or a separate forced air cooling system (i.e. a split system).

The water chiller and steam boiler may be installed as separate units, or a complete package containing both units may be used. Water chiller installations are very rarely used in residential air conditioning. Split system air conditioning dominates this field (i.e. with respect to steam heated residences).

CHAPTER 9

Electrical Heating Systems

There are a variety of different types of heating systems that use electricity as a source of heat. Each of them will employ one or more of the following methods of heat transfer:

1. Radiation,
2. Convection,
3. Forced air.

Radiation is the transmission of heat energy by means of electromagnetic waves. In other words, the heat is transferred directly from the surface of the heating element to the people, objects, and furnishings in the room *without* heating the air. The air will pick up a certain amount of heat from these surfaces (i.e. the bodies of the people, the furnishings, etc.) which then circulates by natural convection.

With *convection,* the temperature of the air is increased by contact with a heated surface. The air rises as it becomes warmer (and lighter), resulting in a circulation of the air in the space. Now note the difference between radiation and convection. In the radiation method of heat transmission, the heat energy passes through the air to the individual or object in much the same way that the sun transfers its heat. The air is *not* used as the medium of heat transfer. The surfaces are warmed and the air picks up heat from these surfaces (convection), not directly from the heat energy passing through it.

In the *forced air* method of heat transmission, the heated or conditioned air is circulated by motor-driven fans integral to the heating unit or included elsewhere in the system. The heated air is forced directly into the room from the heating unit as in the case of a wall heater) or through a duct system from a centrally located furnace (central forced warm air heating systems).

CENTRAL HOT WATER SYSTEMS

Some central hydronic systems (i.e. forced hot water heating systems) use electric-fired boilers as the heat source. These boilers (Fig. 1) are compact units consisting of an insulated steel or cast iron generator with replaceable immersion heating elements. The expansion tank, circulating pump, valves, and pre-wired controls are included within the boiler package, and the entire unit is assembled at the factory (Fig. 2).

The water (sometimes other liquids are used) is heated in the boiler and forced through the pipes by the water circulating pump to baseboard convectors or radiators located in the various rooms.

The controls for an electric boiler are similar to those found on boilers (e.g. oil-fired, gas-fired, etc.) in other types of central heating systems. The basic controls are:

Courtesy *National Better Heating-Cooling Council.*

Fig. 1. Electric-fired boiler.

259

1. A low-voltage thermostat,
2. Pressure and temperature limit controls,
3. Relay and sequence controls,
4. Circulating pump control.

The circulating pump is activated by the low-voltage thermostat. Boiler overheating and pressure build-up is controlled by the boiler high limit controls and the pressure-temperature relief valve.

A more detailed description of electric-fired boilers and their controls is found in other chapters of this book (see, for example, the section on electric-fired boilers in Chapter 15. BOILERS AND BOILER FITTINGS).

For information about the piping system used in electric hydronic heating, read the appropriate sections in Chapter 7 (HOT-WATER HEATING SYSTEMS), and Chapters 8 and 9, Volume 2.

Fig. 2. Interior view of an electric boiler showing the basic components.

Cooling for an electric hydronic system is generally provided by individual room air conditioners (i.e. through-the-wall units) or by chilled water pumped from a central refrigeration unit. The latter type installation is usually found in multifamily structures such as apartment buildings.

CENTRAL FORCED WARM-AIR HEATING SYSTEMS

Central forced warm-air heating systems that use electricity to heat or cool the air rely upon the following heat sources:

1. Electric-fired furnaces,
2. Duct heaters,
3. Heat pumps.

Regardless of the heat source, the air is circulated by fans through a duct system to the various rooms. Variations of these ducted air systems are shown in Fig. 3.

Courtesy *Electric Energy Association.*

Fig. 3. Various duct systems for electric heating and cooling.

The electric-fired furnace is a complete unit designed for zero clearance, and available in several models suitable for either horizontal or vertical installation. The heat is provided by fast activating coiled wire resistance heating elements. When there is more than one coil, the individual coils are energized at intervals in sequence to prevent the type of current overload possible if all were energized simultaneously. The heating coil sequencer is activated and controlled by a low-voltage thermostat mounted on a wall in some convenient location. Overheating of the furnace is prevented by the same type of high-limit control found in other types of furnaces. The warm air is forced through the duct system to outlets in the room by a blower. Evaporator coils can be added to the furnace for cooling. Additional modifications can provide humidity control (humidification and dehumidification), ventilation, and air filtration.

A more detailed description of electric-fired furnaces is found in Chapter 14 (ELECTRIC-FIRED FURNACES). For further information on forced warm-air heating systems, read the appropriate sections in Chapter 6 (WARM-AIR HEATING SYSTEMS). Controls for this type heating system are described in the section ELECTRIC HEATING AND COOLING CONTROLS in this chapter, and in Chapter 4, Volume 2 (THERMOSTATS AND HUMIDISTATS).

Duct heaters (Fig. 4) are factory assembled units that are installed in the main (primary) or branching ducts leading from

Courtesy Vulcan Radiator Co.

Fig. 4. Duct heaters.

the blower unit. Heat is provided by parallel rows of resistance wire formed in the shape of spirals, which may be sheathed or left open. The blower is housed in a specially constructed and insulated cabinet to which the ducts are connected. This unit functions as the power source for circulating the air through the duct heater and the rest of the system. Cooling can be provided by adding a cooling unit to the blower. Humidity control is possible by first supercooling the air in the coil and then reheating it to the desired temperature and moisture content when it passes through the duct heater.

Some duct heaters are designed to be inserted into a portion of the duct through a hole cut in its side, and are more or less permanent units. Other duct heaters are assembled at the factory in a portion of duct which is flanged for easy installation at the site.

Duct heaters may be installed in more than one duct to provide zoned heating. When this is the case, each duct heater is provided with means for independent temperature control. Another common method is to install a single duct heater in the main (primary) duct leading from the blower unit. Controls for either type installation are similar to those used on electric-fired furnaces (see above).

The third type of heat source used in ducted central forced warm-air heating systems is the air-to-air or water-to-air heat pump. These are briefly described in another section of this chapter (see ELECTRIC HEAT PUMPS), and in greater detail in Chapter 12, Volume 3 (HEAT PUMPS).

RADIANT HEATING SYSTEMS

Any electrical conductor that offers resistance to the flow of electricity will generate a certain amount of heat; the amount of heat generated being in direct relation to the degree of resistance. This method of generating heat is employed in *radiant heating systems.*

The type of conductor commonly used in radiant heating systems is an electric heating cable which is imbedded in the floors, walls, or ceilings (Fig. 5). Installation of the cables may be done at the site (as is often the case with new construction), or they may be obtained in the form of prewired, factory-assembled, panel-type units. The heat generated by the cables is transferred to the occupants and surfaces in the room by low intensity radiation.

Site installed heating cables or prewired and assembled panel units are used in the following three types of radiant heating systems:

1. Radiant ceiling panel systems,
2. Radiant wall panel systems,
3. Radiant floor panel systems.

Radiant ceiling heating systems are by far the most commonly used type. The other two have certain disadvantages inherent in their construction. All three types of radiant heating systems are described in greater detail in Chapter 1, Volume 3 (RADIANT HEATING).

Fig. 5. Typical radiant heating system.

The electric heating cables are activated and controlled by wall-mounted low voltage or line thermostats. Cooling can only be accomplished by adding a separate and independent system.

BASEBOARD HEATING SYSTEMS

Baseboard heating systems use electric baseboard units located at floor level around the perimiter of each room (Fig. 6). Each unit consists of a heating element enclosed in a thin metal housing. Zoning is possible with a single, wall-mounted line or low-voltage thermostat located in each room. Heat is provided to the room *primarily* by convection (some radiation is involved) as the room air moves across the heated elements in the baseboard unit. These baseboard units are described more fully in Chapter 2, Volume 3 (RADIATORS, CONVECTORS, AND UNIT HEATERS).

COMES IN DIFFERENT WATTAGES

Fig. 6. Electric baseboard heating units.

As is the case with radiant heating systems, cooling can be provided by installing a separate cooling system (central air or room air conditioners). The baseboard heating system represents the most widely used form of electric heating.

ELECTRIC UNIT VENTILATORS

Electric unit ventilators (similar in design to the heaters, see Fig. 7) are used to heat, ventilate, and cool large spaces that are by nature subject to periods in which there will be a high density of occupancy. They are frequently found in offices, schools, auditoriums, and similar structures.

The basic components of a typical unit ventilator are:

1. The housing,
2. Motor and fans,
3. Heating element,

4. Dampers,

5. Filters,

6. Grilles or diffusers.

Automatic controls activate the unit ventilator and vary the temperature of the air discharged into the room in accordance with room requirements. Outdoor air is drawn through louvers in the wall and into the unit ventilator before being discharged into the interior of the structure. These ventilators are usually floor or ceiling mounted depending on the design of the room.

In addition to electricity, unit ventilators may also be gas-fired, or use steam or hot water as the heat medium. Unit ventilators are described in greater detail in Chapter 2, Volume 3 (RADIATORS, CONVECTORS, AND UNIT HEATERS).

ELECTRIC UNIT HEATERS

Electric unit heaters (Fig. 7) are used primarily for heating large spaces such as offices, garages, warehouses, and similar types of commercial and industrial structures.

As is the case with unit ventilators, the heat conveying medium and combustion source may be other than electricity. For example, the air may be heated by either gas-fired or oil-fired

Fig. 7. Wall or ceiling mounted electric unit heater.

units. Steam or hot water can be substituted for air as the heat conveying medium. Selecting electricity as a medium will depend on such factors as:

1. The availability of cheap electrical power,
2. The need for supplementary heating,
3. The scarcity of other heat conveying mediums.

The typical unit heater contains the following basic components:

1. Fan and motor,
2. Heating element,
3. Directional outlet,
4. Casing or housing.

Unit heaters may be floor mounted or suspended from the ceiling, depending upon the design requirements of the structure.

ELECTRIC SPACE HEATERS

Smaller unit heaters designed and constructed for domestic heating purposes are commonly referred to as *space heaters*. Like the larger units, they are also designed to operate with other heat conveying mediums. However, the advantage of using electric space heaters over the fuel burning types is that they do not have to be vented and require no flue.

Space heaters are available as either portable or permanent types. The latter can be fitted into ducts or mounted on walls or the ceiling.

For further information on space heaters read the appropriate sections in Chapter 2, Volume 3 (RADIATORS, CONVECTORS, AND UNIT HEATERS).

HEAT PUMPS

A *heat pump* (Figs. 8 and 9) is an electrically powered, reversible-cycle refrigeration unit capable of both heating and cooling the interior of a structure.

The heat source is commonly either outside air (in the air-to-air type heat pump) or a closed loop of circulating water (in the water-to-air type heat pump). The former is the most popular type heat pump used for single family dwellings. The water-to-air heat pump system is most often found in multifamily structures. These are essentially central heating systems, with the heat pump replacing the central furnace or boiler as the heat generating unit.

Operating valves in the heat pump unit control the reversal of the refrigeration cycle. It removes heat from the interior of the structure and discharges it outside during hot weather, and supplies heat to the interior spaces during periods of cold weather.

The basic components of a heat pump installation are: (1) the compressor, (2) the condenser, (3) the evaporator, and (4) a

Courtesy *Electric Energy Association.*

Fig. 8. Outdoor section of a split unit heat pump. It is connected to the inside section by refrigerant tubing and electrical wiring.

Courtesy *Electric Energy Association.*

Fig. 9. Outside wall mounted single package heat pump.

low-voltage thermostat. These and other aspects of heat pumps are described in greater detail in Chapter 12, Volume 3.

ELECTRIC HEATING AND COOLING CONTROLS

In an electric heating and cooling system, the best results are obtained by finding a comfortable thermostat setting and leaving it there. Constantly changing the thermostat setting will result in consuming more energy (resulting in increased operating costs) and will create additional wear and tear on the equipment. Experts estimate that for each degree the thermostat is raised above the normal setting there is a three percent increase in heating costs. A comfortable setting for the thermostat will depend upon a number of variables, including: (1) the type of system, (2) environmental conditions, and (3) your personal requirements. For example, 70 degrees is usually adequate for a radiant heating system (i.e. baseboard, panel-type heating installations, etc.) in a properly insulated structure. A comfortable indoor temperature setting for air conditioning is usually 76-78 degrees.

A variety of control methods are used in electric heating and cooling systems to maintain kilowatt demand at a level both economical in operation and suitable in performance. The basic types of controls used for these purposes are:

1. Thermostats,
2. Sequence switching devices,
3. Load-limiting controls,
4. Time-delay relays.

Thermostats can be located in each room or on each heating unit to provide decentralized control. A manual switch can be provided with each thermostat to shut off the current if the room is to be unoccupied for long periods of time.

Sequence switching devices are used to switch electrical current in sequence (rotation) from one room or circuit to the next. A *load-limiting control* is designed to shut off the current to one or more circuits when total electric demand exceeds a preset value. *Time-delay relays* are used to restore service after an interruption *nonsimultaneously* over a period of several minutes so that overloading is avoided. Each of these controls is examined in considerable detail in Chapter 14 (ELECTRIC-FIRED FUR-NACES).

INSULATION FOR
ELECTRICALLY HEATED AND COOLED STRUCTURES

Any structure electrically heated and cooled *must* be properly insulated or operation costs will be unacceptable. Converting an older structure to electric heat is therefore not recommended unless you have no objection to the additional expense of improving the insulation, or you feel you can live with the higher operating costs. Consequently, electric heating is more often considered for new construction.

Insulating a structure that is to be electrically heated and cooled requires greater attention to construction details than do other types of systems; however, the results are well worth the effort because reduced operating costs and other gains (e.g. the reduction of outside noise, quiet operating characteristics, etc.) are soon readily apparent. Because electric heating and cooling systems require an especially well-insulated structure for efficient operation, recommendations for the required minimum levels of insulation are included in this chapter. These recommendations are based on information included in the *Electric Energy Association's* publication "Electric Space Conditioning in Residential Structures".

The recommendations found in the following paragraphs will contain references to U-factors and R-values. The *U-factor* is the overall coefficient of heat transfer and is expressed in Btu's per hour per square foot of surface per degree Fahrenheit difference between air on the inside and air on the outside of a structural section. The *R-value,* on the other hand, is the term used to express thermal resistance of the insulation. The U-factor is the reciprocal of the sum of the thermal resistance values (i.e. R-values) of each element of the structural section. Table 1 gives the insulation recommendations for electrically heated and cooled structures both in terms of R-values and overall coefficients of heat transfer in terms of U-factors.

Table 2 lists heat loss limits for electrically heated structures as recommended by the *Electric Energy Association.* The values

Table 1. Maximum winter heat loss for electrically heated homes.

degree days	maximum heat loss values	
	watts/sq ft	Btuh/sq ft
over 8000	10.0	34
7001 to 8000	9.4	32
6001 to 7000	8.8	30
4501 to 6000	8.2	28
3001 to 4500	7.6	26
2000 to 3000	7.0	24

Courtesy *Electric Energy Association.*

Table 2. Insulation recommendations (R-values) and overall coefficients of heat transfer (U-factors) for major areas of heat loss and heat gain in residential structures.

type of construction	opaque sections adjacent to unheated spaces			opaque sections adjacent to separately heated dwelling units		
	walls	floors	ceilings	walls	floors	ceilings
frame	R-11 U:0.07	R-11 U:0.07	R-19 U:0.05	R-11 U:0.07	R-11 U:0.07	R-11 U:0.07
masonry	R-7 U:0.11	R-7 U:0.11	R-11 U:0.07	R-7 U:0.11	R-7 U:0.11	R-11 U:0.07
metal section	R-11 U:0.07	not applicable	not applicable	R-11 U:0.07	not applicable	not applicable
sandwich	R-* U:0.07	R-* U:0.07	R-* U:0.05	R-* U:0.07	R-* U:07	R-* U:07
heated basement or unvented crawl space	R-7 U:0.11	insulation not required	insulation not required	*Since the thermal resistance of sandwich construction depends upon its composition and thickness, the amount of additional insulation required to obtain the maximum U-factor must be calculated in each case.		
unheated basement or vented crawl space	insulation not required	see "floors"	insulation not required			

Note: Insulation R-values shown above refer to the resistance of the insulation only.

Courtesy *Electric Energy Association.*

271

listed are expressed in watts and Btuh's (1 watt equals 3.413 Btuh) per square foot of floor space measured to the exterior walls. The assumed infiltration rate on which this table is based is approximately three-quarters air change per hour.

The maximum summer heat gain limits for electrically air conditioned homes are indicated in Fig. 10. These figures are adapted from the FHA Minimum Property Standards for single family and multifamily structures.

Frame Walls

Frame walls (Fig. 11), with either wood or metal studs, are commonly insulated with rolls and batts of mineral, glass or wood fiber insulation material. More and more use of sprayed-on or foamed-in-place rigid insulation is also being made. Whatever the type of insulation decided upon, exterior opaque walls and walls between separately heated units should have a maximum U-factor of 0.007, which will require insulation rated at R-11 or greater. A thermal resistance of R-11 can be provided by 3½-inch fiber batts or other equally suitable materials (see Chapter 3. INSULATION PRINCIPLES).

Masonry

A masonry wall is a particularly poor thermal barrier. A typical masonry cavity wall (4-inch face brick on the exterior, an air space, 4-inch concrete block, an air space, and a layer of gypsum wall board) has a thermal resistance equal to approximately 1 inch of rigid insulation. Exterior opaque masonry walls should have a maximum U-factor of 0.11 requiring insulation rated at R-7 or greater. This can be accomplished in the following ways:

1. Place a layer of preformed rigid insulation between the exterior face brick (or other finishing material) and the concrete block during construction (Fig. 12). Another method is to leave the space empty and then to fill it with a foam insulation before capping the wall.

2. If no air space exists between the exterior face brick or other exterior finish and the concrete block, it will be

necessary to add a layer of suitable insulating material (rated R-7 or better) on the inside surface of the block (Fig. 13). The insulating material can be rigid insulation attached to furring strips, foam insulation sprayed between the furring strips, or preformed rigid insulation secured in

NOTE :
1. WHEN THE OUTSIDE DESIGN DRY BULB TEMPERATURE IS LESS THAN 90 F. THE VALUES SHOW FOR 90 F. SHALL BE USED.
2. WHEN THE OUTSIDE DESIGN DRY BULB TEMPERATURE IS MORE THAN 105 F. THE VALUES SHOWN FOR 105 F. SHALL BE USED.

Courtesy *Electric Energy Association.*

Fig. 10. Maximum summer heat gain limits for air conditioned homes.

place with adhesives. An interior wall finish is then applied over the insulating material.

VAPOR BARRIER INSULATION

INTERIOR FINISH

EXTERIOR FINISH

SHEATHING

Fig. 11. Frame wall construction.

RIGID INSULATION PREFORMED OR SPRAYED

INTERIOR FINISH

FACE BRICK

POURED CONCRETE FLOOR

Fig. 12. Masonry wall construction with insulation installed on the outside portion.

If the wall is constructed entirely of concrete, apply preformed rigid insulation to either side by adhesive bonding, and cover the layer of insulation with a wall finish material (Fig. 14). The insulation must be rated R-7 or better.

274

RIGID INSULATION PREFORMED
OR SPRAYED

INTERIOR FINISH

FACE BRICK

POURED CONCRETE
FLOOR

Fig. 13. Masonry wall construction with the insulation installed on the inside portion.

CONCRETE SLAB

INTERIOR FINISH

RIGID INSULATION

Fig. 14. Concrete slab construction.

Metal

Exterior opaque metal walls (Fig. 15) should have a maximum U-factor of 0.07. This will require insulation having a thermal resistance rated at R-11 or bettter, which can be provided by a layer of rigid insulation either preformed or foamed in place.

Fig. 15. Metal sections used for both load-bearing and certain walls in residential construction.

Sandwich Construction

Sandwich or layered construction will vary in composition and thickness, but will typically consist of three layers: two facings and a core. Exterior opaque walls of a sandwich construction should have a maximum U-factor of 0.007. The thermal resistance (i.e. R-value) must be calculated on the basis of its composition and thickness.

Basement Walls

There is no need to insulate the exterior walls of unheated basements because there is no need for reducing heat loss. The exterior walls of *heated* basements (or interior walls separating heated from unheated areas) are a different matter, because heat loss is a concern here. An insulating layer having a thermal resistance rated at R-7 or greater should be applied. Fig. 16 illustrates two methods for insulating these types of walls.

Crawl Space Exterior Walls

The exterior walls of *unvented* crawl spaces should be insulated with a material providing a thermal resistance of at least R-7 or better. A method for doing this is illustrated in Fig. 17. No insulation is required for the exterior walls of *vented* crawl spaces.

FIBER BLANKET INSULATION
VAPOR BARRIER
INTERIOR FINISH
INTERIOR FINISH
RIGID INSULATION

Fig. 16. Exterior walls of heated basements.

RIGID INSULATION

GROUND MOISTURE BARRIER

Fig. 17. Exterior walls of unvented crawl spaces.

Walls Between Separately Heated Dwelling Units

Where walls separating adjacent but independently heated dwelling units occur, insulation with a thermal resistance of at least R-11. The insulation may be placed in either wall or divided proportionally between them (Fig. 18).

Wood or Metal Joist Frame Floors

Wood or metal joist floors over unheated basements or vented crawl spaces should have a maximum U-factor of 0.07 (requiring insulation rated at R-11 or greater). Three methods for providing insulation are illustrated in Figs. 19, 20 and 21.

277

Fig. 18. Walls between adjacent separately heated dwelling units.

Fig. 19. Frame floor construction with insulation between joist.

Concrete Floors

Concrete floors over unheated spaces should have a maximum U-factor of 0.11 requiring insulation rated at R-7 or greater. The following three methods are recommended for applying a layer of insulation:

1. Spray rigid insulation in a suitable thickness on the *under-side* of the floor.

2. Bond preformed rigid insulation to the top or bottom of the surface.

3. Form concrete around a core of rigid insulation.

Fig. 20. Frame floor construction with insulation covering the joist.

Fig. 21. Frame floor construction with rigid insulation panels attached to the bottom of the joist.

Slab-On-Grade Floors

Insulation rated R-5 is required for all slab-on-grade floors. Two methods for insulating slab-on-grade floors is illustrated in Fig. 22. Table 3 lists the maximum heat losses in terms of watts and Btuh per linear foot of exposed slab edge as recommended by the Electric Energy Association. More information can be obtained from the FHA Minimum Property Standards.

Floors of Sandwich Construction

Floors of sandwich construction should be insulated in the same manner as sandwich walls (see above).

279

Fig. 22. Slab-on-grade construction.

Table 3. Slab edge heat loss.

outdoor design temperature (degrees F)	total width of insulation (inches) rated R-5	heat loss per foot of exposed slab edge	
		watts	Btuh
−30 and Colder	24	10.0	34
−25 to −29	24	9.4	32
−20 to −24	24	8.8	30
−15 to −19	24	8.2	28
−10 to −14	24	7.9	27
−5 to −9	24	7.3	25
0 to −4	24	7.0	24
+5 to +1	24	6.4	22
+10 to +6	18	6.2	21
+15 to +11	12	6.2	21
+20 to +16	edge only	6.2	21
+21 and Warmer	none	—	—

Courtesy *Electric Energy Association.*

Frame Ceilings and Roofs

Any ceiling below an unheated space should be insulated with materials having a rating of R-19 or better. The ceilings should have a maximum U-factor of 0.05. Among the various methods for providing such insulation are:

1. Staple blankets or batts to the joists before the finished ceiling is applied (Fig. 23).

2. Blow or manually apply loose insulation between ceiling joists from above.

3. Spray on insulation between the joists from above.

Fig. 23. Ceiling and roof from construction.

Concrete Ceilings

Concrete ceilings below unheated spaces should have a maximum U-factor of 0.07, requiring insulation rated R-11 or greater. The following methods are suggested for insulating concrete ceilings:

1. Attach preformed rigid insulation to either surface with adhesives.

2. Apply sprayed-on rigid insulation to either surface.

3. Make a sandwich constructon with preformed rigid insulation formed around a concrete core.

4. Apply rigid insulation to the *top surface* where the ceiling also serves as the roof of the structure.

Sandwich Ceilings

Ceilings of sandwich construction below unheated spaces should have a maximum U-factor of 0.05, requiring insulation rated at R-19 or better. Sandwich ceilings that also serve as the roof of a structure meet the same requirements.

DOORS AND WINDOWS

Two major heat loss areas for all types of construction are doors and windows. It is particularly important to reduce the rate of heat loss at these points to a minimal level when an electric heating system is used.

An exterior door constructed with a core of rigid insulation and faced with metal or acrylic film will have a very low U-factor (as low as 0.074). Infiltration heat loss around the edges of the door can be effectively reduced with suitable weatherstripping.

A typical 1-inch solid wood door will have a U-factor of approximately 0.64 (compare this to the U-factor for the door described in the preceding paragraph). Doubling the thickness of the door will reduce the U-factor to about 0.43. The addition of metal or glass storm doors will further reduce the U-factor (0.39 and 0.29 respectively).

Windows with double glazing will have a U-factor of 0.58 as compared to 1.13 for single glazing. Double glazing also reduces noise transmission.

FURTHER INFORMATION

Further and more detailed information about insulation, the problems of heat loss and gain, and methods for calculating heat loss can be found by reading the appropriate sections of Chapter 2 (HEATING FUNDAMENTALS), Chapter 3 (INSULATION PRINCIPLES), and Chapter 4 (HEATING CALCULATIONS).

ADVANTAGES OF ELECTRIC HEATING AND COOLING

Among the principal advantages of using electric heating and cooling are:

1. Greater safety,
2. Quiet operation,
3. Economy of space,
4. Reduction of drafts,
5. Reduction of outside noise,
6. Uniform temperature,
7. Structural design flexibility.

Although the chances of an explosion in gas- or oil-fired heating systems are very rare because of the safety features built into these systems, they simply *cannot* occur in total electric heating.

Electric heating units are very compact and therefore utilize very little space. In baseboard-type systems, there is no need for ducts or pipes to carry a heat medium from its source to the space being heated. These factors offer a great degree of structural design flexibility, because duct and pipe arrangements do not have to be taken into consideration. In addition, no chimney or flue arrangement is required.

In a structure properly insulated for electric heating and cooling, there is a marked reduction of drafts and the degree of outside noise penetration. Morever, uniform temperatures will prevail.

Finally, an electric heating system is generally quieter than other types, because fewer mechanical parts are involved. This quietness of operation is particularly characteristic of baseboard-type installations.

DISADVANTAGES OF ELECTRIC HEATING AND COOLING

An electrically heated and cooled structure offers certain inherent disadvantages when compared with other types of heating and cooling systems. Of course, this is true of *any* system regardless of the energy source. When choosing an energy source for a heating and cooling system, it is always necessary to carefully weigh the advantages and disadvantages of the system and its energy source against the requirements you demand from them.

The principal disadvantages of electric heating and cooling are:

1. Generally higher operating costs,
2. Lack of ventilation,
3. Higher insulation costs.

CHAPTER 10

Furnace Fundamentals

The *American Society of Heating, Refrigerating and Air-Conditioning Engineers* (ASHRAE) defines a furnace as being *a complete heating unit for transferring heat from fuel being burned to the air supplied to a heating system.* The *Standard Handbook for Mechanical Engineers* (Baumeister and Marks *Seventh Edition,* 12-104) provides a definition that differs only slightly from the one offered by the ASHRAE. It defines a furnace as being *a self-enclosed, fuel-burning unit for heating air by transfer of combustion through metal directly to the air.* Contained within these closely similar definitions are the two basic operating principles of a furnace; that is to say, some sort of fuel is used to produce combustion and the heat resulting from this combustion is transferred to the air within the structure. Note that *air* and not steam, water, or some other form of liquid is used as the heat conveying medium. This is the feature which distinguishes warm-air heating systems from the other types (see Chapter 6. WARM-AIR HEATING SYSTEMS).

Most modern furnaces are used in warm-air heating systems in which the furnace is a centralized unit, and the heat produced in the furnace is forced or rises by means of gravity through a system of ducts or pipes to the various rooms in the structure. This is what one commonly refers to as a *central heating system.* In other words, the furnace is generally in a centralized location within the heating system in order to obtain the most economical and efficient distribution of heat (although this is not an absolute necessity when a *forced* warm-air furnace is used).

Ductless or pipeless furnaces are also used in some heating applications, but are limited in the size of the area which they can effectively heat. They are installed in the room or area to be heated, but are provided with no means for distributing the heat beyond the immediate adjacent area. This is a far less efficient and economical method of heating than the central heating system, but is found to be adequate for a room, an addition to an existing structure, or a small house or building.

CLASSIFYING FURNACES

There are a number of different ways in which furnaces can be classified. One of the more popular methods is based on the type of fuel used to fire the furnace. Using fuel as classificatory criteria, the following four types of furnaces are recognized:

1. Gas-fired furnaces,
2. Oil-fired furnaces,
3. Coal-fired furnaces,
4. Electric-fired furnaces.

The first three categories of furnaces use a fossil fuel to produce the combustion necessary for heat transfer. The last one, namely electric-fired furnaces, uses electricity. Whether or not electricity can be justifiably called a fuel is not of great importance here, because in this particular instance it functions in the *same* manner as the three fossil fuels: it heats the air being distributed.

Furnaces can also be classified by the means with which the heated air is distributed to the room (or rooms) in the structure. This method of classifying furnaces establishes two broad classes: (1) gravity warm-air furnaces and (2) forced warm-air furnaces. The gravity warm-air furnaces rely *primarily* upon the principle of gravity for circulating the heated air. Because warm air is lighter than cold air, it will rise and pass through ducts or directly into the rooms to be heated. After passing off its heat, the now cooler and heavier air descends through returns to the furnace where it is reheated. Gravity-type furnaces represent the earliest designs used in warm-air heating systems. They were sometimes equipped with fans (integral or booster) to increase the rate of air flow. They have been largely replaced in popularity by forced warm-air furnaces which are equipped with integral fans.

Forced warm-air furnaces are often divided into three principal classes based primarily upon the location on the furnace of the warm air discharge outlet and the return air inlet. Furthermore, there are additional design considerations dependent upon the planned location of the furnace which also enters into the classification of warm-air furnaces. This classificatory system results in three types or classes of warm-air furnaces:

1. Upflow furnaces,
   a. Upflow highboy furnaces.
   b. Upflow lowboy furnaces.
2. Downflow furnaces,
3. Horizontal furnaces.

Upflow Highboy Furnace

A typical *upflow highboy furnace* (also referred to as an *upflow furnace* or a *highboy furnace*) is shown in Fig. 1. These are compact heating units that stand no higher than 5 or 6 feet and occupy a floor space of approximately 4 to 6 sq. ft. (2 ft. x 2 ft. or 2 ft. x 3 ft.).

The heated air is discharged through the top of the upflow furnace (hence the name) and the return air enters the furnace through air intakes in the bottom or sides. Cooling coils can be easily added to the top of the furnace or in the duct system.

Upflow Lowboy Furnace

The *upflow lowboy furnace* (also referred to as an *upflow furnace* or *lowboy furnace*) (Fig. 2) is designed for low clearances and stands only about 4 to 4½ feet high. Although shorter than either the upflow highboy or downflow types (see below), it is longer from front to back.

Both the return air inlet and the warm air discharge outlet are usually located on the top of the lowboy furnace. The lowboy furnace is found in heating installations where the ductwork is located above the furnace.

Downflow Furnace

Fig. 3 shows an example of a *downflow furnace* (also referred to as a *counterflow furnace* or a *downdraft furnace*). It is very similar in size and shape to the upflow furnace, but it discharges its warm air at the bottom rather than the top. The return air intake is located at the top.

Fig. 1. Upflow (highboy) furnace.

A *downflow furnace* is used primarily in heating installations where the duct system is either imbedded in a poured, concrete slab or suspended beneath the floor in a crawl space.

Horizontal Furnace

A *horizontal furnace* (Fig. 4) is designed for installation in a low, cramped space. They are often installed in attics (and re-

287

ferred to as *attic furnaces*) where they are positioned in such a way that a minimum of ductwork is used. This type of furnace is also frequently installed in crawl spaces.

AIR FLOW

Fig. 2. Upflow (lowboy) furnace.

Although dimensions will vary slightly among the various manufacturers, the typical horizontal furnace is about 2 ft. wide by 2 ft. high and 4½ to 5 ft. long.

The terms used in this classificatory system (i.e. *upflow furnace, downflow furnace,* etc.) are commonly employed by furnace manufacturers in the advertising literature describing their products. This is equaly true of their installation and operation manuals. Because of the widespread usage of these terms, they will be employed in the more detailed description of furnaces in the chapters that follow this one.

GRAVITY WARM-AIR FURNACES

Gravity warm-air furnaces rely upon the fact that warm air is lighter than cold air. As a result, the warmer air rises through the ducts or pipes in the structure, gives off its heat to the rooms, and descends to the furnace as it becomes cooler and heavier. It is then reheated and rises once again to the rooms. A continuous circulation path is thus established through the heating and cooling of the air. Sometimes a fan is added to increase the rate of flow, but the primary emphasis is still the effect of gravity on the differing weights of air.

AIR FLOW

Fig. 3. Downflow furnace.

Fig. 4. Horizontal furnace.

AIR FLOW

Depending upon the design, a gravity warm-air furnace will fall into one of the following categories:

1. A gravity warm-air furnace without a fan.
2. A gravity warm-air furnace with an *integral* fan.
3. A gravity warm-air furnace with a *booster* fan.

Each of these three categories represent different types of warm-air furnaces used in *central* heating systems.

Any gravity warm-air furnace not equipped with a fan relies entirely upon gravity for air circulation. The flow rate is very slow and extreme care must be taken in the design and placement of the ducts or pipes. Sometimes an integral fan is added to reduce the internal resistance to air flow and thereby speed up air circulation. A booster fan provides the same function, but is designed not to interfere with air circulation when it is not used.

The round cased gravity warm-air furnace illustrated in Fig. 5 is a coal-fired unit that can be converted to gas or oil (see Chapter 16. BOILER AND FURNACE CONVERSIONS). Depending upon the model, these furnaces are capable of developing up to 108,266 Btu at register and up to 144,319 Btu at bonnet.

Courtesy Oneida Heater Co., Inc.

Fig. 5. Coal-fired gravity warm-air furnace.

Floor, wall, pipeless furnaces, and some unit (space) heaters also operate on the principle of the gravity warm-air furnace. They are distinguished by the fact that the warm air is discharged directly into the room (or rooms) without the use of ducts or pipes.

SELECTING A FURNACE FOR A NEW HOUSE

One of the many decisions involved in building a new house is the selection of the heating, ventilating, and air conditioning system. This is probably one of the most important decisions because the cost of operating and maintaining the system, and its efficiency will be a daily fact of life for as long as you live there. It is therefore in your best interest to select the most efficient type of heating system that you possibly can.

291

The type of heating system selected (i.e. forced warm-air, hydronic, steam, electric baseboard, etc.) will depend upon such factors as the design of the house, the type of climate where the house is located, fuel costs, and personal preference. Comments on the advantages and disadvantages of the various types of heating systems are found in Chapters 6, 7, 8 and 9 (e.g. see Chapter 6. WARM-AIR HEATING SYSTEMS, etc.).

Assuming that you, like the majority of home owners, have decided to install a *forced* warm-air heating system, you should pay particular attention to selecting a suitable furnace. Should this decision be left entirely to your building contractor? Certainly not. For convenience, friendship, or other reasons, he may prefer to subcontract the work to a heating firm with which he customarily does business in all of his jobs. As a result, you may not be getting the best furnace for your needs. For example, John Doe Heating, Ventilating, and Air Conditioning Company may deal in furnace Brand X (or Brands X, Y, and Z). It may be that none of these furnaces is as efficient and economical to operate as the one you might wish to use.

There are many manufacturers of forced warm-air furnaces doing business in the country. The furnaces produced by these manufacturers will differ from one another in a variety of ways, including:

1. Installation cost,
2. Design factors,
3. Type of fuel used,
4. Cost of furnace,
5. Furnace gross output (Btuh),
6. Furnace net output (Btuh),
7. Furnace efficiency.

The *installation cost* will depend upon the area in which you are living, and can vary widely for the same furnace model from the same manufacturer. It is not something you can easily control, because it depends upon local labor costs, delivery charges, and availability of equipment.

Design factors involve development in the heating equipment that improve (or detract from) their performance. For example, some types of electric-fired furnaces use a direct drive blower motor instead of a belt-driven one. The direct-drive results in a smaller, more compact furnace. The belt-driven blower motor, on the other hand, has the advantage of being more easily maintained (it is simply a matter of replacing the standard motor used in the unit). Other design developments include improvement in flame retention (e.g. in oil-fired furnaces), more efficient combustion chambers, or improvements in other components. It may take you a little research, but there are a number of professional journals in the heating field that carry articles reviewing new products. Consumer magazines have also provided this service. Do not take the manufacturer's (or his sales representative's) claim at face value. Go to your local librarian and read what the experts have to say.

The type of fuel used will be an important factor in determining future operating costs. What is the cost of the various fuels in your area? Remember, the cheapest may not necessarily be the best for your purposes. Operating costs will depend on the total amount of fuel necessary to heat your home on a month-to-month basis. As a general rule, electricity is more expensive than the fossil fuels, and gas will be slightly more expensive than oil. Read Chapter 5 (HEATING FUELS) for additional information.

Cost of furnace. The most expensive furnace is not necessarily the best one, and the cheapest is not always that bargain you expect it to be. Your *first* consideration should be furnace performance. Although the initial cost of a suitable furnace may be high relative to others, it will soon pay for itself with lower operating costs.

Furnace efficiency is expressed as a percentage, and is found by subtracting *furnace net output* (the actual amount of heat delivered to the rooms) from *furnace gross output.* For example, an oil-fired furnace with a gross output of 150,000 Btuh and a net output of 120,000 Btuh will have an approximate efficiency of 80%.

By way of review, the three basic steps involved in selecting a suitable furnace for a new house are:

1. Determine the *type* of heating system you want for the house.
2. Estimate the amount of heat loss from the structure (see Chapter 4. HEATING CALCULATIONS).

3. Select a furnace with a heating capacity capable of replacing the lost heat.

SELECTING A FURNACE FOR AN OLDER HOUSE

Sometimes an older house or building will have a furnace that needs replacing. If you are planning to install a newer model of the same kind of furnace, you should have no serious difficulties. However, be sure that the newer model develops a similar Btu rating at both register and bonnet. In a forced warm-air heating system, the ducts are sized in accordance with the overall requirements of the system and a furnace is selected with a capacity to meet these requirements.

Switching from coal to gas or oil also should not present any great difficulties. Conversion burners have been designed and manufactured for just this purpose (see Chapter 16. BOILER AND FURNACE CONVERSIONS).

Installing an electric-fired furnace in an older structure is not generally recommended unless the construction is particularly tight and well insulated.

FURNACE COMPONENTS AND CONTROLS

Most forced warm-air furnaces have the following basic components and controls:

1. Heat exchanger,
2. Blower,
3. Air filter,
4. Primary control,
5. Fan and limit controls,
6. Thermostat control.

The *heat exchanger* (Figs. 6 and 7) is a metal surface (0.05-0.06 inch steel) located between the burning fuel and the circulating air in the furnace. The metal becomes hot and transfers its heat to the air above, which is then circulated through the ducts by the *blower*.

Most forced warm-air furnaces are equipped with either a disposable or permanent (and washable) *air filter* to clean the

BLOWER

Courtesy Meyer Furnace Co.

Fig. 6. Heat exchanger for a horizontal gas-fired furnace.

circulating air. An air filter is not recommended for a gravity warm-air furnace, because it tends to restrict the air flow.

Most warm-air furnaces are thermostatically controlled. In addition, there will be a primary control as well as fan and limit controls. Among the functions of the primary control is the regulation or stoppage of the flow of fuel to the combustion chamber when the fire is out or the thermostat indicates that no further fuel is needed. The fan and limit controls are also actuated by the thermostat, and are designed to start or stop the fan when the temperature in the furnace bonnet reaches predetermined and preset temperatures. In warm-air furnaces, it is suggested that the limit control switch be placed in the warm air plenum with a

FILTER

BLOWER

FAN AND LIMIT
SWITCH

Courtesy Meyer Furnace Co.

Fig. 7. Heat exchanger for an upflow gas-fired furnace.

recommended setting of 200°F for a forced warm-air furnace and 300°F for a gravity warm-air furnace.

These and other controls are offered by furnace manufacturers in a variety of different combinations. There primary function is to provide safe, smooth, and automatic operation of the warm-air furnace. More detailed descriptions of the controls used in furnaces and heating systems are found in Chapter 4 (THERMO-STATS AND HUMIDISTATS), Chapter 5 (GAS AND OIL CONTROLS), Chapter 6 (OTHER AUTOMATIC CONTROLS), and Chapter 9, of Volume 2 (VALVES AND VALVE INSTALLATION). Other chapters will also include sections on automatic controls. Check the index.

Most warm-air furnaces (except hand-fired coal furnaces and electric-fired furnaces) are provided with an apparatus designed to automatically prepare the fuel for combustion or to feed it directly to the fire. These apparatus (oil burners, gas burners, and automatic coal stokers) are described in the appropriate chapters. Electric-fired warm-air furnaces utilize electric resistance heaters.

Cooling coils, electronic air cleaners, and humidifiers are examples of optional equipment that can be added to most *forced* warm-air furnaces to provide total environmental control. It is best to include this optional equipment when the furnace is installed rather than waiting to add them at a later date. By doing so, you avoid complications that might arise in the overall design of your heating system. For example, it is not simply a matter of adding cooling coils to obtain air conditioning. Duct sizes must also be considered and they are not necessarily the same as those used for heating. This optional type equipment is considered in detail in Chapters 7 and 8, Volume 3.

PIPELESS, FLOOR AND WALL FURNACES

A *pipeless furnace* (Fig. 8) is commonly a gravity warm-air furnace installed in a central location beneath the floor. A single grille for the warm air and a return is provided for air circulation. This type of pipeless furnace is sometimes referred to as a *floor furnace*, although the latter is actually a permanently installed room heater and should be distinguished as such. *Wall furnaces* also belong to this category.

Both gas- and oil-fired furnaces are manufactured for installation in recesses cut from the floor. They are available in a wide range of Btu ratings for different types of installations. Some are thermostatically equipped for automatic temperature control and with blowers for forced air circulation (although those that operate on the gravity principle of air circulation are more common). There is also a choice between electric ignition or pilot flame. All gas-or oil-fired floor furnaces *must* be vented.

Another type of pipeless furnace is the gas- or oil-fired vertical furnace installed in a closet or a wall recess. The counterflow types discharge the warm air from grilles located at the bottom of the furnace as shown in Fig. 9.

COLD AIR WARM AIR GRILLE COLD AIR

FLOOR

DOUBLE BAFFLE

FLUE GAS

AIR HOOD

CONNECT FLUE PIPE HERE

FLUE GAS

RADIANT BAFFLE

BURNER

DRAFT HOOD

Fig. 8. Gas-fired gravity floor furnace.

DUCT FURNACES

A *duct furnace* (Fig. 10) is a unit heater designed for installation in a duct system where a blower (or blowers) is used to circulate the air. They are commonly designed to operate on natural or propane gases, although electric duct heaters are also available (see Chapter 7, Volume 2).

FURNACE INSTALLATION

No attempt should be made to install a warm-air furnace until you have consulted the local codes and standards. The *American Gas Association* and the *National Fire Protection Association* have also established codes for the installing of warm-air furnaces which they have made available to the public through their publications (see, for example, NFPA No. 31 *"Installation of Oil Burning Equipment 1972"*, NFPA No. 90B *"Residence Warm Air Heating 1971"*, NFPA No. 204 *"Smoke and Heat Venting*

Courtesy U.S. Department of Agriculture.

Fig. 9. Counterflow type vertical furnace.

1968", etc.). These publications can be obtained free or at a modest price from these organizations. Their addresses appear in Appendix A (PROFESSIONAL AND TRADE ASSOCIATIONS) of this book.

As soon as the heating equipment is shipped to your building site or existing structure, check it for missing or damaged parts. The manufacturer should be notified immediately of any discrepancies so that replacements can be made.

Follow the manufacturer's instructions for installing the heating equipment, but give precedence to local codes and regulations should any conflict appear. Most manufacturers base their installation instructions on existing codes formulated by the *American Gas Association* and the *National Fire Protection Association*.

Locate the furnace so that it has the shortest possible flue run containing as few elbows (turns in the run) as possible. Clearances around the furnace should be kept to the required

299

AIR FLOW

Courtesy *Janitrol*.

Fig. 10. Gas-fired duct furnace.

minimum, but within the regulations established by local codes and standards. Air ventilation must be adequate for efficient operation.

After the furnace has been installed, check all gas or oil lines for possible leaks. Make the necessary repairs if any leaks are found.

More detailed instructions for installing furnaces are found in Chapter 11 (GAS-FIRED FURNACES), Chapter 12 (OIL-FIRED FURNACES), Chapter 13 (COAL-FIRED FURNACES), and Chapter 14 (ELECTRIC-FIRED FURNACES).

FURNACE MAINTENANCE

Furnace maintenance is a very important part of the efficient operation of a warm-air heating system and should never be neglected.

The manufacturer of the heating equipment will provide recommendations for proper maintenance, and these recommendations should be carefully followed. By doing so, you will extend the life of the equipment, improve its efficiency, and reduce operating costs.

Maintenance recommendations specific to gas-fired, oil-fired, coal-fired, and electric-fired furnaces are included in the appropriate chapters (see, for example, Chapter 11. GAS-FIRED FURNACES, etc.). Certain maintenance recommendations are general in nature and will apply to any type of furnace regardless of the fuel used. These general maintenance recommendations are as follows:

1. *Always* read and closely follow the manufacturer's maintenance instructions (if they are available).
2. Check gas and oil pipes for leaks.
3. Check and clean (or replace) air filters.
4. Clean the furnace once a year (*preferably* before the heating season starts).
5. Oil the motor of a gas or oil burner two to three times each year.

GENERAL TROUBLESHOOTING HINTS

Each type of equipment described in this book contains a troubleshooting section in which problems specific to that type of equipment are discussed. A complaint common to all types of heating systems is that the system will not deliver heat. Because this is a troubleshooting problem that is general in nature, recommended corrective procedures will be included in this section.

The first thing that should be done if the system fails to deliver heat is to check the fuse box. Many times the whole problem can be traced to a blown fuse. By simply replacing it, you will save the cost of a service call. By the way, do not look at the manufacturer's name on the room thermostat and call the nearest factory or representative. Thermostat manufacturers do not make furnaces! If the fuse box does not contain the problem, then check the following:

1. Make certain the power is on at the main switch.
2. Make certain the burner motor fuse is not blown.
3. Check the burner on-off switch (if the switch is off, make sure the combustion chamber is free of fuel or fuel vapors, then turn the switch to ON position).
4. Check the fuel supply.

5. Make certain all manual valves are open.

6. Make certain the limit switches are closed.

7. Reset the safety switch and set the thermostat to call for heat.

CHAPTER 11

Gas-Fired Furnaces

Gas-fired furnaces are available in upflow, downflow (counterflow), or horizontal flow models for installation in attics, basements, closet spaces, or at floor level (Figs. 1, 2, and 3). These furnaces are also available in a wide range of heating capacities for different size structures.

This chapter is primarily concerned with a description of the procedures recommended for the installation, operation, servicing, and repair of gas-fired forced warm-air furnaces. Additional information about furnaces is contained in Chapter 10 (FURNACE FUNDAMENTALS).

Only gas-fired furnaces approved by the *American Gas Association* should be used in a heating installation. Not only does A.G.A certification insure a certain standard of quality, it is often required by many local codes and regulations. The furnace should be installed in accordance with procedures outlined by the *National Environmental Systems Contractors Association*.

PLANNING SUGGESTIONS

The first step in planning a heating system is to calculate the maximum heat loss for the structure. This should be done in accordance with procedures described in the manuals of the *National Warm Air Heating & Air Conditioning Association* or by a comparable method. *Correctly* calculating the maximum heat loss for the structure is very important, because the data will be used to determine the size (capacity) of the furnace selected for the installation.

If the heating or heating/cooling installation is to be approved by either the FHA or VA, heat loss and heat gain calculations should be made in accordance with the procedure described in *National Environmental Systems Contractors Manual J*.

Courtesy The Coleman Co., Inc.

Fig. 1. Gas-fired downflow furnace.

FLUE CONNECTION

FAN AND LIMIT
CONTROLS

BLOWER

BLOWER ACCESS
PANEL

BURNER ACCESS PANEL

GAS BURNERS

AUTOMATIC GAS VALVE

GAS MANIFOLD

HEAT EXCHANGER

PILOT LIGHT

Courtesy Lennox Air Conditioning and Heating Co.

Fig. 2. Horizontal gas-fired furnace.

Locations and Clearances

The proper location of a gas-fired furnace will depend upon a careful consideration of the following three factors:

1. Length of heat runs,
2. Chimney location,
3. Clearances.

A gas-fired forced warm-air furnace should be located as near as possible to the center of the heat distribution system and chimney. Centralizing the furnace reduces the need for one or more long supply ducts which tend to lose a certain amount of heat. The number of elbows should also be kept to a minimum for the same reason. Installing the furnace near the chimney will reduce the length of the horizontal run of flue pipe. The flue pipe should always be kept as short as possible.

Sufficient clearance should be provided for access to the draft hood and flue pipe. It is particularly important to locate a gas-fired furnace so that the draft hood is at least 6 inches from any combustible material.

FLUE CONNECTION

BURNER ACCESS PANEL

HEAT EXCHANGER

FAN AND LIMIT CONTROL

MAIN GAS VALVE

PILOT BURNER

MAIN BURNERS

BLOWER MOTOR

HAMMOCK-TYPE AIR FILTER

BELT DRIVE BLOWER

WIRING MAKE-UP BOX

Courtesy Lennox Air Conditioning and Heating Co.

Fig. 3. Gas-fired upflow furnace.

Allowing access for lighting the furnace and servicing it is also an important consideration. Most furnace manufacturers recommend a clearance of 24 to 30 inches in front of the unit for access to the burner and controls.

Clearances between the furnace and any combustible materials are usually provided by the manufacturer in the furnace specifications.

Installation Recommendations

New furnaces for residential installations are shipped preassembled from the factory with all internal wiring completed. In order to install the new furnace, the gas piping from the supply main, the electrical service from the line voltage main, and the

306

low-voltage thermostat must be connected. Directions for making these connections are to be found in the furnace manufacturer's installation instructions.

Make certain you have familiarized yourself with all local codes and regulations which govern the installation of a gas-fired furnace. Local codes and regulations take precedence over national standards. Your furnace installation *must* comply with the local codes and regulations.

Always check a new furnace for damages as soon as it arrives from the factory. If shipping damages are found, the carrier (*not* the factory) should be notified and a claim filed immediately. Heating equipment is shipped FOB from the factory, and it is the responsibility of the carrier to see that it arrives undamaged.

Always place the furnace on a solid level base. This will reduce vibrations from the equipment and keep operating noise to a minimum. In a crawl space installation, the furnace is either supported on a slab or on concrete blocks, or it is suspended from floor joists with ⅜ inch hanger rods.

A furnace installed in an attic should be placed on a fiber board sheeting base to absorb vibrations. Many furnaces designed for attic installations are certified for installation directly on combustible material such as ceiling joists or attic floors. In an attic installation, the furnace should be placed over a load bearing partition for additional support.

Duct Connections

The warm air plenum, warm air duct, and return air duct should be installed in accordance with the recommendations of the *National Environmental Systems Contractors Association*.

Detailed information about the installation of an air duct system is also contained in the following two publications of the *National Fire Protection Association*:

1. "Residence Type Warm Air Heating and Air Conditioning Systems" (NFPA No. 90B).
2. "Installation of Air Conditioning and Ventilating Systems of Other than Residence Type" (NFPA No. 90A).

Additional information about duct connections can be found in Chapter 7, Volume 2 (DUCTS AND DUCT SYSTEMS).

It is important for you to remember the following facts about furnace duct connections and the air distribution ducts:

1. Duct lengths should be kept to a minimum by centralizing the furnace location as much as possible.

2. The duct system must be properly sized. Undersizing will cause a higher external static pressure than the one for which the furnace was designed. This may result in a noisy blower or insufficient air distribution.

3. Duct sizing should include an allowance for the future installation of air conditioning equipment.

4. The furnace should be leveled before any duct connections are made.

5. Seal around the base of the furnace with a calking compound to prevent air leakage if a bottom air return is used.

6. Ducts should be connected to the furnace plenums with tight fittings.

7. Warm air plenum and return air plenum connections should be the same size as the openings on the furnace.

8. Use tapered fittings or starter collars between the ducts and the furnace plenum.

9. Ductwork located in unconditioned spaces (e.g. unheated attics, basements, crawl spaces etc.) should be insulated if air conditioning is planned for some future date.

10. Install canvas connectors between air plenums and the casing of the furnace to insure quiet operation. Check local codes and regulations to make certain canvas connectors are in compliance.

11. Line the first 10 feet or so of the supply and return ducts with acoustical material when extremely quiet operation is necessary.

12. Install locking-type dampers in each warm air run to facilitate balancing the system.

Electrical Wiring

All of the electrical wiring *inside* the furnace is completed and inspected at the factory before the unit is shipped. Wiring instructions will accompany each furnace, and the internal (factory installed) wiring will be clearly marked. The wiring to controls and to the electrical power supply must be completed by the installer, and the wiring should be done in accordance with the diagram supplied by the manufacturer. This wiring is usually indicated on wiring diagrams by broken lines.

All wiring connected to the furnace must comply with the *National Electric Code* and any local codes and regulations. Local codes and regulations will always take precedence over national ones when there is any conflict. In order to validate the furnace warranty, many manufacturers require that a local electrical authority approve all electrical service and connections.

The two types of electrical connections required to field wire a furnace are: (1) line voltage field wiring, and (2) control voltage field wiring. Both of these are shown in the wiring diagram for a *Carrier* heating/cooling unit (Fig. 4).

The line voltage wiring connects the furnace to the building power supply. It runs directly from the building power panel to a fused disconnect switch. From there, the wiring runs to terminals L_1 and L_2 on the power supply terminal block (i.e. the furnace junction box).

The unit must be properly grounded either by attaching a ground conduit for the supply conductors or by connecting a separate wire from the furnace ground lug to a suitable ground. The furnace must be electrically grounded in accordance with the *National Electric Code* (ANSI-C1-1971).

Furnaces equipped with motors in excess of ¾ hp or 12 amps ratings, must be wired to a separate 240 volt service in accordance with the *National Electric Code* and local codes and regulations. A 240 volt transformer should be installed whenever 240 volt service is required and/or a motor larger than ¾ hp is used.

The external control voltage circuitry consists of the wiring between the thermostat and the low voltage terminal block located in the control voltage section of the furnace. Instructions for control voltage wiring are generally shipped with the thermostat.

Ventilation and Combustion Air

A gas-fired furnace requires an unobstructed supply of air for combustion. This combustion air supply is generally supplied

LOW VOLTAGE
TERMINAL STRIP

40-VA TRANSFORMER
(STANDARD COMPONENT)

HUMIDIFIER
RELAY
RECEPTACLE

PLUG IN
BLOWER RELAY
(STANDARD
COMPONENT)

LINE VOLTAGE
CONNECTIONS

Courtesy Carrier Corp.

Fig. 4. Wiring diagram for a Carrier heating and cooling unit.

through natural air infiltration (cracks in the walls and around windows and doors) or through ventilating ducts.

If a gas-fired furnace is located in an open area (basement or utility room) and the ventilation is relatively unrestricted, there should be a sufficient supply of air for combustion and draft hood dilution. On the other hand, if the heating unit is located in an enclosed furnace room or if normal air infiltration is effectively reduced by storm windows or doors, then certain provisions must be made to correct this situation. Fig. 5 illustrates one type of

VENTILATING AIR OPENING
1 SQ. IN. FOR EACH
1000 Btu PER HR. INPUT

AIR INLET OPENING
1 SQ. IN. FOR EACH
1000 Btu PER HR. INPUT

Courtesy Magic Servant Products Co.

Fig. 5. Recommended air openings in furnace room wall.

modification that can be made to provide an adequate supply of combustion and ventilation air to a furnace room. As shown, two permanent grilles are installed in the walls of the furnace room, each of a size equal to one square inch of free area per 1000 Btu/hr of burner output. One grille should be located approximately 6 inches from the ceiling, and the other one near the floor. A possible alternative is to provide openings in the door, ceiling, or floor as shown in Fig. 6.

OPENING FOR
VENTILATION AIR
IN DOOR OR CEILING

OPENING FOR COMBUSTION AIR
IN DOOR OR FLOOR

Courtesy The Coleman Co., Inc.

Fig. 6. Alternative method of providing air supply openings.

If the furnace is located in a tightly constructed building, it should be directly connected to an outside source of air. A permanently open grille sized for at least one square inch of free area per 5000 Btu per hour of burner output should also be provided. Connection to an outside source of air is also recommended if the building contains a large exhaust fan.

Provisions for ventilation and combustion air are described in greater detail in "Installation of Gas Appliances and Gas Piping" (ASA Z21.30—NFPA No. 54). These standards were adopted and approved by the *National Fire Protection Association* and the *National Board of Fire Underwriters*.

Venting

Provisions must be made for venting the products of combustion to the outside in order to avoid contamination of the air in the living or working spaces of the structure. The four basic methods of venting the products of combustion to the outside are:

1. Masonry chimneys,
2. Low heat Type A prefabricated chimneys,
3. Type B gas vents,
4. Type C vents,
5. Wall venting.

Masonry and prefabricated chimneys are described below (see CHIMNEYS and TROUBLESHOOTING CHIMNEYS).

Low heat Type A flues are metal, prefabricated chimneys that have been tested and approved by the *Underwriters' Laboratories.* These chimneys are easier to install than masonry chimneys, and are generally safer under abnormal firing conditions.

Type B gas vents (Fig. 7) are listed by the *Underwriters' Laboratories,* and are recommended for venting all standard gas-fired furnaces. They are not recommended for incinerators, combination gas-oil appliances, or appliances convertible to solid fuel. Check the furnace specifications for the type of vent to use. It will usually be A.G.A. certified for use with a Type B gas vent. A Type C vent is used to vent gas-fired furnaces in attic installations (Fig. 8).

Wall venting involves gas-fired appliances which have their combustion process, combustion air supply, and combustion products isolated from the space which is being heated. These appliances are generally vented through the wall (Figs. 9 and 10).

Flue Pipe

The *flue* is the general term used to describe the passage through which the flue gases pass from the combustion chamber of the furnace (or boiler) to the outside. A flue is also referred to as a *flue pipe, vent pipe,* or *vent connector.*

The term *appliance flue* refers to the flue passages inside a furnace or boiler. A *chimney flue* is the vertical flue passage running up through the chimney. A *vent connector* is specifically the flue passage between the heating unit and the chimney. It is also variously referred to as a *chimney connector* or *smoke pipe.* A

313

BELMONT TOP

STORM COLLAR

ADJUSTABLE ROOF FLASHING
ALSO FLAT AND TALL CONE

ROUND VENT SIZES
FROM 3" TO 12"

SECTION LENGTHS:
6", 12", 18", 3', 5',
AND 12" ADJUSTABLE

ALUMINUM
INNER PIPE

AIR SPACE

GALVANIZED STEEL
OUTER PIPE

45^0 ADJUSTABLE ELBOW
ALSO IN 90^0

FIRESTOP SPACER

PLACEMENT OF REDUCING
TEE AT MINIMUM HEIGHT L_m
ABOVE SMALLER OF TWO OR
MORE APPLIANCES PERMITS
MULTIPLE SYSTEMS.

INCREASER OR REDUCER

DRAFT HOOD CONNECTOR

Courtesy Metalbestos Div. William Wallace Co.

Fig. 7. Double wall gas vent pipe and fittings for single or multiple type B gas vent system.

CLEARANCE - 2" MIN.
BUT NOT LESS THAN $\frac{A}{3}$

A = DIA. OF FLUE
B = A + 4 INCHES
C = A + 4 INCHES
D = NOT LESS THAN 24" OR NOT LESS THAN 6"
    ABOVE ROOF COPING

FLASHING AND ROOF CEMENT

ROOF LINE

ROOF BOARDS

2" OPEN AIR SPACE

Fig. 8. Type C vent.

flue outlet or *vent* is the opening in a furnace or boiler through which the flue gases pass.

The flue pipe (vent pipe) connects the smoke outlet of the furnace with the chimney. It should never extend beyond the inner liner of the chimney, and should never be connected to the flue of an open fireplace. Furthermore, flue connections from two or more sources should never enter the chimney at the same level from opposing sides.

The horizontal run of flue pipe should be pitched toward the chimney with a rise of at least ¼ inch per running foot. Some furnace manufacturers recommend as much as a 1 inch minimum rise per running foot. Do not allow the horizontal run to exceed, in length, 75% of the vertical run. Vent pipe crossovers in an attic must not extend at an angle of less than 60° from the vertical.

At intervals, fasten the horizontal run of flue pipe securely with sheet metal screws and support the pipe with straps or hangers to prevent sagging. Sagging can cause cracks to develop at the joints which may result in releasing toxic flue gases into the living and working spaces of the structure.

Fig. 9. Single package heating and cooling unit with through-the-wall venting.

The point at which the flue pipe enters the chimney should be at least two feet above the clean out opening of the chimney (Fig. 11). The flue pipe must be the same size as the outlet of the flue collector (furnace flue collar). Never install a damper in a flue pipe or reduce the size of the flue pipe. Run the flue pipe from the draft hood to the chimney in as short a distance as possible. All joints in the flue pipe should be made with the length nearest the furnace overlapping the other.

If excessive condensation is encountered, install a drip tee in place of an elbow. Never use dampers or other types of restrictors

Courtesy Climatrol.

Fig. 10. Cutaway of a gas-fired heating/cooling unit with through-the-wall venting design.

in the vent pipe. A minimum distance should be maintained between the vent pipe and the nearest combustible material.

When more than one appliance is vented into a common flue, the area of the common flue should be equal to the area of the largest flue plus 50% of the area of the additional flue.

CHIMNEYS

Most chimneys are constructed of brick or metal. If a brick chimney is used, it should be lined with a protective material to prevent damage from water vapor. The most commonly used liner is smooth faced tile. Prefabricated factory built chimneys are also used, but only those listed by *Underwriters' Laboratories* are suitable for use with fuel burning equipment.

FLUE PIPE

2 FT. MINIMUM

CLEANOUT DOOR

Fig. 11. Entry point of flue pipe into chimney.

The standard chimney must be at least 3 feet higher than the roof or 2 feet higher than any portion of the structure within 10 feet of the chimney flashing in order to avoid downdrafts (Fig. 12).

DOWN DRAFT BECAUSE OF PRESSURE DIFFERENCE

WIND

MINIMUM CLEARANCE 2 FEET

WIND

Courtesy Dunham-Bush.

Fib. 12. Examples of correct and incorrect chimney design.

TROUBLES

TOP OF CHIMNEY LOWER THAN
SURROUNDING OBJECTS

CHIMNEY CAP OR VENTILATOR

COPING RESTRICTS OPENING

OBSTRUCTION IN CHIMNEY

JOIST PROJECTING
INTO CHIMNEY

BREAK IN CHIMNEY LINING

COLLECTION OF SOOT AT
NARROW SPACE IN
FLUE OPENING

OFFSET

TWO OR MORE OPENINGS IN
SAME CHIMNEY

SMOKE PIPE EXTENDS
INTO CHIMNEY

LOOSE-SEATED PIPE
IN FLUE OPENING.

FAILURE TO EXTEND THE
LENGTH OF FLUE PARTITION
DOWN TO THE FLOOR

LOOSE-FITTED CLEANOUT DOOR

Fig. 13. Common chimney problems and their corrections.

An existing chimney should always be checked to make sure it is smoke tight and clean. Any dirt or debris must be cleaned out before the furnace or boiler is used.

TROUBLESHOOTING THE CHIMNEY

The chimney must give sufficient draft for combustion or the furnace will not operate efficiently. It must also provide a means for venting the products of combustion to the outside. Although a chimney produced draft is not as important for the combustion process in gas-fired appliances as it is in other types of fuel burning equipment, the venting capacity of the chimney is extremely important. The chimney must be of a suitable area and height to vent all the products of the combustion process.

Fig. 13 illustrates some of the common chimney problems that can cause insufficient draft and improper venting. Some of these problems are detectable by observation; others require the use of a draft gauge.

Problem	Remedy
Top of chimney is lower than surrounding objects	Extend chimney above all objects within 20 feet
Chimney cap or ventilator	Remove
Coping restricts opening	Make opening as large as inside of chimney
Obstruction in chimney	Use rod or weight on string or wire to break and dislodge
Joist projecting into chimney	Change support for joist so that chimney will be clear. Should be handled by a competent brick contractor
Break in chimney lining	Rebuild chimney with a course of brick between flue tiles
Collection of soot at narrow space in flue opening	Clean out with weighted brush or bag of loose gravel on end of line. May be necessary to open chimney

Offset	Change to straight or long offset
Two or more openings in same chimney	Least important opening must be closed using some other chimney flue.
Smoke pipe extends into chimney	Shorten pipe so that end is flush with inside of tile
Loose fitted cleanout door	Leaks should be eliminated by cementing all pipe openings
Failure to extend the length of flue partition down to the floor	Extend partition to floor level
Loosely fitted cleanout door	Close all leaks with cement

Sometimes a chimney will be either too small or too large for the installation. When this is the case, the chimney should be rebuilt with the necessary corrections made in its design.

DRAFT HOOD

A gas-fired furnace should be equipped with a draft hood attached to the flue outlet of the appliance. The draft hood used on the appliance should be certified by the *American Gas Association*. Only gas conversion furnaces equipped with power-type burners and conversion burner installations in large steel boilers with inputs in excess of 400,000 Btu/hr are not required to have draft hoods.

A draft hood is a device used to insure the maintenance of constant low draft conditions in the combustion chamber. By this action, it contributes to the stability of the air supply for the combustion process. A draft hood will also prevent excessive chimney draft and downdrafts that tend to extinguish the gas burner flame. Because of this last function, a draft hood is often referred to as a *draft diverter*.

Draft hoods may be either internally or externally mounted depending upon the design of the furnace. Never use an external type draft hood with a furnace already equipped with an *internal* draft hood. Either vertical or horizontal discharge from the draft hood is possible (Fig. 14). Locations of draft hoods in conversion burner installations are illustrated in Fig. 15.

321

BUILT-IN DRAFT HOOD WITH
VERTICAL DISCHARGE

BUILT-IN DRAFT HOOD WITH
HORIZONTAL DISCHARGE

EXTERNAL DRAFT HOOD
WITH VERTICAL DISCHARGE

EXTERNAL DRAFT HOOD
WITH HORIZONTAL DISCHARGE

Courtesy American Gas Association.

Fig. 14. External and internal (built-in) draft hoods with vertical and horizontal discharge.

NEUTRAL PRESSURE POINT ADJUSTER

In some installations, a neutral pressure point adjuster is installed in the flue pipe between the furnace and the draft hood. The procedure for making a neutral pressure point adjuster is

Courtesy Magic Servant Products Co.

Fig. 15. Location of draft hoods in conversion burner installation.

illustrated in Fig. 16. *Always* leave the neutral pressure point adjuster in wide open position until after the burner rating has been established.

BASIC COMPONENTS

The basic components of a gas-fired forced warm-air furnace are:

1. Automatic controls,
2. Heat exchanger,
3. Gas burner,
4. Gas pilot assembly,
5. Blower and motor,
6. Air filter(s).

323

SAW SLOT

SUGGESTED FORM OF NEUTRAL PRESSURE POINT ADJUSTER. INSERT IN SLOT CUT IN FLUE PIPE.

WHEN ADJUSTMENT IS COMPLETED, SCRIBE A LINE ON ADJUSTER SO IT CAN BE RELOCATED.

SCRIBE LINE AROUND FLUE PIPE

TRIM OFF EXCESS STOCK

1/2"

SCRIBED LINES

REMOVE ADJUSTER, TRIM OFF EXCESS, SLIT REMAINDER VERTICALLY AND BEND SEGMENTS IN ALTERNATE DIRECTIONS.

REPLACE IN FLUE PIPE, RECHECK ADJUSTMENTS TO INSURE UNCHANGED CONDITIONS, AND FASTEN ADJUSTER IN PLACE WITH SHEET METAL SCREWS.

Courtesy Magic Servant Products Co.

Fig. 16. Suggested construction of a neutral pressure point adjuster.

Each of these components is described in the sections that follow. Additional information is contained in Chapter 10 (FURNACE FUNDAMENTALS) and the various chapters in which furnace controls and fuel burning equipment are described.

AUTOMATIC CONTROLS

The automatic controls used in a heating system are designed to insure its safe and efficient operation. Detailed descriptions of these controls are found in Chapter 4 (THERMOSTATS AND HUMIDISTATS); Chapter 5 (GAS AND OIL CONTROLS); and Chapter 6 of Volume 2 (OTHER AUTOMATIC CONTROLS). Additional information can be found in the appropriate sections of Chapter 16 (BOILER AND FURNACE CONVERSIONS) and Chapter 2, Volume 2 (GAS BURNERS). This section is primarily concerned with outlining the operating principles of the automatic controls used with a gas-fired furnace. These controls include:

1. Room thermostat,
2. Fan and limit controls,
3. Pilot safety valve,
4. Automatic main gas valve,
5. Gas pressure regulator.

The operation of the furnace is controlled by the room thermostat. The thermostat senses air temperature changes in the space or spaces being heated and sends an electrical signal to open or close the automatic main gas valve. The thermostat is generally wired in series with the pilot safety valve, the automatic main gas valve, and the limit control. See Chapter 4, Volume 2 (THERMOSTATS AND HUMIDISTATS) for a detailed description.

The fan and limit control is a safety device which operates on a thermostatic principle. This combined control is designed to govern the operation of the furnace within a specified temperature range (usually 80°F to 150°F), and is located in the furnace plenum where it responds to outlet air temperatures.

One of the functions of this control is to prevent the furnace from overheating by shutting off the gas supply when the plenum air temperature exceeds the upper temperature setting (e.g. 180°F) on the control. The limit control switch (and pilot safety relay) responds to excessive plenum temperatures by closing the automatic main gas valve and thereby shutting off the flow of gas to the burner or burner assembly.

Another function of the fan and limit control is to stop the fan when the gas burner or burner assembly has been turned off.

When the plenum air drops below the lower setting (e.g. 80°F) on the control, the fan is automatically shut off. For additional information about fan and limit controls read the appropriate sections in Chapter 6, Volume 2 (OTHER AUTOMATIC CONTROLS).

The function of the automatic pilot safety valve is to shut off the main gas supply when the pilot flame goes out. There are several types of these pilot safety devices on the market, but the most commonly used ones are those which employ a thermo-couple lead in conjunction with a valve switch.

The automatic main gas valve controls the flow of gas to the burner manifold (in the case of a burner assembly) or to the gas burner. It is located between the burner (or burners) and the gas pressure regulator. Chapter 6, Volume 2 (OTHER AUTOMATIC CONTROLS) contains additional information about both pilot safety and main gas valves.

The gas in the supply main is frequently subjected to pressure variations. A gas pressure regulator installed in the piping between the supply riser and the automatic main gas valves insures a constant gas pressure in the burner manifold.

HEAT EXCHANGER

Each gas-fired furnace has a burner compartment containing the gas burner and pilot. The gas and air necessary for combustion are mixed in the venturi or mixing tube prior to ignition. This mixture is then ignited by the pilot flame as it leaves the burner. The burner flame is encased in a combustion chamber or fire box located directly below the furnace heat exchanger. The heat of the combustion process is transferred to the metal walls of the heat exchanger, and then to the air rising through it (Fig. 17).

The air flowing through a heat exchanger is subject to temperatures which frequently approach the 700-800°F range. Because of these high temperatures, the air flow must be uniform across all sections of the heat exchanger, and must move at a sufficient rate of speed to keep these temperatures from building any higher. A low volume of air flowing across one section of the heat exchanger will result in overheating and the failure of the entire heat exchanger assembly (Fig. 18).

Overheating can also be caused by a portion of the gas flame impinging on the inner surface of the heat exchanger (Fig. 19).

Fig. 17. Uniform air flow across all sections of the heat exchanger.

Courtesy The Trane Co.

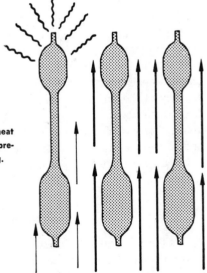

Fig. 18. Low air flow across one heat exchanger section may result in premature failure due to overheating.

Courtesy The Trane Co.

327

Fig. 19. Flame impingement in one section of the heat exchanger.

This results in the generation of extremely high temperatures. Under normal operating conditions, the gas flame should be directed upward through the center of each section.

Another cause of heat exchanger failure is over firing. This condition is usually traced to an excessively high manifold pressure or an orifice that is too large. The result is a firing rate that exceeds the design limits.

If the heat exchanger becomes so covered with soot, carbon deposits, and other contaminants that it reduces the operating efficiency of the furnace, it should be thoroughly cleaned. Gaining access to the heat exchanger is not easy, because it requires removing the burners and manifold assembly, vent connector, draft hood (diverter) and flue baffles. For this reason, you are advised to call a local service repairman if you want to clean the heat exchanger.

GAS BURNERS

Gas burners and gas burner assemblies are designed to provide the proper mixture of gas and air for combustion purposes. The gas burners used in residential heating installations operate primarily on the same principle as the *Bunsen* burner. See Chapter 2, Volume 2 (GAS BURNERS).

Gas-fired furnaces require both primary and secondary air for combustion process. The gas passes through the small orifice in the mixer head which is shaped to produce a straight flowing jet moving at high velocity (Fig. 20). As the gas stream enters the throat of the mixing tube, it tends to spread and induce air in through the opening of the adjustable shutter. This is the *primary air* which

Fig. 20. Schematic of air movements in a typical gas burner.

mixes with the gas *before* the air-gas mixture is forced through the burner ports (Fig. 20). The energy in the gas stream forces the mixture through the mixing tube into the burner manifold casting, from which it issues through ports where additional air must be added to the flame to complete combustion. This *secondary air* supply flows into the heat exchanger and around the burner. Its function is to mix with unburned gas in the heat exchanger in order to complete the combustion process.

The primary air is admitted at a ratio of about 5 parts primary air to 1 part gas for manufactured gas, and at a 10 to 1 ratio for natural gas. These ratios are generally used as theoretical values of air for purposes of complete combustion. Most burners used in gas-fired furnaces operate efficiently on 40 to 60 percent of the theoretical value. Primary air is regulated by means of an adjustable shutter. When burning natural gas, the air adjustment is gen-

329

erally made to secure as blue a flame as possible. Burner manufacturers provide a number of different ways to field adjust primary air. For example, shutters for primary air control are part of the main burners. Factory adjusted, angled orifices are also used on some burners.

The secondary air is drawn into the burner by natural draft. No provision is made by manufacturers to field adjust secondary air. Secondary air is controlled primarily by the design of the flue restrictors and draft diverter. The general configuration of the heat exchanger and burners are also important factors in secondary air control. Excess secondary air constitutes a loss and should be reduced to a proper minimum (usually not less than 25 to 35 percent).

As shown in Fig. 21, a typical gas flame consists of a well-defined inner cone surrounded by an outer envelope. Combustion of the primary air and gas mixture occurs along the *outer* surface of the inner cone. Any unburned gases combine with the secondary

OUTER ENVELOPE - COMBUSTION COMPLETED HERE

SECONDARY AIR

INNER CONE - COMBUSTION OF PRE-MIXED AIR AND GAS BURNS HERE

PERCENTAGE OF MAXIMUM VISIBLE CONE HEIGHT

UNBURNED GAS-AIR

AIR-GAS MIXTURE

Courtesy The Trane Co.

Fig. 21. Typical gas flame.

air and complete the combustion process in the outer envelope of the flame.

As Fig. 21 illustrates, the major portion of the inner cone consists of an unburned air-gas mixture. It is along the outer surface of the inner cone that combustion takes place. The height of the inner cone can be reduced by increasing the amount of primary air contained in the air-gas mixture. The extra oxygen increases the burning rate which causes the inner cone to decrease in size.

A burner is generally considered to be properly adjusted when the height of the inner cone is approximately 70% of the maximum visible cone height. The gas should flow out of the burner ports fast enough so that the flame cannot travel or flash back into the burner head. The velocity must not be so high that it blows the flame away from the port(s).

CROSSOVER

CROSSOVER ORIFICE

CROSSOVER IS FED FROM THE BURNER MANIFOLD. AIR FOR COMBUSTION IS SUPPLIED THROUGH THE PRIMARY AIR OPENING (NOT SHOWN) NEAR THE ORIFICE. IN THIS DESIGN THERE ARE 2 ORIFICES: ONE ON EITHER END OF THE CROSSOVER

BURNER MANIFOLD

Courtesy The Trane Co.

Fig. 22. Crossover located perpendicular to and across top of the main burners near the port area.

331

CROSSOVER BURNER

Most modern gas-fired furnaces consist of several individual heat exchanger sections with a gas burner inserted in each section. In this arrangement, all burners must fire almost simultaneously or there will be the danger of unburned gas flowing into the combustion chamber. Simultaneous ignition can be provided by installing a crossover burner across the top of the main burners near the port area. Figs. 22 and 23 show the types of crossovers used in *Trane* gas-fired furnaces.

MUST FIT SNUGLY AGAINST THE BURNER

CROSSOVER

GAS-AIR MIXTURE FED TO THIS SECTION OF CROSSOVER BY BURNER #3

BURNER #1   BURNER #2   BURNER #3

BURNER #2 FEEDS THIS SECTION OF CROSSOVER.
SCOOP IN BURNER DIRECTS FLOW UP INTO CROSSOVER

Courtesy The Trane Co.

Fig. 23. Crossover arrangement in which each burner feeds a portion of the gas-air mixture into the crossover where it is ignited by the pilot flame as it leaves the crossover port.

GAS PILOT ASSEMBLY

The gas pilot assembly consists of a pilot burner and a pilot safety device (Fig. 24). The most commonly used pilot safety is the thermocouple lead operating in conjunction with a companion valve or switch. The thermocouple itself is actually a miniature generator which converts heat (from the pilot flame) into a small electrical current. The upper half of the thermocouple is located in the pilot flame. As long as the pilot flame is operating, an electrical circuit is maintained between the pilot burner and the gas valve. If the pilot flame goes out, the circuit between the thermocouple and the gas valve is broken. As a result, the gas valve closes and shuts off the gas flow to the pilot burner.

Other types of pilot safety devices used with gas-fired furnaces are described in Chapter 5, Volume 2 (GAS AND OIL CONTROLS).

PILOT BURNER ADJUSTMENT

It is sometimes necessary to adjust the pilot flame. On some combination gas controls a pilot adjustment screw is provided for

Fig. 24. Pilot burner assembly.

333

this purpose. On units using natural gas, the screw should be adjusted until the pilot flame completely envelopes the thermocouple. If propane gas is used, the screw should be wide open.

Sometimes a pilot gas regulator is used on a furnace to control pressures in areas where gas pressure variation are great. The pilot valve is placed in wide open position and pilot flame adjustments are made by adjusting the gas pressure regulator.

GAS INPUT ADJUSTMENT

The gas input of a furnace must not be greater than that specified on the rating plate. However, it is permissible to reduce the gas input to a *minimum* of 80 percent of the rated input in order to balance the load.

The gas input can be checked by timing the flow through the meter. The procedure for clocking the meter is as follows:

1. Turn off all gas units connected through the meter except the furnace. The main burners of the furnace must be on for the timing check.

2. Time one complete revolution of the hand (or disc) on the meter. The hand should be the one with the smallest cubic footage per revolution (usually ½ or 1 cubic foot per revolution on small meters and 5 cubic feet on large meters).

3. Using the size of the test meter dial (½, 1, 2, or 5 cubic feet) and the number of seconds per revolution, determine the cubic feet per hour from the appropriate column in Table 1.

Table 1. Gas rate in cubic feet per hour.

Seconds For One Revolution	Size of Test Meter Dial				Seconds For One Revolution	Size of Test Meter Dial			
	One-Half Cu. Ft.	One Cu. Ft.	Two Cu. Ft.	Five Cu. Ft.		One-Half Cu. Ft.	One Cu. Ft.	Two Cu. Ft.	Five Cu. Ft.
	Cubic Feet Per Hour					Cubic Feet Per Hour			
10	180	360	720	1800	50	36	72	144	360
11	164	327	655	1636	51	35	71	141	353
12	150	300	600	1500	52	35	69	138	346
13	138	277	555	1385	53	34	68	136	340
14	129	257	514	1286	54	33	67	133	333

15	120	240	480	1200	55	33	65	131	327
16	112	225	450	1125	56	32	64	129	321
17	106	212	424	1059	57	32	63	126	316
18	100	200	400	1000	58	31	62	124	310
19	95	189	379	947	59	30	61	122	305
20	90	180	360	900	60	30	60	120	300
21	86	171	343	857	62	29	58	116	290
22	82	164	327	818	64	29	56	112	281
23	78	157	313	783	66	29	54	109	273
24	75	150	300	750	68	28	53	106	265
25	72	144	288	720	70	26	51	103	257
26	69	138	277	692	72	25	50	100	250
27	67	133	267	667	74	24	48	97	243
28	64	129	257	643	76	24	47	95	237
29	62	124	248	621	78	23	46	92	231
30	60	120	240	600	80	22	45	90	225
31	58	116	232	581	82	22	44	88	220
32	56	113	225	562	84	21	43	86	214
33	55	109	218	545	86	21	42	84	209
34	53	106	212	529	88	20	41	82	205
35	51	103	206	514	90	20	40	80	200
36	50	100	200	500	94	19	38	76	192
37	49	97	105	486	98	18	37	74	184
38	47	95	189	474	100	18	36	72	180
39	46	92	185	462	104	17	35	69	173
40	45	90	180	450	108	17	33	67	167
41	44	88	176	440	112	16	32	64	161
42	43	86	172	430	116	15	31	62	155
43	42	84	167	420	120	15	30	60	150
44	41	82	164	410	130	14	28	55	138
45	40	80	160	400	140	13	26	51	129
46	39	78	157	391	150	12	24	48	120
47	38	77	153	383	160	11	22	45	113
48	37	75	150	375	170	11	21	42	106
49	37	73	147	367	180	10	20	40	100

Courtesy Dunham-Bush, Inc.

4. Determine the actual Btu per hour input of the burner by multiplying the cubic feet per hour rate (obtained in Step 3) by the Btu per cubic feet of gas (obtained from the local gas company).

By the way of example, let's assume that you have contacted the local gas company and they have given you a value of 1040 Btu per cubic foot. When you clock the meter, you find that it takes exactly 18 seconds for the hand to make one complete revolution on the ½ cubic foot meter dial. With this information, turn to Table 1 and find 18 seconds in the extreme left hand column. Opposite the value 18 seconds under the ½ cubic foot column heading read 100 cubic feet. The rest is simple multiplication:

$$100 \text{ cu. ft. per hour} \times 1040 \text{ Btu per cu. ft.}$$
$$= 104,000 \text{ Btu per hour.}$$

If the Btu per hour rate is not within 5% of the desired value, change the burner orifices or adjust the pressure regulator. Pressure regulator setting changes are *only* used for *minor adjustments* in Btu input. Major adjustments are made by changing the burner orifices.

When making adjustments, remember the following: *Btu input may never be less than the minimum or more than the maximum on the rating plate.*

There is no meter to clock on furnaces fired with propane. The burner orifices on these units are sized to give the proper rate at 11 inches W.C. (water column) manifold pressure.

After the correct Btu input rate has been determined, adjust the burner for the most efficient flame characteristics (see COMBUSTION AIR ADJUSTMENTS).

HIGH ALTITUDE ADJUSTMENT

For elevations *above* 2000 feet, the Btu input should be reduced (derated) 4% for each 1000 feet above sea level. This adjustment is required because of the increase in air volume at higher elevations. Air expands at approximately 4% per 1000 feet of elevation. At elevations above 2000 feet, the volume of air required to supply enough oxygen for complete combustion exceeds the maximum air inducing ability of the burners. Because the primary air volume cannot be increased, the gas flow (input) must be decreased (i.e. derated).

CHANGING BURNER ORIFICES

A gas-fired furnace is supplied with standard burner orifices for the gas shown on the rating plate. The orifice size will depend on the calorific value of the gas (Btu per cubic foot) and the manifold pressure (3.5 inches W.C. for natural gas and 11 inches W.C. for LP gas). The burner orifice *size* partially determines the firing rate by allowing only a predetermined volume of gas to pass. An orifice that is too large will allow too much gas to pass through into

Table 2. Orifice capacity table.

Drill Size	*Approximate Orifice Capacity in Btu Per Hour	
	Natural Gas @ 3½" W.C.	Propane Gas @ 11" W.C.
60	4650	13,000
59	4900	13,750
58	5150	14,450
57	5400	15,150
56	6300	17,600
55	7900	22,000
54	8800	24,750
53	10,400	28,800
52	11,750	33,000
51	13,000	36,700
50	14,250	40,000
49	15,500	43,500
48	16,750	47,200
47	17,900	50,200
46	19,000	53,500
45	19,500	55,000
44	21,500	60,050
43	23,000	
42	25,250	
41	26,750	
40	27,800	
39	28,600	
38	29,900	
37	31,100	
36	33,000	
35	35,000	

(*Values determined using the following data: Propane at 1.52 Specific Gravity/ 2500 Btu Per Cubic Foot and Natural Gas at .62 Specific Gravity/1000 Btu Per Cubic Foot).

Courtesy *The Trane Company*

the burner. This condition sometimes results in overheating the heat exchanger, or restricting the air inducing ability of the burner.

The gas burner orifices supplied with a furnace are suitable for *average* calorific values of gas at the listed manifold pressures. Table 2 lists the various orifice capacities for different drill sizes. The procedure for determining the suitability of a particular drill size is as follows:

1. Divide the total Btu input of the furnace by the *total* number of burner orifices in the unit.

2. From Table 2, select the burner orifice size closest to the calculated value (plus or minus 5%) for the gas used.

This procedure can be illustrated by determining the most suitable drill size for a 150,000 Btu input gas-fired furnace. This furnace has 4 burners with 2 orifices per burner, and uses natural gas.

$$\text{capacity of each burner orifice} = \frac{150,000}{4 \times 2}$$

$$= \frac{150,000}{8}$$

$$= 18,750$$

In Table 2, the figure closest to 18,750 in the left hand column (Natural Gas) is 19,000. The drill size opposite this figure is #46.

MANIFOLD PRESSURE ADJUSTMENT

A suitable manifold pressure is important to efficient furnace operation. If the gas pressure is too low, it will cause rough ignition, incomplete and inefficient combustion, and incorrect fan control response. An excessively high manifold pressure may cause the burners to overfire the heat exchanger. Overfiring the heat exchanger not only reduces the life of this component, it also may result in repeated cycling of the burner on the high limit control.

Manifold pressure should be set at 3.5 inches W.C. for natural gas. Control manufacturers normally preset the pressure regulator in natural gas valves at the time of manufacture so that natural gas units are fired at this 3.5 inches W.C. rate. Only small variations in the gas flow should be made by adjusting the pressure regulator. This adjustment should never exceed plus or minus 0.3 inches W.C. Major changes in the gas flow should be made by changing the size of the burner orifice (see below).

The manifold gas pressure can be tested by using a U-tube water manometer (Fig. 25). The manometer is connected to the manifold through an opening covered by a ⅛ inch plug cap. The pressure test *must* be run while the unit is operating. On *Dunham-Bush* gas-fired furnaces, the test is run through an opening in the gas valve (Fig. 26). The furnace manufacturer's installation literature should contain instructions for testing the manifold pressure.

Once you have determined the manifold pressure, you may find it necessary to adjust it for better operating characteristics. Manifold pressure can be increased by turning the adjusting screw clockwise, and decreased by turning it counterclockwise.

Courtesy The Trane Co.

Fig. 25. Applying a V-tube manometer to gas manifold.

339

PRESSURE TAP –
REMOVE 1/8" PIPE PLUG IN
GAS VALVE AND INSTALL
"B" COCK FOR TUBING
CONNECTION. REPLACE
1/8" PLUG AFTER
CHECKING PRESSURE

Courtesy Dunham-Bush, Inc.

Fig. 26. Applying a V-tube manometer to gas valve.

LP gas units are fired at 11 inches W.C. manifold pressure. LP gas (propane and butane) is heavier than natural gas and has a higher heating value. As a result, LP gas needs more primary air for combustion. Thus, a higher manifold pressure is required to induce the greater volume of primary air into the burner than is the case in natural gas units.

An LP furnace does not have a pressure regulator in the gas valve. Pressure adjustments are made by means of the regulator at the supply tank. These adjustments *must* be made by the installer, and regularly checked by the serviceman for possible malfunctions.

COMBUSTION AIR ADJUSTMENT

Primary air shutters are provided on all furnaces to enable adjustment of the primary air supply. The purpose of this adjustment is to obtain the most suitable flame characteristics. Experience has shown that the most efficient burner flame for natural gas has a soft blue cone without a yellow tip. For propane gas, there should be a little yellow showing in the tips of the flame.

The procedure for adjusting the primary air on a *Coleman* gas-fired furnace may be summarized as follows:

1. Light the pilot burner.
2. Turn up the setting on the room thermostat until the main burners come on.
3. Allow the main burners 10 minutes to warm up.
4. Loosen lock nut on adjusting screw (Fig. 27).

Courtesy The Coleman Co., Inc.

Fig. 27. Position of the lock nut and adjusting screw on a Coleman gas-fired furnace.

5. Turn adjusting screw IN (clockwise) until yellow tip appears in the flame.
6. Turn adjusting screw OUT (counterclockwise) until the yellow tip *just* disappears.
7. Hold adjusting screw and tighten lock nut.
8. Repeat Steps 4-7 on each of the other burners.

On a *Fedders* gas-fired furnace, a locking screw is located above the primary air opening on each gas burner (Fig. 28). The primary air adjustment procedure is as follows:

1. Loosen the locking screw at the base of the burner.
2. Adjust the air shutter opening to a position that gives a slight yellow tip on the end of the flame.

341

LOCKING SCREW

PRIMARY
AIR OPENING

AIR SHUTTER
(ADJUSTABLE)

BURNER

Courtesy Fedders Corp.

Fig. 28. Primary air adjustment on a *Fedders* gas-fired furnace.

3. Open the air shutter until the yellow tip just disappears.

4. Tighten the locking screw.

5. Repeat Steps 1-4 for each of the burners.

A primary air shutter assembly is provided on all *Carrier* gas-fired furnaces in order to simplify the adjustment procedure. As shown in Fig. 29, adjustment of the end burner results in the simultaneous adjustment of all burners.

After the primary air adjustments have been made, check to see that the burners are level. The burner flames should be uniform and centered in the heat exchanger (see HEAT EXCHANGER in this chapter).

OUTBOARD BURNER ADJUSTS ALL BURNERS SIMULTANEOUSLY

Courtesy Carrier Corp.

Fig. 29. Primary air shutter assembly on *Carrier* furnace.

342

Before allowing the furnace to continue operating, you must check to see that it has the proper draft. This can be done by passing a match along the draft hood opening. If the vent is drawing properly, the match flame will be drawn inward (i.e. into the draft hood). If the furnace is not receiving proper draft, the products of combustion escaping the draft hood will extinguish the flame. The draft must be corrected before operating the furnace.

GAS SUPPLY PIPING

The *gas supply piping* (also referred to as the *gas service piping)* must be sized and installed in accordance with the recommendations contained in the latest edition of "American National Standards for Installation of Gas Appliances and Gas Piping" (ANSI Z21.30-1964) and "Piping for Gas Appliances" (NFPA No. 54-1964). The location of the main shutoff valve will usually be specified by local codes or regulations. Consult the local codes or regulations for recommended sizes of pipe for the required gas volumes. Local codes or regulations *always* take precedence over national standards when there is a conflict between the two.

Generally 1 inch supply pipe is adequate for furnace inputs up to 125,000 Btu/hr. A 1¼ inch pipe is recommended for higher inputs. The inlet gas supply pipe size should never be smaller than ½ inch. The gas line from the supply should serve only a single unit.

Table 3. Capacity of pipe (Cu. Ft./Hr.)
(Pressure Drop 0.3 — Specific Gravity 0.60)

Length of Pipe (Ft.)	Diameter of Pipe (Inches)			
	¾	1	1¼	1½
15	172	345	750	1220
30	120	241	535	850
45	99	199	435	700
60	86	173	380	610
75	77	155	345	545
90	70	141	310	490
105	65	131	285	450
120		120	270	420

Courtesy Janitrol.

A *drip leg* (also called a *dirt leg* or a *dirt trap*) should be installed at the bottom of the gas supply riser to collect moisture or

Fig. 30. Location of manual valve on horizontal gas supply line.

impurities carried by the gas. A manual main shutoff valve should be installed either on the gas supply riser or on the horizontal pipe between the riser and ground union joint Figs. 30 and 31). The location of the main shutoff valve will usually be specified by local codes and regulations.

The *ground union joint* should be installed between the manual main shutoff valve and the gas control valve on the furnace or boiler (Fig. 32). The ground union joint provides easy access to the gas controls on the unit for servicing or repair.

SIZING GAS PIPING

The gas supply pipe should be sized per volume of gas used and allowable pressure drop. The pipe must also be of adequate size to prevent undue pressure drop. The diameter of the supply pipe must *at least* equal that of the manual shut off valve in the supply riser. Minimum acceptable pipe sizes are listed in Table 3.

The volume rate of gas in cubic feet per hour is determined by dividing the Btu per hour required by the furnace by the Btu per

MANUAL GAS VALVE

DIRECTION OF FLOW

Fig. 31. Location of union and drip leg for connecting furnace to main gas supply line.

TEE

UNION

PIPE CAP

GAS VALVE

UNION

GAS SUPPLY

MANUAL GAS VALVE

DIRT LEG

Courtesy The Coleman Co.

Fig. 32. Location of ground union joint.

345

Table 4. Multiplier table.

Specific Gravity	Multiplier	Specific Gravity	Multiplier
.40	.813	1.30	1.47
.50	.910	1.40	1.53
.60	1.00	1.50	1.58
.70	1.08	1.60	1.63
.80	1.15	1.70	1.68
.90	1.22	1.80	1.73
1.00	1.29	1.90	1.77
1.10	1.35	2.00	1.83
1.20	1.42		

Courtesy *Janitrol*

cubic foot of gas being used. The cubic feet per hour value can then be used to determine pipe size for gas of a specific gravity other than 0.60. The data used for these calculations is listed in Tables 3 and 4. The procedure is as follows:

1. Multiply the cubic feet per hour required by the multiplier (Table 4).

2. Find recommended pipe size for length of run in Table 3.

Tables 3 and 4 apply to the average piping installation where four or five fittings are used in the gas piping to the furnace. Existing pipe should be converted to the proper size of pipe where necessary. In no case should pipe less than ¾ inch diameter be used. Any extensions to existing piping should conform to Tables 3 and 4.

INSTALLING GAS PIPING

The principal recommendations to be followed when installing gas piping are:

1. A single gas line must be provided from the main gas supply to the furnace.

2. The gas supply pipe must be sized according to the volume of gas used and the allowable pressure drop.

3. The supply riser and drop leg must be installed adjacent to the furnace on either side. Never install it in front of the furnace where it might block access to the unit.

4. A manual main shutoff valve or plugcock should be installed on the gas supply piping (usually on the riser) outside the unit in accordance with local codes and regulations.

5. Install a tee for the drop leg in the pipe at the same level as the gas supply (inlet) connection on the furnace.

6. Extend the drip leg and cap it.

7. Install a ground union joint between the tee and the furnace controls.

8. Use a pipe compound resistant to LP gas on all threaded pipe connections in an LP gas installation.

9. Test for pressure and leaks. Use a solution of soap or detergent and water to detect any leaks. *Never* check for leaks with a flame.

START UP INSTRUCTIONS

The start up procedure for most gas-fired furnaces is very similar, particularly insofar as use of the thermostat is concerned. The major difference is in the lighting of the pilot burner, and this will depend upon the type of gas valve used.

The start up procedure for a gas-fired furnace may be summarized as follows:

1. Open all warm air registers.

2. Shut off main gas valve and pilot gas cock, and wait at least 5 minutes to allow gas which may have accumulated in the burner compartment to escape.

3. Set the room thermostat at the lowest setting.

4. Open the water supply valve if a humidifier is installed in the system.

5. Turn off the line voltage switch if one is provided in the furnace circuit.

6. Light the pilot (see below).

7. Turn electrical power on.

8. Replace all access doors.

9. Light main burners by setting the room thermostat *above* the room temperature.

10. Set the room thermostat at the desired temperature after the main burners are lighted.

Lighting The Pilot

The pilot lighting instructions should be included with the furnace and prominently displayed. Lighting instructions for the 24 volt *Honeywell* V8280, V8243, and *Robertshaw (Unitrol)* 7000 combination gas valves basically summarize the pilot lighting procedure (Fig. 33). The procedure is as follows:

1. Depress gas cock dial and turn it counterclockwise to "pilot" position.
2. Keep gas cock dial depressed and light pilot with a match.
3. Keep dial depressed until pilot remains lighted when released.
4. Depress dial and turn counterclockwise to "on" position.

BLOWERS AND MOTORS

A forced warm-air furnace should be equipped with a variable speed electrical motor designed for continuous duty. The motor

Courtesy Robertshaw Controls Co.

Fig. 33. *Robertshaw Unitrol* 7000 series combination gas valve.

should be provided with overcurrent protection in accordance with the *National Electrical Code* (ANSI C1-1971).

Blower motors are available as either direct-drive or belt-driven units. On direct-drive units, the blower wheel is mounted directly on the motor shaft and runs at motor speed. The blower wheel of a belt-driven unit is operated by a V-belt connecting a fixed pulley mounted on the blower shaft with a variable pitch drive pulley mounted on the motor shaft.

The blower motors used in residential furnaces are single-phase electrical motors. Depending upon the method used for starting, they may be classified as:

1. shaded pole,
2. permanent split capacitor,
3. split-phase,
4. capacitor-start motors.

A direct-drive blower generally uses a shaded pole (SHP) or permanent capacitor (PSC) type motor. Belt-driven blower motors are usually split-phase (SPH) types up to the ⅓ hp size. Larger motors are capacitor-start split-phase motors.

Blower motors are fractional horsepower motors commonly available in the following sizes: ¹⁄₁₂, ¹⁄₁₀, ⅙, ¼, ⅓, ½, and ¾ horsepower. A fractional horsepower motor should be protected with temperature or current-sensitive devices to prevent motor winding temperatures from exceeding those allowed in the specifications.

Provisions should be made for the periodic lubrication of the blower and motor. Use only the type and grade of lubricant specified in the operating instructions. Do not over lubricate. Too much oil can be just as bad as too little. A direct-drive blower should be capable of operating within the range of air temperature at the static pressures for which the furnace is designed and at the voltage specified on the motor name plate.

Motors using belt drives must be supplied with adjustable pulleys. The only exception is a condenser motor used on a forced warm-air furnace with a cooling unit. *Always* provide the means for adjusting a belt-driven motor in an easily accessible location.

Most forced warm-air furnaces are shipped with the blower motors factory adjusted to drive the blower wheel at a correct speed for heating. This is a *slower* speed than the one used for air conditioning, but it is relatively easy to alter the air delivery to meet the requirements of the installation see AIR CONDITIONING in this chapter).

Air Delivery and Blower Adjustment

The furnace blower is used to force air through the heat exchanger in order to remove the heat produced by the burners. This removal of heat by the blower serves a twofold purpose: it distributes the heat through the supply ducts to the various spaces in the structure, and (2) it protects the surfaces of the heat exchanger from overheating.

Most blowers are factory set to operate with a specific air temperature rise. For example, *Janitrol* Series 37 gas-fired furnaces are certified by the American Gas Association to operate within a range of 70°-100°F air temperature heat rise. In other words, the temperature of the air within the supply ducts may be 70°-100°F higher than the air temperature in the return duct. A temperature rise of 85°-90°F with approximately 0.1 inch external pressure in the area beyond the unit and filter is generally considered adequate for most residential applications.

The furnace air delivery rate should be adjusted at the time of installation to obtain a temperature rise within the range specified on the furnace rating plate. Many furnace manufacturers recommend that the temperature rise be set **below** the maximum to insure that temperature rise will not exceed maximum requirement should the filter become excessively dirty.

Although the blower and motor are factory set to operate within a specific temperature rise range, the unit can be field adjusted to deliver more or less air as required. The air delivery rate is adjusted by changing the blower speed. On direct-drive motors, this is accomplished by moving the line lead to the desired terminal. Belt-driven motors are adjusted by changing the effective diameter of the driver pulley. (see below: DIRECT DRIVE BLOWER ADJUSTMENT and BELT DRIVE BLOWER ADJUSTMENT).

Direct Drive Blower Adjustment

The blower (fan) of a direct drive unit is operated by a drive shaft connected to the motor. Three common types of direct drive motors used with blowers are:

1. Two speed direct drive shaded pole motor.
2. Three speed direct drive PSC motor.
3. Four speed direct drive PSC motor.

These blower motors are shipped from the factory set for low speed heating operation. If a higher operating speed is desired, the red blower motor wire must be moved from the low speed red terminal to a higher speed terminal (Fig. 34).

Courtesy *The Coleman Co., Inc.*

Fig. 34. Blower motor speed terminals.

Belt Drive Blower Adjustment

Air temperature rise can be decreased by increasing the blower speed. A belt drive unit can be adjusted for a higher speed as follows:

1. Loosen the set screw in the movable outer flange of the motor pulley.
2. Reduce the distance between the fixed flange and the movable flange of the pulley.
3. Tighten the set screw.

The procedure for decreasing the blower speed (and thereby increasing the air temperature rise) is similar except that in Step 2 above the distance between the fixed flange and the movable flange of the pulley are *increased*.

The motor and blower pulley should be aligned to minimize belt wear. The belt should be parallel to the blower scroll, as shown in Fig. 35. Adjustments can be made by moving the motor or adjusting the pulley on the shaft.

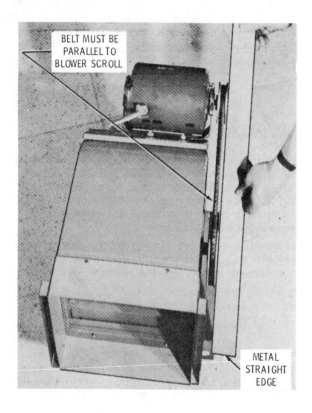

BELT MUST BE PARALLEL TO BLOWER SCROLL

METAL STRAIGHT EDGE

Courtesy The Coleman Co., Inc.

Fig. 35. Checking belt alignment.

Correct belt tension is also important. Belt slippage occurs when there is too little tension. Too much tension will cause motor overload and bearing wear. Correct belt tension will usually be indicated by the amount of depression in the belt midway between the blower and motor pulleys. Usually this will be a 1 inch depression at the center of the belt. On the *Coleman* belt driven unit shown in Fig. 36 it is only ½ inch.

Fig. 36. Checking belt tension.

AIR FILTERS

A *forced* warm-air gas-fired furnace is supplied with either a disposable air filter or a permanent (washable) one. The type of filter will be indicated on a label attached to the filter. *Never* use a filter with a gravity warm-air furnace, because it will obstruct the air flow.

Proper maintenance of the air filter is very important to the operating efficiency of the furnace. A dirty, clogged air filter reduces the air flow through the furnace and causes air temperatures inside the unit to rise. This increase in air temperature results in a reduced life for the heat exchanger and a lower operating efficiency for the furnace.

If a disposable air filter is used, it should be inspected on a monthly basis. If it is dirty, it should be replaced. This is usually done at the beginning and middle of the heating season. The same schedule should be followed during the summer months when air conditioning is used.

If a permanent filter is provided, it must be removed and cleaned from time to time. Cleaning the filter involves the following procedure:

353

1. Shake it to remove dust and dirt particles.
2. Vacuum it.
3. Wash it in a solution of soap or detergent and water.

Courtesy Fedders Corp.

Fig. 37. Location of air filter in upflow furnace.

RETURN AIR PLENUM

FLEXIBLE DUCT CONNECTOR
RECOMMENDED

BLOWER CABINET

AIR FILTERS

FLUE PIPE

USE APPROPRIATED
FLOOR BASE LISTED
IN TABLE V WHEN
INSTALLED ON
COMBUSTIBLE FLOORING

FIBERGLASS SEAL
SUPPLIED BY INSTALLER

NONCOMBUSTIBLE FLOOR

DISCHARGE AIR DUCT

Courtesy Fedders Corp.

Fig. 38. Location of air filter in a counterflow furnace.

355

4. Allow time for it to dry.

5. Replace it in the furnace.

Read the label on the filter for any special instructions that may apply to cleaning it. The location of filters in upflow and counter-flow furnaces is shown in Figs. 37 and 38. Note that in each case the air filter is placed directly in the path of the air entering the furnace. An external filter unit for a horizontal gas-fired furnace is shown in Fig. 39.

AIR FLOW

Courtesy Carrier Corp.

Fig. 39. External air filter unit in a Carrier horizontal gas-fired furnace.

A filter should be installed in the return plenum or duct when air conditioning is added to the system. Because air conditioning normally requires a greater volume of air than heating, the filter should be sized for air conditioning rather than heating.

Access to internally installed filters is generally provided through a service panel on the furnace. Instructions for cleaning or replacing the filters should be placed on the service panel. It would be a good idea to include the dimensions of the replacement filter(s) in the information.

An adequate filter should provide at least 50 square inches of area per 100 cfm of air circulated. For an air velocity of 1000 cfm, this would mean a filter area of 400 square inches. Filters sized on 50 square inches for each 100 cfm meet the requirements of air conditioning, which requires a greater volume of air than heating.

Never use a smaller size filter than the one required in the specifications. To do so would create excessive pressure drop at the filter which would reduce the operating efficiency of the furnace.

Additional information about air filters can be found in Chapter 14, Volume 3 (AIR CLEANERS AND FILTERS).

AIR CONDITIONING

If you intend to install air conditioning equipment at some future date, the duct sizing should include an allowance for it. Air conditioning involves a greater volume of air than heating. Quite often the blowers and motors of the furnace will be sized for the addition of cooling with matching evaporator coil and condensing unit. Check the specifications to be sure.

All ductwork located in unconditioned areas (e.g. attics, crawl spaces, etc.) *downstream* from the furnace should be insulated. A furnace used in conjunction with a cooling unit should be installed in parallel or on the upstream side of the evaporator coil to avoid condensation on the heating element. In a parallel installation, dampers or comparable means should be provided to prevent chilled air from entering the furnace.

INSTALLATION AND MAINTENANCE CHECK LIST

A properly installed and maintained furnace will operate efficiently and economically. The following installation check list is offered as a guide for the installer.

1. Adjust primary air shutter.
2. Provide sufficient space for service accessibility.
3. Check all field wiring.
4. Supply line fuse or circuit breaker must be of proper size and type for furnace protection.
5. Line voltage must meet specifications while furnace is operating.
6. Air flow must be sufficient.
7. Ventilation and combustion air must be sufficient.
8. Duct work must be checked for proper balance, velocity, and quietness.

9. Measure the manifold pressure.

10. Check all gas piping connections for gas leaks (USE SOAP AND WATER NOT A FLAME).

11. Cycle the burners.

12. Check limit switch.

13. Check fan switch.

14. Adjust blower motor for desired speed.

15. Make sure air filter is properly secured.

16. Make sure all access panels have been secured.

17. Pitch air conditioning equipment condensation lines towards a drain.

Courtesy Robertshaw Controls Co.

Fig. 40. Installation and maintenance check list.

18. Check thermostat heat anticipator setting.

19. Check thermostat for normal operation.

20. Clear and clean the area around the furnace.

A general check list for both furnace installation and maintenance is shown in Fig. 40. Other maintenance recommendations are contained in various sections of this chapter.

TROUBLESHOOTING A GAS-FIRED FURNACE

The list that follows contains the most common operating problems associated with gas-fired furnaces. Each problem is given in the form of a symptom, the possible cause, and a suggested remedy. This list is intended to provide the operator with a quick reference to the cause and correction of a specific problem.

Symptom and Possible Cause	Possible Remedy
Flame Too Large	
(a) Pressure regulator set too high	(a) Reset using manometer
(b) Defective regulator	(b) Replace
(c) Burner orifice too large	(c) Replace with correct size
Noisy Flame	
(a) Too much primary air	(a) Adjust air shutters
(b) Noisy pilot	(b) Reduce pilot gas
(c) Burr in orifice	(c) Remove burr or replace orifice
Yellow Tip Flame	
(a) Too little primary air	(a) Adjust air shutters
(b) Clogged burner ports	(b) Clean ports
(c) Misaligned orifices	(c) Realign
(d) Clogged draft hood	(d) Clean
Floating Flame	
(a) Blocked venting	(a) Clean
(b) Insufficient primary air	(b) Increase primary air supply

359

Delayed Ignition

(a)	Improper pilot location	(a)	Reposition pilot
(b)	Pilot flame too small	(b)	Check orifice, clean, increase pilot gas
(c)	Burner ports clogged near pilot	(c)	Clean ports
(d)	Low pressure	(d)	Adjust pressure regulator

Failure to Ignite

(a)	Main gas supply off	(a)	Open manual valve
(b)	Burned out fuse	(b)	Replace
(c)	Limit switch defective	(c)	Replace
(d)	Poor electrical connections	(d)	Check, clean and tighten
(e)	Defective gas valve	(e)	Replace
(f)	Defective thermostat	(f)	Replace

Burner Will Not Turn Off

(a)	Poor thermostat location	(a)	Relocate
(b)	Defective thermostat	(b)	Check calibration. Check switch and contacts. Replace.
(c)	Limit switch maladjusted	(c)	Replace
(d)	Short circuit	(d)	Check operation at valve. Check for short and correct.
(e)	Defective or sticking automatic valve	(e)	Clean or replace

Rapid Burner Cycling

(a)	Clogged filters	(a)	Clean or replace
(b)	Excessive anticipation	(b)	Adjust thermostat anticipator for longer cycles
(c)	Limit setting too low	(c)	Readjust or replace limit
(d)	Poor thermostat location	(d)	Relocate

360

Rapid Fan Cycling

(a) Fan switch differential too low

(a) Readjust or replace

(b) Blower speed too high

(b) Readjust to lower speed

Blower Will Not Stop

(a) Manual fan ON

(a) Switch to automatic

(b) Fan switch defective

(b) Replace

(c) Shorts

(c) Check wiring and correct

Noisy Blower and Motor

(a) Fan blades loose

(a) Replace or tighten

(b) Belt tension improper

(b) Readjust (usually allow 1 inch slack)

(c) Pulleys out of alignment

(c) Realign

(d) Bearings dry

(d) Lubricate

(e) Defective belt

(e) Replace

(f) Belt rubbing

(f) Reposition

Burner Will Not Turn On

(a) Pilot flame too large or too small

(a) Readjust

(b) Dirt in pilot orifice

(b) Clean

(c) Too much draft

(c) Shield pilot

(d) Defective automatic pilot valve

(d) Replace

(e) Defective thermocouple

(e) Replace

(f) Improper thermocouple position

(f) Properly position thermopile

(g) Defective wiring

(g) Check connections, tighten and repair shorts

(h) Defective thermostat

(h) Check for switch closure and repair or replace

(i) Defective automatic valve

(i) Replace

Blower Will Not Run

(a) Power not on

(a) Check power switch. Check fuses and replace if necessary

(b) Fan control adjustment too high

(b) Readjust or replace

(c) Loose wiring

(c) Check and tighten

(d) Defective motor overload, protector, or motor

(d) Replace motor

Not Enough Heat

(a) Thermostat set too low

(a) Raise setting

(b) Lamp or some other heat source too close to thermostat

(b) Move heat source away from thermostat

(c) Thermostat improperly located

(c) Relocate thermostat

(d) Dirty air filter

(d) Clean or replace

(e) Thermostat out of calibration

(e) Recalibrate or replace

(f) Limit set too low

(f) Reset or replace

(g) Fan speed too low

(g) Check motor and fan belt and tighten if too loose

Too Much Heat

(a) Thermostat set too high

(a) Lower setting

(b) Thermostat out of calibration

(b) Recalibrate or replace

(c) Short in wiring

(c) Locate and correct

(d) Valve sticks open

(d) Replace valve

(e) Thermostat in draft or on cold wall

(e) Relocate thermostat to sense average temperature

(f) By-pass open

(f) Close by-pass

CHAPTER 12

Oil-Fired Furnaces

Oil-fired furnaces are available in upflow, downflow (counterflow), or horizontal flow models and in a wide range of heating capacities for installations in attics, basements, closet spaces, or at floor level (Figs. 1 and 2). The furnace, oil burner, and controls are generally available from the manufacturer as a complete package for residential installations. All internal wiring is done at the factory.

Oil-fired furnaces should be listed by the *Underwriters' Laboratories* for construction and operating safety. Furnaces approved by the agency are marked *UL Approved*. Furnaces not made under UL standards should not be installed.

Another source of information concerning standards for oil-fired furnaces is the publication "Installation of Oil Burning Equipment 1972" (NFPA No. 31) from the *National Fire Protection Association*. See also "American Standard Performance Requirements for Oil-Powered Central Furnaces" (ANSI Z91.1-1972), a publication of the *American National Standards Institute*.

PLANNING SUGGESTIONS

The first step in planning a heating system is to calculate the maximum heat loss for the structure. This should be done in accordance with procedures described in the manuals of the *National Warm Air Heating and Air Conditioning Association* or by a comparable method. This is very important, because the data will

FLUE CONNECTION

FAN AND LIMIT CONTROLS

HEAT EXCHANGER

FLAME
OBSERVATION
PORT

PRIMARY SAFETY
CONTROL AND
RESET LEVER

FUEL PUMP

OIL BURNER

BURNER MOTOR

HAMMOCK-TYPE
AIR FILTER

BLOWER MOTOR

BLOWER AND FILTER
ACCESS DOOR
(LIFT UP AND OUT
TO REMOVE)

BELT DRIVE
BLOWER UNIT

Courtesy Lennox Air Conditioning and Heating Co.

Fig. 1. Lennox upflow oil-fired furnace.

be used to determine the size (capacity) of the furnace selected for the installation.

Location and Clearance

An oil-fired warm-air furnace should be located as near as possible to the center of the heat distribution system and chimney. A centralized location for the furnace usually results in the best operating characteristics, because long supply ducts and the heat loss associated with them are eliminated. The furnace should be installed as close as possible to the chimney to reduce the horizontal run of flue pipe.

An oil-fired furnace should be located a safe distance from any combustible material. Observe the minimum clearances sug-

ATTRACTIVELY STYLED CABINET

SLIPSTREAM HEAT EXCHANGER

FAN AND LIMIT CONTROL

ACCESS PORT

STAINLESS STEEL
COMBUSTION CHAMBER

PRIMARY OIL CONTROL

FUEL PUMP
(LOCATED BEHIND BURNER ASSEMBLY)

MINI BURNER

AIR BLOWER

BLOWER HOUSING AND FAN MOTOR

Courtesy Carrier Corp.

Fig. 2. Cutaway of a *Carrier* upflow oil-furnace.

gested by the manufacturer for the furnace except where they may come into conflict with local codes and regulations (Table 1). Local standards *always* take precedence over the manufacturer's specifications or national standards. Where there are no local standards specifying clearances, use those listed in "Installation of Oil Burning Equipment 1972" (NFPA No. 31-1972). Contents of the 1972 edition have been approved by the *American National Standards Institute* (ANSI).

Allow access for servicing the furnace and oil burner. Sufficient clearance should also be provided for access to the barometric draft control assembly.

Installation Recommendations

The installation of an oil-fired furnace should comply with all local codes and regulations, and "Installation of Oil Burning Equipment 1972".

New furnaces for residential installations are shipped preassembled from the factory with all internal wiring completed. The fuel tank must be installed and the fuel line connected to the furnace by the installer. The electrical service from the line voltage main, and the low-voltage thermostat will also have to be connected. Directions for making these connections are included in the furnace manufacturer's installation literature.

Make certain you have familiarized yourself with all local codes and regulations which govern the installation of an oil-fired furnace. Local codes and regulations *always* take precedence over national standards. Read the section INSTALLATION RECOMMENDATIONS in Chapter 11 (GAS-FIRED FURNACES) for additional information.

Table 1. Minimum clearances specified for *Carrier* oil-fired upflow (58HV), loboy upflow (58HL), downflow (58HC), and horizontal (58HH) furnace models.

From:	Furnace				
	58HV	58HL	58HC	58HH	
			Clearance (in.)		
Top of furnace or plenum	2	2	2	6	*2
Bottom of furnace	0	0	—	6	*2
Front of furnace	15	24	15	24	*48
Rear of furnace	1	6	2	6	*2
Sides of furnace	1	6	1	6	*4
Top of horizontal warm air duct within 6 ft of furnace	2	—	4	6	*3
Flue pipe: Horizontally or below pipe	9	—	9	—	—
Vertically	18	18	15	18	*9
Any side of supply plenum and warm air duct within 6 ft of furnace	2	—	6	—	—

*Alcove installation

Courtesy Carrier Corp.

Duct Connections

Detailed information concerning the installation of an air duct system is contained in the following two publications of the *National Fire Protection Association:*

1. "Installation of Air Conditioning and Ventilating Systems of Other Than Residence Type" (NFPA No. 90A).
2. "Residence Type Warm Air Heating and Air Conditioning Systems" (NFPA No. 90B).

Additional information about duct connections can be found in Chapter 7, Volume 2 (DUCTS AND DUCT SYSTEMS). The specific recommendations concerning furnace duct connections and the air distribution ducts contained in Chapter 11 (GAS-FIRED FURNACES) also apply to gas-fired furnaces.

Electric Wiring

Always follow all national and local codes when wiring a unit. The local code should take precedence when there is a conflict between the two. If there is no local electrical code in force, do all wiring in accordance with the *National Electric Code.*

Closely follow the wiring diagrams provided by the furnace manufacturer and those who manufacture the automatic controls. Failure to do this may result in losing the factory warranty.

A typical wiring diagram for an oil-fired furnace is shown in Fig. 3. The comments made in Chapter 11 (GAS-FIRED FUR-NACES) concerning line voltage and control voltage wiring connections also pertain to the wiring of oil-fired units.

Ventilation and Combustion Air

An oil-fired furnace should be located where a sufficient supply of air is available for combustion, proper venting, and the maintenance of suitable ambient temperature. In buildings of exceptionally tight construction, an outside air supply should be introduced for ventilation and combustion purposes.

If the furnace is located in a confined space such as a furnace room, closet, or utility room, the enclosure should be provided with two permanent openings for the passage of the air supply. One opening should be located approximately 6 inches from the top of the enclosure and the other opening approximately 6 inches from the bottom. Each opening should have a free area

Fig. 3. Wiring diagram for a *Trane* oil-fired furnace.

of at least 1 square inch per 1000 Btu per hour (minimum size 100 square inches) of the total input rating, or one square foot per gallon of oil per hour.

The size of the openings must correspond to the bonnet capacity of the furnace. Furnace manufacturers generally recommend suitable sizes for air openings in their equipment specifications (Table 2).

Air from an attic or a crawl space can also serve as a ventilation and combustion air supply to a confined furnace. Suitable

Table 2. Recommended sizes of ventilation openings for
Trane counterflow oil-fired furnace.

Bonnet Capacity 1,000 B.T.U./Hr.	Length In.	Height In.
85	24	12
100	24	12
125	26	13

Courtesy *The Trane Co.*

openings near (or on) the floor and near (or on) the ceiling must be provided for the passage of the air.

When an oil furnace is installed in an unconfined area such as a full basement, the full amount of air necessary for combustion is generally supplied by infiltration. Only when the construction is exceptionally tight is it necessary to provide openings to the outdoors.

Combustion Draft

An oil-fired furnace requires sufficient draft for the proper burning of oil. The average combustion draft is usually 0.02 inches wg over the fire (the so-called *overfire draft*), and between 0.04 and 0.06 inches wg draft in the stack (the *flue draft*) for units having inputs up to 2.0 gph. For higher inputs, 0.06 to 0.08 inches wg will be the rule. These figures are *average* ones and do not apply to all installations. For example, a *XXth Century* Zeph-Air Low Boy furnace only requires 0.01 inches wg draft over the fire and a flue draft of 0.02 to 0.03 inches wg. *Always* consult the furnace manufacturer's operating manual first for recommended drafts.

Venting

Oil-fired appliances use the same methods for venting the products of combustion to the outside air as do gas-fired units, except when the venting method is specifically limited to a particular fuel (e.g. Type B gas vents). Additional information about venting methods is contained in Chapter 11 (GAS-FIRED FURNACES). The type of venting to be used with a furnace is generally recommended by the manufacturer in the installation literature.

Flue Pipe

Use 24 gauge galvanized flue pipe (or its equivalent) for the flue connection between the furnace and the chimney. The horizontal run of flue pipe should slope upward at least ¼ inch per running foot. The pitch should not exceed 75% of the vertical vent. Locate the unit so that the furnace flue outlet is no more than *ten* flue pipe diameters from the chimney connection. If the distance is greater, use the next size larger flue pipe.

Insulate the flue pipe when it passes near combustible material. The *Underwriters' Laboratories* require that any uninsulated flue pipe be installed with a *minimum* clearance of 6 inches from any combustible materials.

Draft Regulator

A *draft regulator* (or *barometric draft regulator*) is a device designed to maintain a proper draft in an oil-fired furnace by automatically reducing the chimney draft (i.e. the flue draft) to the desired value (Fig. 4).

Draft regulators are recommended on all installations for *each* oil-fired appliance connected to a chimney *unless* the appliance is listed for use without one.

Do not use a small draft regulator, because they will not do a satisfactory job. The draft regulator should be installed in the

FRICTION DAMPER

BAROMETRIC DRAFT CONTROL

8" DIA. VENT CONNECTOR TEE

Courtesy Hydrotherm, Inc.

Fig. 4. Barometric draft control assembly.

horizontal flue pipe as close as possible to the chimney. When this flue pipe is too short for a satisfactory installation, the draft regulator should be installed in the chimney either above or below the flue pipe entry into the chimney. Some authorities oppose installing the draft regulator in the chimney, but experience has shown that this can be a very satisfactory installation.

CHIMNEYS

The chimney used with an oil-fired furnace should be of sufficient height and large enough in its cross-sectional area to meet the requirements of the furnace. The chimney area should be at least 20% greater than the area of the flue outlet. To insure the adequate removal of flue gases, the height and size of the chimney must be sufficient to create 0.02 to 0.04 inches wg draft over the fire for inputs up to 2.0 gph. For higher inputs, 0.06 to 0.08 wg is recommended. If the stack height is limited, it may be necessary to apply mechanically induced draft to the installation.

Ideally, *only* the furnace should be connected to the chimney. Read the section CHIMNEYS in Chapter 11 (GAS-FIRED FURNACES) for additional information.

Chimney Troubleshooting

Read the section CHIMNEY TROUBLESHOOTING in Chapter 11 (GAS-FIRED FURNACES) for a description of common chimney problems and suggested remedies. The chimneys used with oil-fired furnaces are essentially identical to those used with gas-fired furnaces.

BASIC COMPONENTS

The basic components of an oil-fired forced warm air furnace are:

1. Automatic controls,
2. Combustion chamber,
3. Heat exchanger,
4. Oil burner,
5. Fuel pump,

6. Blower and motor,

7. Air filter(s).

Each of these components is described in the sections that follow. Additional information is contained in Chapter 10 (FURNACE FUNDAMENTALS) and the various chapters in which furnace controls and fuel burning equipment are described.

AUTOMATIC CONTROLS

The automatic controls for an oil-fired furnace can be divided into the following three principal categories:

1. Room thermostat,

2. Fan and limit control,

3. Primary control.

The thermostat and limit controls are essentially identical to those used with gas-fired furnaces. The thermostat controls the furnace operation by signalling the primary control to turn the burner on and off in response to temperature changes in the space or spaces being heated. The fan and limit control is designed to govern the operation of the furnace within a specified temperature range.

Until comparatively recently, the primary control consisted of the oil burner relay and a stack switch to shut off the oil flow to the burner if the oil failed to ignite or the flame went out. Stack switches have largely been replaced by burner-mounted cadmium detection cells.

A typical automatic control system for an oil-fired furnace is shown in Fig. 5. Additional and more detailed information about the various types of controls can be found in Chapter 4 (THERMOSTATS AND HUMIDISTATS), Chapter 5 (GAS AND OIL CONTROLS), and Chapter 6 of Volume 2 (OTHER AUTOMATIC CONTROLS).

COMBUSTION CHAMBER

The combustion chamber is that portion of the furnace which surrounds the oil burner flame and within which the combustion process takes place.

ROOM THERMOSTAT

CONDUIT OR BX

FLOOR

TO FUSE BOX

DRAFT REGU-LATOR

STACK

FURNACE

LINE VOLTAGE LIMIT CONTROL

STACK MOUNTED PROTECTORELAY PRIMARY CONTROL

ON-OFF SWITCH

IGNITION TRANSFORMER

BURNER MOTOR AND BLOWER

COMBUSTION CHAMBER

Courtesy Honeywell Tradeline Controls

Fig. 5. Typical automatic control system for an oil-fired furnace.

The typical combustion chamber is made of stainless steel or some sort of refractory material such as firebrick (Figs. 6 and 7). Round combustion chambers are probably the most common, although square, rectangular, and other shapes do occur.

HEAT EXCHANGER

The heat exchanger is a device used to transfer the heat of the combustion process to the air being circulated through the ducts of the heating system. The heat exchanger is generally constructed

373

of heavy gauge steel (14-16 gauge). The wrap around radiator type heat exchanger is one of the most commonly used designs. It consists of an upper and lower chamber, each with an extension or pouch which must be aligned when assembled.

Courtesy Bard Mfg. Co.

Fig. 6. Stainless steel combustion chamber and heat exchanger on a *Bard* oil-fired furnace.

FIRE BRICK COMBUSTION CHAMBER

HEAT EXCHANGER

Courtesy Dorback Furnace and Foundry Co.

Fig. 7. Heat exchanger and combustion chamber on a *Dorback* oil-fired furnace.

OIL BURNER

An oil burner is a mechanical device used to combine fuel oil with the proper amount of air for combustion and deliver it to the point of ignition. Oil burners are described in considerable detail in Chapter 1, Volume 2 (OIL BURNERS).

When installing the burner nozzle, make sure that the proper size nozzle for the installation has been selected. Electrodes are adjusted at the time of manufacture, but they should be checked when the unit is installed to make sure they are set to the burner manufacturer's specifications.

FUEL PUMPS AND FUEL UNITS

Almost all oil burners used in residential heating installations are equipped with fuel units. It is very important to distinguish between a fuel unit and a fuel pump, because the two are not identical. A fuel unit is an oil pumping device which consists of a pump, a pressure regulating valve, and a filter. A pump, on the other hand, is *only* an oil pumping device. A pressure regulating valve and a filter can be attached to the mountings on the pump if so desired. Fuel pumps and fuel units are described in Chapter 1, Volume 2 (OIL BURNERS).

Manufacturers produce single-stage and two-stage pumps. The single-stage pumps are recommended for installations where the fuel tank is located at a level *above* the burner (Fig. 8). Two-stage

Table 4. Types of pumps recommended for different vacuums.

Total Vacuum	Type of Pump
Up to 3 inches vacuum	Single-stage pump
4-13 inches vacuum	Two-stage pump
14 inches vacuum or more	Single-stage pump for the burner and a separate lift pump with a reservoir.

Courtesy National Fuel Oil Institute.

pumps are used in installations with underground tanks where the combination of lift, horizontal run, fittings and filters does not exceed the manufacturer's rating in inches of vacuum.

AIR VENT

1-1/4" MIN

FILL PIPE

OIL TANK

SHUT-OFF VALVE

FUEL UNIT

INLET

Courtesy Sundstrand Corp.

Fig. 8. Fuel tank located above burner.

The capacity of the integral pump on an oil burner should be sufficient to handle the total vacuum in the system. The vacuum is expressed in inches, and can be determined by the following procedure:

1. One inch of vacuum for each one foot of lift.
2. One inch of vacuum for each 90° elbow in either the suction or return lines.
3. One inch of vacuum for each ten feet of horizontal run (⅜ inch OD line).
4. One inch of vacuum for each twenty feet of horizontal run (½ inch OD line).

After you have calculated the total vacuum, you can use this data to select the most suitable pump for the burner. Table 4 lists various vacuums and suggests appropriate pump capacities.

377

TROUBLESHOOTING FUEL PUMPS AND FUEL UNITS

The following troubleshooting list contains the most common operating problems associated with fuel pumps and fuel units. Each problem is given in the form of a symptom, the possible cause, and a suggested remedy. The purpose of this list is to provide the operator with a quick reference to the cause and correction of a specific problem. Much of the information used in this troubleshooting list was obtained from the "Sundstrand Technical Fuel Unit Manual," courtesy the *Sundstrand Corporation*.

Symptom and Possible Cause	Possible Remedy
No Oil Flow at Nozzle	
(a) Clogged strainer or filter.	(a) Remove and clean strainer (Fig. 9). Repack filter element.
(b) Air binding in two pipe system.	(b) Check and insert by-pass plug.
(c) Frozen pump shaft.	(c) Remove unit and return to factory for repair.
Oil Leak	
(a) Loose plugs or fittings.	(a) Dope with good quality thread sealer.
(b) Leak at pressure adjusting end cap nut.	(b) Fibre washer may have been left out after adjustment of valve spring. Replace washer.
(c) Blown seal	(c) Replace fuel unit.
(d) Seal leaking	(d) Replace fuel unit.
Noisy Operation	
(a) Air inlet line.	(a) Tighten all connections and fittings in the intake line and unused intake port plugs (Fig. 10).

Courtesy Sundstrand Corp.

Fig. 9. Cleaning the strainer with clean fuel oil or kerosene.

Courtesy Sundstrand Corp.

Fig. 10. Tightening connections and fittings.

(b) Bad coupling alignment.

(b) Loosen mounting screws and shift fuel unit to a position where noise is eliminated. Retighten mounting screws.

(c) Pump noise

(c) Work in gears by continued running or replace.

Pulsating Pressure

(a) Air leak in intake line.

(a) Tighten all fittings and valve packing in intake line.

379

(b) Air leaking around strainer cover.

(b) Tighten strainer cover screws.

(c) Partially clogged strainer.

(c) Remove and clean strainer.

(d) Partially clogged filter.

(d) Replace filter element.

Low Oil Pressure

(a) Nozzle capacity is greater than fuel unit capacity.

(a) Replace fuel unit with unit of correct capacity.

(b) Defective gauge.

(b) Check against another and replace if necessary.

Improper Nozzle Cutoff

(a) Filter leaks.

(a) Check face of filter cover and gasket for damage. Replace if necessary.

(b) Partially clogged nozzle strainer.

(b) Clean strainer or change nozzle.

(c) Air leak in intake line.

(c) Tighten intake fittings and packing nut on shutoff valve. Tighten unused intake port plug.

(d) Strainer cover loose.

(d) Tighten screws.

(e) Air pockets between cutoff valve and nozzle.

(e) Start and stop burner until smoke and after-fire disappears.

The cause of improper cutoff can be determined by inserting a pressure gauge in the nozzle port of the fuel unit (Fig. 11). Allow the fuel unit to operate for a few minutes and then shut off the burner. If the pressure drops to 0 psi on the gauge, the fuel unit should be replaced. A gauge reading above 0 psi indicates that the fuel unit is operating properly. The probable cause of improper cutoff in those cases where a gauge reading is obtained is usually air in the system.

FUEL TANK AND LINE

The fuel tank and line must be installed in accordance with the requirements of the *Underwriters' Laboratories* and any local codes

Courtesy Sundstrand Corp.

Fig. 11. Checking fuel unit pressure with a gauge.

and regulations. Local codes and regulations take precedence over national standards. Additional information can be obtained from "Installation of Oil Burning Equipment 1972" (NFPA No. 31), a publication of the *National Fire Protection Association*. Use only an approved tank and line for the installation.

A one-pipe system is recommended for an oil tank installation when the following conditions are present:

1. Tank is installed in the basement or similar location and/or below the oil pump inlet port.
2. Vacuum at pump does not exceed 2 inches.

A two-pipe (return line) system is recommended for a tank installation when the following conditions exist:

1. Tank is located outside and/or below the oil pump inlet port.
2. Vacuum at pump does not exceed 10 inches.

Use a two-stage pump for installations having especially long lines or high lifts *if* the vacuum at the inlet port does not exceed 20 inches.

A shutoff valve should be installed in the suction line close to the oil burner. This valve enables the operator to disconnect the line without draining it. Do *not* place a shutoff valve in the return line. Closing a shutoff valve on the return line while the oil burner is operating could damage the pump. All joints in the fuel line should be sealed with a suitable oil piping compound.

OIL FILTER

An oil filter should be included in the fuel line inside the building and as close to the fuel tank as possible (Fig. 12). One filter is adequate on oil burners using more than 1.00 gph.

Fig. 12. Typical location of oil filter.

Change the filter cartridge at least once a year. The filter body should be thoroughly cleaned before installing a new cartridge.

PRIMING THE FUEL PUMP

On occasion, the oil burner may fail to pump oil. When this occurs, check the oil supply line to the furnace for leaks. If there are no leaks, it may be necessary to prime the fuel pump. Pumps are self priming for single-stage, two-pipe systems and two-stage pumps. A single-stage pump (one-pipe system) should be primed as follows:

1. Turn off the electrical power supply to the unit.
2. Read and follow the priming instructions provided by the manufacturer.
3. Prime the pump until the oil is free from bubbles.

When a *new* pump fails to prime, it may be due to dry pump conditions. This can be corrected by removing the vent plug and filling the pressure cavity slowly so that the fuel oil 'wets" the gears (Fig. 13).

TO VENT PUMP
ATTACH 1/4" ID
PLASTIC TUBE.
USE 3/8" WRENCH
TO OPEN VENT 1/8
TURN MAXIMUM.

RETURN PORT
1/4" PIPE THREAD. USED
AS RETURN ON TWO
PIPE INSTALLATION.
MAY BE USED AS
OPTIONAL INLET OR
VACCUM TEST PORT
ON SINGLE PIPE
INSTALLATION.

TO ADJUST PRESSURE
REMOVE COVER PLUG OF
PRESSURE ADJUSTING
PORT LOCATED BEHIND
INLET PORT. INSERT
STANDARD SCREWDRIVER.
TURN COUNTER-
CLOCKWISE TO BELOW
PRESSURE DESIRED.
TURN CLOCKWISE TO SET
TO-DESIRED PRESSURE.

FOR USE AS
GAGE PORT
REMOVE PLUG TO
INSTALL 1/4" PIPE
THREADED GAGE.
REMOVE INSERT TO
INSTALL 1/8" PIPE
THREADED GAGE.

NOZZLE PORT
1/8" PIPE THREAD.

NOZZLE ◀
RETURN

WEBSTER ELECTRIC
RACINE, WISCONSIN
A STA-RITE INDUSTRY

SINGLE STAGE
FUEL UNIT
1725 RPM

INSERT BY-PASS
PLUG FOR TWO
PIPE SYSTEM
THRU BOTTOM
INLET

INLETS ▶

INLET PORT

BYPASS PLUG
1/16" PIPE THREAD.
9/64" ALLEN.
ACCESS THROUGH
BOTTOM INLET.

INLET PORT
1/4" PIPE THREAD.

COLOR OF PRINTING
DENOTES OPERATING SPEED
BLUE - 1725 RPM
GREEN - 3450 RPM

INLET PORT
1/4" PIPE THREAD.

Courtesy Webster Electrical Co., Inc.

Fig. 13. *Webster* **model M series fuel pump.**

Table 5. Maximum recommended pressures.

Bonnet Capacity (1000 Btuh)	Firing Rate (gph)	Standard Nozzle Size	Pump Pressure (psig)
85	.76	.75	103
100	.90	.90	100
125	1.12	1.10	104
150	1.35	1.35	100
200	1.80	1.75	112
250	2.25	2.25	100
335	3.00	1.50	100

Courtesy Carrier Corp.

ADJUSTING OIL PUMP PRESSURE

The oil pressure regulator on the fuel pump is generally factory set to give nozzle oil pressures of 100 psig. The firing rate is indicated on the nameplate, and can be obtained with standard nozzles by adjusting the pump pressures as follows (Fig. 13).

1. Turn the adjusting screw clockwise to increase pressure.
2. Turn the adjusting screw counterclockwise to decrease pressure.
3. Never exceed the pressures indicated in Table 5.

STARTING THE BURNER

The procedure for starting an oil burner may be summarized as follows:

1. Open all warm air registers.
2. Check to be sure all return air grilles are unobstructed.
3. Open the valve on the oil supply line.
4. Reset the burner primary relay.
5. Set the thermostat above the room temperature.
6. Turn on the electric supply to the unit by setting the main electrical switch to the ON position.
7. Change the room thermostat setting to the desired temperature.

The oil burner should start after the electric power has been switched on (Step 6). There is no pilot to light as is the case with gas-fired appliances. The spark for ignition is provided automatically on demand from the room thermostat.

Allow the burner to operate at least 10 minutes before making any final adjustments. Whenever possible use instruments to adjust the fire.

COMBUSTION TESTING AND ADJUSTMENTS

The instruments recommended for combustion testing and adjustments are:

1. Draft gauge,

Figure 6. Smoke scale.

BACHARACH SMOKE SCALE NO.	RATING	SOOTING PRODUCED
1	EXCELLENT	EXTREMELY LIGHT IF AT ALL
2	GOOD	SLIGHT SOOTING WHICH WILL NOT INCREASE STACK TEMPERATURE APPRECIABLY
3	FAIR	MAY BE SOME SOOTING BUT WILL RARELY REQUIRE CLEANING MORE THAN ONCE A YEAR
4	POOR	BORDERLINE CONDITION SOME UNITS WILL REQUIRE CLEANING MORE THAN ONCE A YEAR
5	VERY POOR	SOOT RAPIDLY AND HEAVILY

Courtesy Bacharach Instrument Co.

2. Smoke tester,
3. Carbon dioxide tester,
4. 200/1000°F. stack thermometer,
5. 0/150-psig pressure gauge,
6. 0/30-inch mercury vacuum gauge.

Smoky combustion indicates poor burner performance. The amount of smoke in the flue gas can be measured with a smoke tester (Fig. 14). The tube of the smoke tester is inserted through a ⅜ inch hole drilled in the flue pipe, and the test is run as shown in Fig. 15. Any smoke in the air drawn into the smoke tester will register on a filter paper inserted in the device. The results are interpreted according to the smoke scale in Table 6.

One of the most common causes of smoky combustion is soot formation on the heating surfaces. This is easily corrected by cleaning. *Other* possible causes of smoky combustion include:

1. Insufficient draft,
2. Poor fuel supply,
3. Fuel pump malfunctioning,
4. Defective firebox,
5. Incorrectly adjusted draft regulator,
6. Defective oil burner nozzle,

7. Wrong size oil burner nozzle,

8. Improper fan delivery,

9. Excessive air leaks in boiler or furnace,

10. Unsuitable fuel-air ratio.

Net stack temperatures in excess of 700°F for conversion units and 500°F for packaged units are considered abnormally high. The *net* stack temperature is the difference between the temperature of the flue gases inside the pipe and the room air temperature outside. For example, if the flue gas temperature is 600°F and the room temperature is 75°F, then the net stack temperature is 525°F (600°F − 75°F = 525°F).

A 200/1000°F stack thermometer is used to measure the flue gas temperature. The thermometer stem is inserted through a hole drilled in the flue pipe (Fig. 16). A high stack temperature may be caused by any of the following:

1. Undersized furnace.

2. Defective combustion chamber.

3. Incorrectly sized combustion chamber.

4. Lack of sufficient baffling.

5. Dirty heating surfaces.

6. Excessive draft.

Fig. 14. *Bacharach true-spot smoke tester.*

Courtesy Bacharach Instrument Co.

| Insert filter test paper into TRUE-SPOT | Withdraw gas sample from flue pipe by 10 pump strokes | Grade soot spot on test paper by comparison with shadings on scale |

Courtesy Bacharach Instrument Co.

Fig. 15. A smoke tester in use.

Courtesy Bacharach Instrument Co.

Fig. 16. Running a stack gas temperature test with a stack thermometer.

7. Boiler or furnace overfired.

8. Unit unsuited to automatic firing.

9. Draft regulator improperly adjusted.

When the carbon dioxide (CO_2) content of the flue gas is too low (less than 8%), heat is lost up the chimney and the unit operates inefficiently. This condition is usually caused by one of the following:

1. Underfiring the combustion chamber.

2. Burner nozzle is too small.

3. Air leakage into the furnace or boiler.

When the carbon dioxide content is too high, the furnace operation is generally characterized by excess smoke and/or pulsations and other noises. A high carbon dioxide content is usually caused by insufficient draft or an overfired burner.

The carbon dioxide reading is also taken through a hole drilled in the flue pipe with a CO_2 indicator (Fig. 17). The CO_2 indicator is used as shown in Fig. 18. The results are indicated by a test liquid on a scale calibrated in percent CO_2.

A correct draft is essential for efficient burner operation. Insufficient draft can make it almost impossible to adjust the oil burner for its highest efficiency. Excessive draft can reduce the percentage of carbon dioxide in the flue gases and increase the stack temperature.

For the most efficient operating characteristics, the overfire draft generally should be not less than 0.02 inches wg. Smoke and odor often occur when the overfire draft falls below 0.02 inches wg.

It may be necessary to adjust the barometric draft regulator to obtain the correct overfire draft. If it is not possible to adjust the overfire draft for a minus 0.01 to 0.02 inches wg., install a mechanical draft inducer between the chimney and the barometric draft regulator.

The primary air band should be adjusted to a O+ smoke or until a hard clean flame is visible. A clean flame is preferred to one with high carbon dioxide. Adjust the overfire draft for a minus 0.01 to a minus 0.02 inches wg. An excessive overfire draft condition will cause high stack temperature and inefficient operation. A

Fig. 17. Bacharach fyrite CO$_2$ indicator.

Courtesy Bacharach Instrument Co.

1 Push rubber connector down — then squeeze bulb 18 times

2 Turn bottom side up and back again

3 Read fluid level on scale for percentage of gas

Courtesy Bacharach Instrument Co.

Fig. 18. CO$_2$ test.

389

too low or positive draft over the fire will usually cause the flue gases and fumes to seep into the space upon start up or shutdown.

The flue pipe draft in most residential oil burners is between 0.04 and 0.06 inches water. This is sufficient to maintain a draft of 0.02 inches in the firebox.

The furnace or boiler draft is measured with a draft gauge as shown in Fig. 19. A hole is drilled in either the fire door (overfire

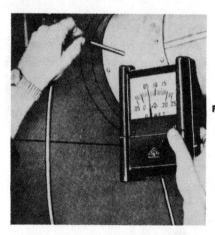

Fig. 19. Bacharach model MZF draft gauge.

Courtesy Bacharach Instrument Co.

draft measurement) or flue pipe (flue pipe draft measurement) and the unit is run for approximately 5 minutes. The draft tube is then inserted into the test hole and the gauge is read (Fig. 20).

BLOWERS AND MOTORS

Both direct-drive and belt-driven blowers are used with oil-fired furnaces. These blowers and motors are identical to those used in gas-fired furnaces. Read the section BLOWERS AND MOTORS in Chapter 11 (GAS-FIRED FURNACES) for additional information.

AIR DELIVERY AND BLOWER ADJUSTMENT

It is sometimes necessary to adjust the blower speed to produce a temperature rise through the furnace that falls within the limits

Overfire Draft

Flue Pipe Draft
Courtesy Bacharach Instrument Co.

Fig. 20. Overfire and flue pipe draft tests.

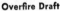

stamped on the furnace nameplate. Blower adjustment procedures are described in Chapter 11 (GAS-FIRED FURNACES).

AIR FILTER

A *forced* warm-air oil-fired furnace is supplied with either a disposable air filter or a permanent (washable) one. *Never* use a filter with a gravity warm-air furnace, because it will obstruct the air flow.

A disposable air filter should be inspected on a regular basis and replaced when dirty. Always replace the air filter with one of the same size and type of filter media. This information is usually found on a label attached to the filter.

A permanent air filter must also be regularly inspected. When it is dirty, it must be removed and cleaned. The usual method is to vacuum it and then to wash it in a soap or detergent and water solution.

Additional information about air filters is found in Chapter 11 (GAS-FIRED FURNACES) and Chapter 14, Volume 3 (AIR CLEANERS AND FILTERS).

AIR CONDITIONING

A furnace used in conjunction with a cooling unit should be installed in parallel or on the upstream side of the evaporator coil to avoid condensation on the heating element. In a parallel installation, dampers or comparable means should be provided to prevent chilled air from entering the furnace.

Duct sizing should include an allowance for air conditioning even though it may not be initially installed with the furnace. Air conditioning involves a greater volume of air than heating.

All ductwork located in unconditioned areas (attics, crawl spaces, etc.) *downstream* from the furnace should be insulated to prevent unwanted heat gain.

INSTALLATION AND MAINTENANCE CHECK LIST

A properly installed and maintained furnace will operate efficiently and economically. The following installation check list is offered as a guide to the installer.

1. Be sure there is at least 0.01 inches wg draft over the fire.
2. Check for sufficient combustion and ventilation air.
3. Eliminate downdraft or back draft.
4. Provide sufficient space for service accessibility.
5. Check all field wiring.
6. Supply line fuse or circuit breaker must be of proper size and type for furnace.
7. Line voltage must meet specifications while furnace is operating.
8. Ductwork must be checked for proper balance, velocity, and quietness.
9. Check the fuel line for leaks.
10. Cycle the burner.
11. Check the limit switch.
12. Check the fan switch.
13. Make final adjustments to the fire.
14. Adjust the blower motor for desired speed.
15. Make sure air filter is properly secured.
16. Make sure all access panels have been secured.
17. Pitch air conditioning equipment condensate lines toward a drain.

18. Check thermostat heat anticipator setting.

19. Check thermostat for normal operation. Observe at least five ignition cycles before leaving the installation.

20. Clear and clean the area around the furnace.

Other installation and maintenance recommendations are found in various sections of this chapter.

TROUBLESHOOTING AN OIL-FIRED FURNACE

The list that follows contains the most common operating problems associated with oil-fired furnaces. Each problem is given in the form of a symptom, the possible cause, and a suggested remedy. This list is intended to provide the operator with a quick reference to the cause and correction of a specific problem.

Symptom and Possible Cause	Possible Remedy
Change in Size of Fire	
(a) Dirty nozzle	(a) Clean or replace
(b) Low pressure	(b) Adjust at pump
(c) Plugged strainer	(c) Clean
(d) Cold oil	(d) Adjust pressure
No Oil Flow	
(a) Oil level below intake line in the supply tank	(a) Fill tank
(b) Clogged strainer	(b) Remove and clean strainer
(c) Clogged nozzle	(c) Clean or replace
(d) Air leak in intake line	(d) Tighten fittings and plugs, check valves
(e) Restricted intake line (high vacuum reading)	(e) Replace kinked tubing, check valves, filters
(f) Two pipe system air bound	(f) Check bypass plug

393

(g) Simple pipe system air bound

(g) Loosen gauge port and drain oil until foam is gone

(h) Slipping or broken coupling

(h) Tighten or replace coupling

(i) Frozen pump shaft

(i) Replace

Oil Spray But No Ignition

(a) Dirty electrodes

(a) Clean or replace

(b) Improper spacing

(b) Reset

(c) Cracked porcelain

(c) Replace

(d) Dead transformer

(d) Replace

(e) Loose connection

(e) Tighten

(f) Faulty relay

(f) Replace

Burner Motor Does Not Start

(a) Defective thermostat

(a) Replace

(b) Fuse burned out

(b) Replace

(c) Limit control open

(c) Check setting and correct

(d) Contact dirty or open on primary relay

(d) Clean or replace relay

(e) Relay transformer burned out

(e) Replace relay

(f) Motor stuck or burned out or overload protector out

(f) Replace if burned out

(g) Primary relay off on safety

(g) Push reset button

Noisy Operation

(a) Bad coupling alignment

(a) Loosen fuel unit or motor

(b) Loose coupling

(b) Tighten set screws

(c) Air in oil line

(c) Bleed oil line, look for leaks

(d) Pump noise

(d) Continued running sometimes works in gears. If not, replace.

(e) Hum vibration

(e) Isolate pipes from structural members

(f)	Combustion noise	(f)	Adjust noise
(g)	Furnace too small	(g)	Check heat loss to be sure furnace properly sized
(h)	Burner noisy	(h)	Check mounting and position Adjust air
(i)	Blower noisy	(i)	Oil bearing Tighen shaft collars Adjust belt tension Align and tighten pulleys Position rubber isolators

Burner Will Not Run Continuously

(a)	Lockout timing too short	(a)	Replace primary control
(b)	Poor flame due to too much air; too little oil	(b)	Check nozzle, air adjustment, oil pressure, size of nozzle
(c)	Water or air in oil	(c)	Look for leak in supply
(d)	Control wired wrong	(d)	Check and rewire

Pulsation

(a)	Air adjustment	(a)	Readjust air
(b)	Pressure over fire	(b)	Correct draft to 0.02″ W.C negative
(c)	Dirty or improperly set electrodes	(c)	Clean and reset Wire primary control for continuous ignition
(d)	Too much oil impingement	(d)	Check nozzle and pump pressure Check nozzle size and angle and position of drawer assembly

Short Cycling of Fan

(a)	Fan control setting	(a)	Set lower turn on (115°F)
(b)	Input too low	(b)	Check burner input

(c) Temperature rise too low due to excessive speed of blower

(c) Slow blower down and check ventilation

Short Cycling on Limit Control

(a) Limit setting low

(a) Reset to maximum

(b) Input too high

(b) Check burner input

(c) Temperature rise too high due to blower running too slow

(c) Increase blower speed

(d) Temperature rise too high due to restricted returns or outlets

(d) Open dampers or add additional outlets or returns

(e) Fan control setting too high

(e) Reset lower (115°F)

(f) Control out of position

(f) Place cad cell in proper position

High Fuel Consumption

(a) Input too high

(a) Check burner input

(b) Flue loss too great

(b) Measure CO_2 and flue gas temperature. If loss is more than 25%, reset air, check input, speed up blower.
Check static pressure in return and outlet plenum and correct to recommended values.

Not Heating

(a) Low input

(a) Check nozzles and input

(b) Insufficient air circulating

(b) Speed up blower. Check size and location of ducts and outlets. Set fan control and blower for continuous air circulation.

CHAPTER 13

Coal-Fired Furnaces

Coal-fired furnaces burn coal or coke. Some are designed to burn coal in combination with one or more other types of fuels such as gas, oil, and/or wood (see MULTI-FUEL FURNACES in this chapter). Both gravity and forced warm-air furnaces are available for use with solid fuels (Figs. 1 and 2).

Coal furnaces are either hand fired or fired with a coal stoker. A stoker is an automatic coal feeding device that carries the coal from the storage bin to the furnace as needed.

The fire in a coal furnace must be kept burning throughout the heating season. The fire *must* be attended at *regular* intervals. The supply of fuel, method of feeding and regulation is always based on weather conditions.

Solid fuel forced warm-air furnaces are rated in accordance with equations provided by the *National Warm Air Heating and Air Conditioning Association*. Check the latest edition of their publication "Commercial Standard for Solid-Fuel-Burning Forced Air Furnaces."

PLANNING SUGGESTIONS

The first step in planning a heating system is to calculate the maximum heat loss for the structure. This should be done in accordance with procedures described in the manuals of the *National Warm Air Heating and Air Conditioning Association* or by

Fig. 1. Oneida coal-fired forced warm-air furnace.

a comparable method. This is very important because the data will be used to determine the size (capacity) of the furnace selected for the installation.

Location and Clearance

A coal-fired furnace should be located a safe distance from any combustible materials. Consult the local codes and regulations for the required clearances.

Never obstruct the front of the furnace. Access must be provided to the fire and ashpit doors in order to operate the furnace.

A centralized location for the furnace usually results in the

Fig. 2. Oneida coal-fired gravity warm-air furnace.

best operating characteristics, because long supply ducts and the heat loss associated with them are eliminated.

Installation Recommendations

Make certain you have familiarized yourself with all local codes and regulations which govern the installation of coal-fired furnaces and coal stokers. Local codes and regulations take precedence over national standards.

Coal-fired furnaces are shipped disassembled with complete assembly instructions. The installer must assemble the furnace in accordance with these instructions. If the furnace is fired by a coal stoker or contains some automatic controls, the installer must connect the electrical service from the line voltage main. The low-voltage thermostat must also be connected.

Duct Connections

Detailed information concerning the installation of an air duct system is contained in the following two publications of *The National Fire Protection Association:*

1. "Installation of Air Conditioning and Ventilating Systems of Other Than Residence Type" (NFPA No. 90A).
2. "Residence Type Warm Air Heating and Air Conditioning Systems" (NFPA No. 90B).

Additional information about duct connections can be found in Chapter 7, Volume 2 (DUCTS AND DUCT SYSTEMS). The comments in Chapter 11 (GAS-FIRED FURNACES) about furnace duct connections and air distribution ducts also apply to ducts used with coal-fired forced warm-air furnaces.

Electrical Wiring

A wiring diagram specifying the electrical connections between the various controls should be sent by the stoker manufacturer.

As shown in Fig. 3, a room thermostat operating through the stoker time relay starts and stops the stoker in response to temperature conditions. The time relay operates the stoker to keep the fire alive during periods when heat may not be required by the room thermostat. A stoker time relay should include a device to shut down the stoker immediately after a shut down call from the thermostat. This action eliminates fuel waste by preventing the overshooting of room temperature.

A high limit control is used to protect the system against excessive temperatures. This control takes the form of an air switch in a warm air heating system.

A snap switch should be installed in the wiring between the stoker time relay and the stoker motor to open the stoker motor circuit when the fire is being cleaned (Fig. 3).

Fig. 3. Snap action switch located between stoker time relay and stoker motor. The snap switch is used to open the stoker motor circuit when cleaning the fire.

A wiring diagram for a combination wood-oil furnace is shown in Fig. 4. Note that the wood controls and oil controls are wired as separate systems. This is a common practice on furnaces designed to burn more than one fuel.

Ventilation and Combustion Air

The total draft requirement of a coal-fired furnace is greater than those which burn gas or oil, because the chimney draft must overcome the resistance of the fuel bed.

Both a primary and secondary air supply are necessary for the combustion of solid fuels. The primary air passes through the fuel bed generally on an upward path from the ashpit. The secondary air is usually admitted through slots in the furnace fire door and passes *over* the fire to complete the combustion process (Fig. 5).

Courtesy Oneida Heater Co., Inc.

Fig. 4. Wiring diagram for an Oneida wood-oil furnace.

Fig. 5. Entry points for primary and secondary air.

On residential type furnaces, the secondary air slots in the fire door should be kept *closed*. There is usually enough air leakage to admit a suitable amount of secondary air without having to open the fire door. An excessive amount of secondary air (particularly after the fire has taken hold) will actually reduce the efficiency of the furnace.

The draft will determine the rate at which the fuel is burned. The draft in furnaces using coal or coke will depend upon the following factors:

1. Type of fuel (anthracite or bituminous; coke, etc.).
2. Size of fuel.
3. Fuel bed thickness.
4. Ash and clinker accumulation.
5. Flue resistance to the flow of gas.
6. Soot accumulation in flue.
7. Grate area.

The rate at which the fuel is burned will depend upon the heating load demand of the specific location. The burning rate can be varied by regulating the draft. One of the most effective methods of doing this is by placing a damper in the flue outlet (Fig. 6). This method proportionally reduces both primary and secondary air.

The use of an ashpit damper is required for low combustion rates. A cold air check damper is necessary when there is excessive chimney draft.

Venting

Provisions must be made for venting the products of combustion to the outside in order to avoid contamination of the air in the living or working spaces of the structure. Masonry chimneys and low heat Type A prefabricated chimneys are the most common methods of venting coal-fired furnaces.

Flue Pipe

Coal-fired furnaces require flue passages that are larger than those used with other types of fuels. This is due to the far greater volume of smoke, soot and other products of combustion associated with burning solid fuels.

Fig. 6. Damper placed in flue outlet.

CHIMNEY

The chimney used with a coal-fired furnace should be of sufficient height and area to meet the requirements of the furnace. For best results, only the furnace should be connected to the chimney. Read the section CHIMNEYS in Chapter 11 (GAS-FIRED FURNACES).

Chimney Troubleshooting

The chimneys used with coal-fired furnaces are basically the same as those used with gas-fired and oil-fired units. Read the section CHIMNEY TROUBLESHOOTING in Chapter 11 (GAS-FIRED FURNACES) for a description of common chimney problems and suggested remedies.

BASIC FURNACE COMPONENTS

The basic components of a coal-fired forced warm-air furnace are:

1. Automatic controls,
2. Furnace grate,
3. Heat exchanger,
4. Coal stoker,
5. Blower and motor.

Each of these components is described in the sections that follow. Additional information is contained in Chapter 10 (FURNACE FUNDAMENTALS) and the various chapters in which furnace controls and fuel burning equipment are described.

Automatic Controls

The automatic controls of a coal-fired furnace which uses a stoker to feed the coal are shown in Fig. 7. This control system is very similiar to the one used for an oil burner except for the automatic timer included with the stoker relay and transformer. During the heating season, the fire of a coal furnace must burn continually. The timer is a device designed to operate the coal stoker for a few minutes every hour or half hour in order to keep the fire alive during those periods when little heat is required.

A stack thermostat or similar device is recommended for use with stokers in areas subject to electric power failures. An electric power failure will shut down the stoker. If the shutdown period is long enough, the fire will die for lack of fuel. When the electricity comes on again, the stoker will feed coal to a cold fire pot. A stack thermostat or light sensitive electronic device will monitor the stack heat or fire, and prevent stoker operation when the fire is out.

FURNACE
THERMOSTAT
(High-Limit and
Fan Control)

ROOM THERMOSTAT

JUNCTION BOX

To 110-volt
Supply Line

STOKER
MOTOR

STOKER RELAY, TIMER
AND TRANSFORMER

FAN MOTOR

Courtesy U. S. Dept. of Agriculture.

Fig. 7. Automatic controls for a coal-fired furnace fed by a stoker.

A blower fan control is generally included with most forced warm-air furnaces. Many manufacturers will also provide a hand operated draft control with a coal-fired furnace. Electrically operated dampers for draft control are generally available, but at extra cost.

The operation of a furnace is controlled by the room thermostat. In heating systems in which a coal-fired furnace is used, the thermostat opens or shuts dampers to increase or decrease the supply of air to the fire. When the supply of air is increased, the fire becomes hotter and the amount of heat generated by the furnace is increased. Decreasing the supply of air to the fire has the opposite effect. Older coal-fired furnaces are not controlled by a room thermostat. Hand operated dampers were used instead.

Furnace Grate

An essential part of any coal-fired furnace is the metal grate on which the fuel is burned (Fig. 8). The grate should be designed to allow sufficient primary air to pass upward through the fuel bed for the combustion process. The ashes will drop to the ashpit below through the same openings when the grate is shaken.

There should be a metal heat transfer surface (heat exchanger) of sufficient size above the fire to transfer the heat from the combustion process to the water or air in the heating system.

Coal Stoker

A coal stoker is a device used to automatically feed coal to a coal-fired furnace or boiler. See Chapter 3, Volume 2 (COAL FIRING METHODS) for additional information about coal stokers.

Courtesy Oneida Heater Co., Inc.

Fig. 8. Cutaway of a coal-fired furnace showing grate and combination chamber details.

Hand Firing Methods

The method used to hand fire a furnace will depend largely upon the type of solid fuel used in it. Coke and the various types of coals each have their own special hand firing methods. Some of these firing methods are described in Chapter 3, Volume 2 (COAL FIRING METHODS).

Blower and Motor Assembly

The blowers used on coal furnaces are the same centrifugal types found on gas, oil, and electric forced, warm-air furnaces. The blowers are usually installed on either side of the furnace (Fig. 1) or at the back (Fig. 9). The latter arrangement is especially recommended for wood burning and multi-fuel furnaces. A detailed description of blowers and blower motors is included in Chapter 11 (GAS-FIRED FURNACES).

Blower Adjustment

Instructions for making blower adjustments can usually be obtained from the furnace manufacturer. A brief description of methods used to make blower adjustments is included in Chapter 11 (GAS-FIRED FURNACES).

Air Filter

The number, size, and type of air filter will be recommended by the furnace manufacturer in the specifications. See also Chapter 14, Volume 3 (AIR CLEANERS AND FILTERS).

Air Conditioning

One method of adding summer air conditioning to a coal-fired heating installation is to install a separate and independent air conditioning system. This is expensive, because it includes the equipment and separate ducts, but it avoids many complications.

If you are considering the idea of adding air conditioning at some future date, you should select a solid fuel furnace capable of meeting your cooling needs. For example, *Oneida* All-Fuel and Two-in-one furnaces are equipped with blowers large enough to handle up to four tons of air conditioning and electronic air cleaning.

MULTI-FUEL FURNACES

Furnaces designed for use with more than one fuel are available from some manufacturers for use in residential warm-air heating systems. Both gravity and forced warm-air units are manufactured.

The *Oneida* All-Fuel furnace will burn any type of combustible material (coal, oil, gas, wood). It is designed to burn coal and wood, oil and wood, or gas and wood at the same time. The cast iron grates allow the burning of anthracite coal, bituminous coal, or wood.

WOOD-FIRED FURNACES

Furnaces are available that burn wood in combination with other fuels (see above), or which burn *only* wood (Figs. 9 and 10). One standard cord of good hard wood is approximately equal to the heating equivalent of 200 gallons of fuel oil or one ton of coal.

The burning of wood requires a furnace built specifically for this purpose. The *Daniels* gravity warm-air furnaces illustrated in Figs. 9 and 10 are constructed of steel with fire doors and rectangular firepots big enough to burn large chunks of wood.

The so-called "chunk" furnace in Fig. 9 has a plate grate with round holes and slotted holes. Each time the furnace is fired, the wood coals are pulled to the front of the grate and stirred up so that there is a draft up through the holes in the plate. By closing the draft door located in the large fire door and opening the draft door in the ash pit door, a draft is forced up through the wood coals to complete their burning. For general firing the ash pit draft door is closed and the fire is regulated by the draft door in the fire (feed) door.

Forced warm-air wood-fired furnaces are also available for use in residential heating systems. The *Daniels* furnace shown in Fig. 11 is available in capacities ranging from 100,000 to 400,000 Btu (at bonnet). These units contain three large filters and the air is circulated by means of a large, low velocity fan. One advantage of this type of furnace is the provision for gravity heat. When a power failure occurs, the furnace can provide gravity heat by removing the cold air panels.

Courtesy Sam Daniels Co.

Fig. 9. Blower installation on a wood-fired forced warm-air furnace.

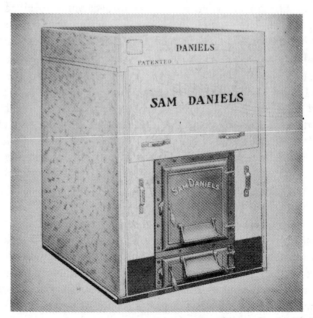

Courtesy Sam Daniels Co.

Fig. 10. Wood-fired chunk furnace.

Fig. 11. Pipeless chunk furnace.

MAINTENANCE AND OPERATING INSTRUCTIONS

The following recommendations are offered as a guide to the maintenance and operation of a coal-fired furnace:

1. Carry a deep or high fire by keeping the pot full. Let coals come up to the feed door and higher, and then slope back. Do not disturb the fire by frequent feeding in driblets, or poking or shaking through the day. Feed, shake or clear grate at regular intervals.

2. When feeding coal to the furnace or shaking the fire, keep the choke damper open and the draft door closed.

3. In severe weather, give the furnace most careful attention late at night. Clear the grate until it is bright underneath, and fill the pot full.

411

4. In mild weather (running with checked fire), feeding the fire in the morning and at night is usually enough. In extreme weather, it may be necessary to feed more often.

5. Do not leave the feed door open in order to check the fire (in mild weather or at night). Correct adjustment of the damper control will regulate the fire to supply just the right amount of heat needed.

6. Do not allow ashes to bank up under the grate in the ash pit. Grate bars are hardy, but it is possible to warp them by carelessness. Taking up ashes once a day is the best rule, even if few ashes have fallen into the pit. If convenient, spray the ashes with a water hose before taking them up.

7. Keep the heating surfaces and flues clean. Soot will reduce heating efficiency and the furnace will require much more fuel to produce the required heat than would be the case if it were clean.

8. Do not overshake or poke the fire in mild weather. Shake it enough to make room for a little more fuel. Never poke a hard coal fire from above.

9. Clean the fire regularly especially in cold weather. This will result in fuel savings and increased heating efficiency. Do not slice or poke the fire from the top. Use the feed door for removing large clinkers. Do all cleaning through the clinker door, or by shaking the dumping grate.

10. If the fire does not burn evenly over the entire grate or if it seems sluggish in starting up after the grates have been thoroughly shaken, a thorough cleaning is required. This is accomplished by allowing the fire to burn down until a thin layer of fuel is left on the grate. Shake well and remove all the slate and clinkers left on the grate with a slice bar and hoe. To start the fire, first add a thin layer of fresh coal and feed the full charge of fuel only after this layer is burning briskly.

11. The grate will sometimes get covered with slate and clinkers massed together when a poor quality coal is burned. The best way to correct the situation is to dump the grate and build an entirely new fire.

12. Hard clinkers lodged between the grate bars should be removed with a poker or slice bar.

TROUBLESHOOTING A COAL-FIRED FURNACE

The list that follows contains the most common operating problems associated with coal-fired furnaces. Each problem is given in the form of a symptom, the possible cause, and a suggested remedy. This list is intended to provide the operator with a quick reference to the cause and correction of a specific problem.

Symptom and Possible Cause	Possible Remedy
Inadequate Fire	
(a) Insufficient draft	(a) Clean ash pit and remove obstruction to primary air supply. Adjust damper control.
(b) Dirty furnace	(b) Clean furnace
(c) Dirty flue	(c) Clean flue
(d) Poor quality fuel	(d) Replace with better quality fuel
(e) Grate clogged with slate and clinkers	(e) Dump grate and rebuild fire. Dislodge by poking gently with poker.
Coal Stoker Stops	
(a) Power off	(a) Check main power switch, fuses and correct
(b) Obstruction in feed screw	(b) Remove obstruction
(c) Dirty fire	(c) Clean fire
Stoker Motor Fails to Start	
(a) No electrical power	(a) Check main power switch, fuses and correct

(b) Overload

(b) Push reset button on transmission.
Push reset button on stoker.

(c) Blown fuses

(c) Replace fuses

(d) Limit control contacts open

(d) Let furnace cool off

Stoker Operates Continuously

(a) Controls out of adjustment

(a) Readjust controls or call local sales representative for service

(b) Dirty fire

(b) Clean fire

(c) Fire out

(c) Rebuild fire

(d) Dirty furnace

(d) Clean furnace

Excessive Coal in Firebox

(a) Stoker feeding too much coal

(a) Reduce coal feed rate

(b) Insufficient air

(b) Open manual draft

(c) Stoker windbox full of siftings

(c) Clean out windbox

(d) Accumulation of clinkers in fire

(d) Clean fire

Excessive Fuel Use

(a) Dirty furnace

(a) Clean furnace

(b) Dirty flue

(b) Clean flue

(c) Improper draft

(c) Adjust dampers for correct rate of combustion

Not Enough Heat

(a) Thermostat set too low

(a) Raise setting

(b) Thermostat improperly located

(b) Relocate thermostat

(c) Thermostat out of calibration

(c) Recalibrate or replace

(d) Lamp or other heat source too near to the thermostat

(d) Remove heat source

414

(e) Dirty air filter

(e) Clean or replace

(f) Limit set too low

(f) Reset or replace

(g) Fan speed too slow

(g) Check motor and fan belt, and tighten if too loose.

Blower Won't Run

(a) Power not on

(a) Check power switch. Check fuses and replace if necessary.

(b) Fan control adjustment too high

(b) Readjust or replace

(c) Loose wiring

(c) Check and tighten

(d) Defective motor overload, protector or motor

(d) Replace motor

Blower Won't Stop

(a) Manual fan ON

(a) Switch to automatic

(b) Fan switch defective

(b) Replace

(c) Short in wiring

(c) Check wiring and correct

Rapid Fan Cycling

(a) Fan switch differential too low

(a) Readjust or replace

(b) Blower speed too high

(b) Readjust to lower speed

Noisy Blower and Motor

(a) Fan blades loose

(a) Replace or tighten

(b) Belt tension improper

(b) Readjust to allow 1 inch slack

(c) Pulleys out of alignment

(c) Realign

(d) Bearings dry

(d) Lubricate

(e) Defective belt

(e) Replace

(f) Belt rubbing

(f) Reposition

Electric-Fired Furnaces

Electric heating is the only heat that is produced almost as fast as the thermostat calls for it. It is almost instantaneous, because there are no heat exchangers to warm up. The heating elements start producing heat the moment the thermostat calls for it.

Because there is no flame with electric heat, there is no need to vent smoke or flue gases to the outside. Furthermore, there is no chimney loss with electric heat. It is 100% efficient, compared with an efficiency of 80% in furnaces using other fuels.

Electric-fired furnaces are available in upflow, downflow, or horizontal flow models, and in a wide range of sizes. For example, *Carrier* electric furnaces are available in 15 standard models from 5 to 35 kW in 5 kW increments. Other manufacturers offer a similar range of models (Figs. 1 and 2).

An electric-fired furnace should be listed by the *Underwriters' Laboratories* for construction and operating safety. Furnaces approved by the agency are marked UL Approved.

Electric-fired furnaces should be installed in accordance with local codes and regulations; *The National Electrical Code;* and recommendations made by *The National Fire Protection Association.*

ELECTRICAL POWER SUPPLY

Contact the local power company and make certain adequate electrical service is available for the furnace load *plus* all other appliances that will be on the line.

CONTINUOUS AIR CIRCULATION SWITCH

ELECTRIC HEATING ELEMENTS

ACCESS PANEL, PULL OUT AND LIFT UP TO REMOVE

FUSE BLOCKS (ELEMENT HEATING FUSES)

FUSE ACCESS DOOR

HAMMOCK AIR FILTER

BLOWER MOTOR

BLOWER WHEEL

Courtesy Lennox Air Conditioning and Heating.

Fig. 1. Lennox model E10 upflow electric furnace.

Check *The National Electrical Code* and the local code requirements. All wiring (including sizing) *must* comply with the requirements of these codes. When there is any conflict, the local codes and regulations will take precedence. The manufacturer's requirements are also important. For example, *Janitrol* discourages the use of aluminum wire although it is acceptable to *The National Electrical Code.* Use of aluminum wire in this case could jeopardize the furnace warranty.

Be sure to consult the manufacturer's wiring diagrams for electrical power requirements. Read the instructions carefully and be sure you understand them thoroughly before you begin work.

PLANNING SUGGESTIONS

If an electric-fired furnace is being planned for a new structure, the maximum heat loss for each heated space must be calculated in accordance with procedures described in the manuals of the National Warm Air Heating & Air Conditioning Association

HEATING ELEMENT

FUSE BLOCK

GROUND TERMINAL

FAN SPEED
TERMINAL BOARD

LOW-VOLTAGE BOARD

FAN CAPACITOR

BLOWER

FILTER

Courtesy Carrier Corp.

Fig. 2. Carrier forced warm-air electric furnace.

or by a comparable method. This is very important because this data will be used to determine the size (capacity) of the furnace selected for the installation.

Do not consider an electric heating system unless the structure is properly insulated. This type of insulating will be more extensive than that used with other types of heating systems. For example, ceilings should have a *minimum* of 6 inches of blanket or loose type insulation, and cavities between studs in exterior walls should be completely filled with insulation. A description of the insulation requirements for a structure in which an electric heating system is used is included in Chapter 9 (ELECTRIC HEATING SYSTEMS).

If the heating or heating/cooling installation is to be approved by either the FHA or VA, heat loss and heat gain calculations should be made in accordance with the procedures described in *National Environmental Systems Contractors Manual J.*

Location and Clearance

An electric-fired forced warm-air furnace should be located as near as possible to the center of the heat distribution system. Centralizing the furnace eliminates the need for one or more exceptionally long supply ducts. Long supply ducts are uneconomical because they are subject to a certain amount of heat loss. The number of elbows should be kept to a minimum for the same reason.

Electric furnaces are not vented because electric heat is not produced by the combustion process. No flue gases or other toxic products of the combustion process occur with electric heat. As a result, a chimney and flue pipe are not required and it is not necessary to consider these factors when locating the furnace.

A clearance of 24-30 inches in front of the furnace access panel should be provided for servicing and repairs. There is no minimum clearance requirement for ductwork and combustible materials. Electric furnaces may be installed with zero clearance between the cabinet and combustible materials, because the heat from the furnace is not produced by a flame.

INSTALLATION RECOMMENDATIONS

New electric furnaces for residential installation are shipped from the factory with all internal wiring completed. These furnaces are also generally shipped as a preassembled unit. In order to install the new furnace, the electric service from the line voltage main and the low voltage thermostat must be connected. Di-

HEATING AND VENTILATING

rections for making these connections are found in the furnace
manufacturer's installation instructions.

Familiarize yourself with all local codes and regulations which
govern the installation of an electric-fired furnace. Local codes
and regulations take precedence over national standards.

Check the insulation of the structure to determine if it is properly insulated for electric heat. The insulation should be installed
in accordance with recommendations in "All Weather Comfort
Standard of Electrically Heated and Air Conditioned Homes"
(Electric Heating Association).

The furnace should be mounted on a level surface. If the unit
is not level, it may develop serious vibrations. An insulating material can be placed under the furnace in most installations to
reduce sound vibrations when the unit is operating. A noncombustible base is recommended for counterflow models.

Duct Connections

Detailed information concerning the installation of an air
duct system is contained in the following two publications of the
National Fire Protection Association:

1. "Installation of Air Conditioning and Ventilating Systems of Other Than Residence Type" (NFPA No. 90A).
2. "Residence Type Warm Air Heating and Air Conditioning Systems" (NFPA No. 90B).

Additional information about duct connections can be found
in Chapter 7, Volume 2 (DUCTS AND DUCT SYSTEMS). The
comments made in Chapter 11 (GAS-FIRED FURNACES) about
furnace duct connections and air distribution ducts apply for the
most part to ducts used with electric-fired forced warm-air
furnaces.

BASIC COMPONENTS

An electric-fired forced warm-air furnace will generally consist of the following basic components (Fig. 3):

1. Automatic controls,
2. Safety controls,

420

3. Heating elements,

4. Blower and motor,

5. Air filter(s).

Each of these components is described in the sections that follow. Additional information is contained in Chapter 10 (FURNACE FUNDAMENTALS) and the various chapters in which furnace controls are described.

AUTOMATIC CONTROLS

The automatic controls used in an electric heating system are designed to insure its safe and efficient operation. Detailed descriptions of these controls are found in Chapter 4, Volume 2 (THERMOSTATS AND HUMIDISTATS). This section is primarily concerned with outlining the operating principles of the automatic controls used with an electric furnace. These controls include:

1. Room thermostat,

2. Thermostat heat anticipator,

3. Timing sequencers.

In a central heating system, the wall mounted room thermostat is the control that governs the normal operation of the furnace. The operating principle is simple. The temperature selector on the thermostat is set for the desired temperature. When the temperature in the room falls below this setting, the thermostat will call for heat and cause the first heating circuit in the furnace to be turned on. There is generally a delay of about 15 seconds before the furnace blower starts. This prevents the blower from circulating cool air in the winter. After about 30 seconds, the second heating circuit is turned on. The other circuits are turned on one by one in timed sequence.

When the temperature reaches the required level, the thermostat opens. After a short period of time, the first heating circuit is shut off. The others are shut off one by one in timed sequence. The blower will continue to operate until the air temperature in the furnace drops below a specified temperature.

A typical room thermostat will have a fan switch, a system switch, and a temperature selector (Fig. 4). The temperature

UPPER
ACCESS
DOOR

UNIT WIRING
DIAGRAM

HEATING ELEMENT

HIGH LIMIT
CONTROLS

TRANSFORMER

FUSES

LATCH

AIR FILTER

TERMINAL
BLOCKS

FAN AND
MOTOR

Courtesy The Trane Co.

Fig. 3. Principal components of an electrical furnace.

selector (a dial or lever device) on the thermostat is used to select the desired temperature. The actual operation of the heating system is governed by the positions of the fan and system switches. The switch positions and their functions are listed in Tables 1 and 2.

Most room thermostats contain a *heat anticipator*. This is a device designed to assist the thermostat in controlling closer to the desired temperature range (Fig. 5).

When timing sequencers are used (see below), the current flowing through the first time delay sequencer (relay) must also

422

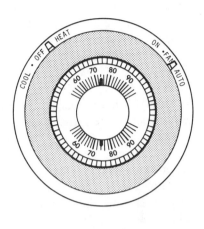

Courtesy Lennox Air Conditioning and Heating

Fig. 4. Typical room thermostats used in central electric heating systems.

HEAT ANTICIPATOR ADJUSTMENT

Courtesy The Trane Co.

Fig. 5. Thermostat heat anticipator.

flow through the heat anticipator. In order to obtain satisfactory operation, the heat anticipator setting must be equal to the current draw of the sequencer.

The furnace manufacturer will generally recommend the setting for the heat anticipator adjustment for each size unit. For example, the setting recommended for a *Trane* Model EUADH 07 electric furnace is 0.45. This thermostat adjustment will vary depending upon the type of time delay sequencer used, the furnace manufacturer, and the size of the furnace. This may be illustrated by the recommended heat anticipator settings given by *Coleman* for its 10 kW, 15 kW, and 20 kW furnace models (Table 3). All *Coleman* 25 kW models require a heat anticipator setting of 0.60.

After you have adjusted the heat anticipator to the suggested setting, operate the furnace several hours and observe the results. If there is *insufficient* heat, it may be caused by short furnace cycles. This can be corrected by moving the heat anticipator pointer to a slightly higher setting. If there is too much heat, then long furnace cycles are overheating the structure. This can be corrected by moving the heat anticipator pointer to a slightly lower setting. After making these thermostat adjustments, allow the furnace to operate several hours to determine if further adjustment will be required. Additional information about the thermo-

HEATING ELEMENT SUPPORT ROD

HEATING
ELEMENTS

FAN CONTROL   HIGH LIMIT CONTROL

Courtesy The Trane Co.

Fig. 6. Heating elements and controls.

stat heat anticipator is contained in Chapter 11 (GAS-FIRED FURNACES).

HEATING ELEMENTS

The heating elements in an electric furnace are resistance coils made from high temperature, chrome nickel heat generating wire. Most manufacturers design their furnaces so that the entire heating element assembly can be removed for easy maintenance or repair (Fig. 6).

TIMING SEQUENCERS

When all the heating elements in an electric furnace turn on simultaneously, there is a momentarily excessive demand on the power supply. The result is a temporary interruption in the electrical service. This problem of power drains and surges can be eliminated by timing the heating elements so that they start one at a time in predetermined increments. Sequencing relays are used for this purpose.

SAFETY CONTROLS

Most electric-fired furnaces are equipped with a variety of different safety controls to protect the appliance against current overloading or excessive operating temperatures. These safety controls are:

1. Temperature limit controls.
2. Secondary high limit control.
3. Furnace fuses.
4. Circuit breakers.
5. Control voltage transformer.
6. Thermal overload protector.

TEMPERATURE LIMIT CONTROLS

In the line control diagram of a *Trane* EUADH 07A model electric furnace shown in Fig. 7, each heating element is shown

HEATING AND VENTILATING

TD = TIME DELAY RELAY
FC = FAN CONTROL
F = FUSE
HLC = HIGH LIMIT CONTROL
H = HEATER
CR = CONTROL RELAY
TS = TERMINAL STRIP

OUTSIDE OR 2 STAGE
THERMOSTAT MAY BE
CONNECTED BETWEEN 1 & 2
FOR STAGING.

Courtesy *The Trane* Co.

Fig. 7. Line control wiring diagram of a *Trane* electric furnace.

426

with a high limit control device located between the heating element and the time delay relay. These high limit control devices are designed to limit the outlet air temperature on *Trane* electric furnaces to 200°F. If the temperature of the outlet air should exceed 200°F., the high limit control will open and interrupt the supply of electric power to the heating element.

SECONDARY HIGH LIMIT CONTROL

Secondary high limit protection is provided by a fusible link in each electric heating element (Fig. 8). This device is designed to shut off the current when temperatures in the furnace become excessive. It functions as a back up system in case of limit switch failure.

Courtesy International Heating and Air Conditioning Corp.

Fig. 8. Fusible link connection in electric heating element.

FURNACE FUSES

Furnace fuses are used to provide protection against possible overload conditions, and to insure correct, safe field wiring.

Each heating element circuit is protected by two branch fuses as shown in Fig. 9. These are sized to limit the current draw of each heater and are designed to open on a short circuit or an overloaded circuit condition. For overload protection, the blower motor and relay are also safeguarded with a separate fuse.

HEATING   FAN CONTROL   HIGH
ELEMENT   IN BACK OF   LIMIT
TERMINAL   PANEL   CONTROL

FAN
CONTROL
RELAY

HEATING
ELEMENT

CONTROL
POWER
TRANSFORMER

MOUNTING
SCREWS

TIME RELAY
SEQUENCE

CONTROL
POWER
FUSES

LINE POWER FUSES
(TWO FOR EACH
HEATING ELEMENT)

LINE POWER   MAIN POWER
FUSES   TERMINAL BLOCKS

Courtesy The Trane Co.

Fig. 9. Electric furnace heating section control panel showing location of line power fuses.

CIRCUIT BREAKERS

Some electric furnaces are equipped with circuit breakers and a terminal board. The wiring diagram for the *Coleman* 25 kW electric furnace illustrated in Fig. 10 indicates that both the terminal board and circuit breakers are located in the power supply feed line to the furnace.

When it is desirable to run branch circuits to the circuit breakers (bypassing the terminal board), the jumper wires between the circuit breakers and the wiring is connected as shown in Figs. 11 and 12.

TRANSFORMER

A control voltage transformer is used to limit the amount of output current. Limiting the amount of output current permits the use of open type control wiring.

Courtesy The Coleman Co., Inc.

Fig. 10. Wiring diagram for the Coleman 25 kW electric furnace.

429

Courtesy The Coleman Co., Inc.

Fig. 11. Electrical wiring for a Coleman 15 kW or 20 kW electric furnace with circuit breakers.

THERMAL OVERLOAD PROTECTOR

The fan motor is protected against locked rotor or overheated conditions by thermal overload devices. When these conditions occur, the fan motor circuit is automatically opened and the motor is shut off.

Fig. 12. Electric wiring for a Coleman 25 kW electric furnace with circuit breaker.

ELECTRICAL WIRING

All internal furnace wiring is done at the factory before it is shipped. At the site, the following two types of electrical connections are required to field wire the unit:

1. Line voltage field wiring.
2. Control voltage field wiring.

A typical example of line voltage field wiring is shown in Fig. 13. Line voltage wiring involves the connection of the furnace to the building power supply. Line voltage wiring runs directly from the building power panel to a fused disconnect switch. From there, the wiring runs to terminals L_1 and L_2 on the power supply terminal block.

Courtesy Fedders Corp.

Fig. 13. Typical line voltage field wiring shown for a *Fedders* electric furnace.

The unit must be properly grounded either by attaching a grounded conduit for the supply conductors (knockouts in the side panels of *Fedders* electric furnaces are provided for this purpose) or by connecting a separate wire from the furnace ground lug to a suitable ground.

The external control voltage circuitry consists of the wiring between the thermostat and the low voltage terminal block located in the control voltage section of the furnace. Instructions for control voltage wiring are generally shipped with the thermostat.

Some typical examples of control voltage field wiring used with *Fedders* electric furnaces are shown in Figs. 14 and 15. Control voltage field wiring connections used with *Coleman* electric furnaces are shown in Fig. 16.

Fig. 14. Control voltage wiring for two-stage heating, single-stage cooling applications.

Fig. 15. Control voltage wiring for single-stage heating, single-stage cooling applications.

Courtesy Fedders Corp.

The National Electric Code requires that furnaces larger than 10 kW be supplied with branch circuit fusing. Power connections on units of this size are usually made to lugs on the fuse blocks.

434

10, 15 or 20 kW Model

25 kW Model

20 or 25 kW Model

Courtesy The Coleman Co., Inc.

Fig. 16. Three connections for indoor and outdoor heating (only) thermostats for a Coleman model 6806 electric furnace.

BLOWERS AND MOTORS

The blowers and motors that are used with electric-fired forced warm-air furnaces are identical to those used in gas-fired furnaces.

435

You should read the section BLOWERS AND MOTORS in Chapter 11 (GAS-FIRED FURNACES) for additional information.

AIR DELIVERY AND BLOWER ADJUSTMENT

It is sometimes necessary to adjust the blower speed to produce a temperature rise through the furnace that falls within the limits stamped on the furnace nameplate. Blower adjustment procedures are described in full detail in Chapter 11 (GAS-FIRED FURNACES).

AIR FILTER

The air filters used in electric-fired forced warm-air furnaces are either permanent types that can be removed and cleaned on a periodic basis, or replaceable, throw away type filters.

More detailed information concerning furnace air filters is contained in Chapter 14, Volume 3 (AIR CLEANERS AND FILTERS). Also read the section on air filters in Chapter 11 (GAS-FIRED FURNACES). The filters used in the various types of forced warm-air furnaces are essentially the same regardless of the type of fuel used to fire the unit.

AIR CONDITIONING

A furnace should be installed in parallel or on the upstream side of the cooling unit to avoid condensation in the heating section. Parallel installation will require dampers or some other means to prevent cool air from entering the furnace.

The air conditioning component of a typical electric heating and cooling system generally consists of an outdoor condensing unit, indoor coils, and a cabinet to house the cooling coils. Most furnace manufacturers provide detailed instructions for adding air conditioning to the heating unit. The important thing to remember is to size the ducts for the larger volume of air used in air conditioning.

Additional information about air conditioning can be found in Chapters 7-8, Volume 3.

MAINTENANCE AND OPERATING INSTRUCTIONS

Maintenance and operating instructions will normally be provided for the furnace by the manufacturer. If no instructions are available, try contacting a field representative or writing the company for a duplicate copy of the owners manual.

Always shut off the electrical power supply to the furnace before attempting to service it. This is very important, because there is the possibility of a fatal electric shock. Be sure to open all furnace fused disconnect switches before servicing.

A clean air filter is important to the efficient operation of the furnace. Air filters should be removed and inspected on a periodic basis (preferably every six months). Clean or replace the filters *as* required.

Inspect the heating element and heating control wiring to make certain connections are tight and clean. Check for any breaks or cracks in the wire insulation. These can cause shorting. All terminal block wiring connections should also be tight and clean.

Check the furnace wiring diagram to make certain the fuses are of the correct type and amperage.

Some fan motors are permanently lubricated and will not require further attention. Others have ports for oiling and require periodic lubrication. See BLOWERS AND MOTORS in Chapter 11 (GAS-FIRED FURNACES).

A combined heating and cooling system is designed to operate year around without being shut down. Only changeover that is required is that of the room thermostat and periodic maintenance.

When an electric furnace is installed *without* an air conditioning unit, the furnace should be shut down at the end of the heating season. The procedure for doing this will be found in the furnace owner's manual. It is a simple operation which generally consists of opening the main fused disconnect switch or switches in the power supply lines serving the furnace.

TROUBLESHOOTING AN ELECTRIC FURNACE

Any appliance may sometimes fail to operate efficiently because of a malfunction somewhere in the equipment. The problems most commonly associated with electric-fired furnaces are listed below.

Symptom and Possible Cause	Possible Remedy
Unit Fails to Operate	
(a) Defective furnace transformer	(a) Replace
(b) Blown or defective transformer fuse	(b) Replace fuse
(c) Defective thermostat	(c) Replace thermostat
(d) Improperly set room thermostat	(d) Change thermostat to proper setting
(e) Open fused disconnect switch	(e) Correct problem and close furnace circuit
(f) Blown or faulty furnace fuse	(f) Replace fuse
Fan Operates With Low Or No Heat	
(a) Blown or faulty heater element fuse	(a) Replace fuse
(b) Defective time delay sequencer	(b) Replace sequencer
Heats Without Fan Operation	
(a) Faulty fan control relay	(a) Replace
(b) Defective fan motor	(b) Repair or replace
(c) Faulty fan motor wiring or loose connections	(c) Repair or replace wiring
(d) Defective fan motor run capacitor	(d) Repair or replace
Individual Heater Fails To Operate	
(a) Blown or faulty heater circuit fuse	(a) Replace fuse
(b) Defective high limit control	(b) Replace

438

(c) Defective time delay sequencer

(c) Replace

(d) Faulty heater element

(d) Replace

Fan Operates On Heating Not On Cooling

(a) Defective cooling cycle control relay

(a) Replace

(b) Improperly connected or faulty room thermostat

(b) Make proper connections to thermostat

(c) Defective or improper fan motor connections

(c) Repair or replace

Table 1. Thermostat operation in single stage heating/two stage cooling; two stage heating/single stage cooling; two stage heating/two stage cooling.

Thermostat Switch Setting		Function
Fan	System	
Auto	Off	System completely shut down.
On	Off	Blower only, continuous operation. Provides air circulation when no cooling or heating is desired.
Auto	Cool	Blower and cooling system cycle on and off as thermostat demands.
On	Cool	Blower runs continuously, cooling system cycles on and off as thermostat demands.
Auto	Heat	Blower and furnace cycle on and off as thermostat demands.
On	Heat	Blower runs continously, furnace cycles on and off as thermostat demands.
Auto	Auto	Unit cycles for either cooling or heating as per demand of thermostat.
On	Auto	Blower runs continuously, cooling and heating cycle as per demand of thermostat.

Courtesy *Fedders Corporation*

Table 2. Thermostat operation in single stage heating/cooling.

Thermostat Switch Setting		Function
Fan	System	
Auto	Off	System completely shut down.
On	Off	Blower only, continuous operation. Provides air circulation when no cooling or heating is desired.
Auto	Cool	Blower and cooling system cycle on and off as thermostat demands.
On	Cool	Blower runs continuously, cooling system cycles on and off as thermostat demands.
Auto	Heat	Blower and furnace cycle on and off as thermostat demands.
Heat	On	Blower runs continuously, furnace cycles on and off as thermostat demands.

Courtesy Fedders Corporation

Table 3. Recommended heat anticipator settings for Coleman 10 kW, 15 kW, and 20 kW electric furnaces.

Sequencer	10 kW	15 kW	20 kW
White-Rogers	.15	.15	.1
Klixon	.15	.30	
Honeywell	.4	.4	

Courtesy The Coleman Company, Inc.

CHAPTER 15

Boilers and Boiler Fittings

A boiler is a device used to supply steam or hot water for heating, processing or power purposes. This chapter is primarily concerned with a description of the low-pressure steam and hot-water boilers used in the heating systems of residences and small buildings.

The basic construction of both low-pressure steam and hot-water space heating boilers fired by fossil fuels consists of an insulated steel jacket enclosing a lower chamber in which the combustion process takes place and an upper chamber containing cast iron sections or steel tubes in which water is heated or converted to steam for circulation through the pipes of the heating system (Figs. 1, 2, 3, and 4).

The design and construction of the lower chamber will depend upon the type of fuel used to fire the boiler. It serves as a combustion chamber for coal- and oil-fired boilers, and as a compartment for housing the gas burner assembly on gas-fired boilers. These gas burner assemblies are generally designed for easy removal so that they can be periodically cleaned or serviced (Fig. 5).

Oil burners are externally mounted with the burner nozzle extending into the combustion chamber. This is also true of gas conversion burners. Gas burner assemblies, on the other hand, are located inside the lower chamber of the boiler.

The cast iron sections or steel tubes in the upper chamber of the boiler contain water which circulates through the pipes in the heating system either in the form of steam or hot water. The heat

CAST IRON SECTIONS

JACKET

CONTROLS

BURNERS

FACTORY-ASSEMBLED
MANIFOLD

Courtesy American Standard

Fig. 1. Gas-fired boiler for a forced hot-water heating system.

from the combustion process in the lower chamber of the boiler is transferred through the metal surface of the cast iron sections or steel tubes to the water contained in them, causing a rise in temperature. The amount of water contained in these passages is one of the ways in which steam and hot water space heating boilers are distinguished from one another. In hot water space heating boilers, these passages are completely filled with water; whereas in low-pressure steam boilers only the lower two thirds is filled. In the latter, the water is heated very rapidly causing steam to form in the upper one third. The steam, under pressure, rises through the supply pipes connected to the top section of the boiler.

442

Fig. 2. Basic components of a *Bryan* steel-tube gas-fired boiler used in hot-water heating systems.

A boiler jacket contains a number of different openings for pipe connections and the mounting of accessories. The number and type of openings on a specific boiler jacket will depend upon the type of boiler (e.g. steam or hot water). Among the different types of openings to be found on a boiler jacket are the flue connection, water feed (supply) connection, inspection and cleanout tapping, blow down tapping, relief valve tapping, control tapping, drain tapping, expansion tank tapping and return tapping. There

443

DRAFT
HOODS

LOW WATER
CUTOFF

FLUE
CANOPY

STEEL
JACKET

CAST
IRON
SECTIONS

BASE
FRAME

GAS
CONTROL
TRAIN

MANIFOLD

CAST IRON
BURNERS

COMBUSTION
CHAMBER

Courtesy American Standard

Fig. 3. Gas-fired steam heating boiler.

are also gas and oil burner connections. Fig. 6 illustrates the arrangement of control tappings for *American Standard* gas-fired steam and hot-water boilers.

Most (but not all) of the controls on low-pressure steam and hot-water space heating boilers fired by the *same* fuel are similar in design and function; however, there are exceptions. For example, a few boiler controls and fittings are designed to be specifically used on steam boilers; whereas others are found only on hot-water space heating boilers. The various types of boiler con-

444

CONTROLS

CAST IRON
SECTIONS

FRONT
CLEANOUT
PANEL

WATER
HEATER

JACKET

INSULATED
STEEL BASE

OIL BURNER

Fig. 4. Oil-fired boiler for a forced hot-water heating system.

trols and fittings are described in the appropriate sections of this chapter.

BOILER RATING METHODS

The construction of low-pressure steel and cast-iron heating boilers is governed by the requirements of the ASME Boiler and Pressure Vessel Code. This is a nationally recognized code used by boiler manufacturers, and any boiler used in a heating installation should clearly display the ASME stamp. State and local codes are usually patterned after the ASME Code.

Courtesy H.B. Smith Co., Inc.

Fig. 5. Gas burner boiler heads and support frame. Support frames are mounted on casters to facilitate removal of burner for cleaning.

The location of the identification symbols used by the ASME is specified by the code and determined by the type of boiler. For

location on boiler	size (inches)	steam boilers	water boilers
A	6	supply	supply
B	4	return	return
C	3	safety valve	relief valve
D	1¼	blow-off	not required
E	1	water feeder	not required
F	¾	pressure limit	temperature limit
G	¾	drain	drain
H	½	pressure gauge	temperature and altitude gauge
J	½	gauge glass and low water cut-off	not required
K ■	⅜	try-cocks	not required

■ tappings available on special order only.

Fig. 6. Arrangement of control tappings.

front of boiler

layout of tappings
(left end shown)
Courtesy American Standard

example, on a water-tube boiler, it appears on a head of the steam-outlet drum near and above the manhole opening. On vertical fire-tube boilers, the stamp bearing the identification symbol should appear on the shell above the fire door and handhole opening. Other types of boilers (e.g. scotch marine, superheaters,

447

etc.) have their own specified location for the identification symbol stamp.

The *ASME Boiler and Pressure Vessel Code* only applies to boiler construction, specifically to maximum allowable working pressures, and not to its heating capacity. A number of different methods are used to rate the heating or operating capacity of a boiler. The boiler manufacturers have developed their own ratings, but these are generally used along with rating methods available from the following organizations:

1. Institute of Boiler and Radiator Manufacturers (IBR).
2. Steel Boiler Institute (SBI).
3. American Gas Association (AGA).
4. Mechanical Contractors Association of America (MCAA).

The Institute of Boiler and Radiator Manufacturers has adopted a code for rating most types of sectional cast-iron boilers. Steel boilers are rated by the Steel Boiler Institute rating code. The Mechanical Contractors Association of America has devised a method for rating boilers not covered by either the SBI or IBR Codes. Finally, gas-fired boilers are rated in accordance with methods developed by the *American Gas Association.*

Other organizations which rate boilers include the Packaged Fire-tube Branch of the American Boiler and Affiliated Industries (ratings for modified scotch marine boilers) and numerous international groups.

In terms of its heating capacity, the rating of a boiler can be expressed in square feet of equivalent direct radiation (EDR) or thousands of Btu per hour. Sometimes a boiler horsepower rating is also given, but this has proven to be misleading.

For steam boilers, 1 square foot of direct radiation (EDR) is equal to the emission of 240 Btu per hour. For a water boiler, 1 square foot EDR is considered equal to the emission of 150 Btu per hour.

A boiler horsepower (BHP) is the evaporation of 34.5 pounds of water into dry steam from and at 212°F. For rating purposes, 1 BHP is considered as the heat equivalent of 140 square feet of steam radiation per hour. BHP ratings are in some cases obtained by dividing steam SBI Ratings by 140.

A boiler is rated according to its operating or heating capacity, but this rating will vary in accordance with the *type* of load used as the basis for the rating. The three types of connected loads used to determine the rating of a boiler are:

1. Net load,
2. Design load,
3. Gross load.

Net load refers to the actual connected load of the heat emitting units in the steam or hot-water heating system. *Design load* includes the net load rating plus an allowance for piping heat loss. Finally, *gross load* will equal the net load and the piping heat loss, plus an additional allowance for the pickup load.

BOILER HEATING SURFACE

The boiler *heating surface* (expressed in square feet) is that portion of the surface of the heat transfer apparatus in contact with the fluid being heated on one side and the gas or refractory being cooled on the other side. The *direct* or *radiant* heating surface is the surface against which the fire strikes. The surface which comes in contact with the hot gases is called the *indirect* or *convection* surface.

The *heating capacity* of any boiler is influenced by the amount and arrangement of the heating surface, and the temperatures on either side. The arrangement of the heating surface refers to the ratio of the diameter of each passage to its length, also its contour (straight or curved), cross sectional shape, number of passes, and other design variables.

BOILER EFFICIENCY

The *boiler efficiency* is the ratio of the heat output to the calorific value of the fuel. Boiler efficiency is determined by a number of different factors, including the type of fuel used, the method of firing, and the control settings. For example, oil- and gas-fired boilers have boiler efficiencies ranging from 70 to 80 percent. On the other hand, a hand fired boiler in which anthracite coal is used will have a boiler efficiency of 60 to 75 percent.

TYPES OF BOILERS

The boilers used in low-pressure steam and hot-water space heating systems can be classified in a number of different ways. Some of the criteria used in classifying them will include:

1. Construction material,
2. Construction design,
3. Boiler position,
4. Number of passes of the hot gases,
5. Length of travel of the hot gases,
6. Type of heating surface,
7. Type of fuel used.

Most boilers are either cast iron or steel construction. *Very few* are constructed from nonferrous materials. Cast-iron boilers

Fig. 7. Coal-fired round vertical boiler.

generally display a greater resistance to the corrosive effects of water than steel ones do, but the degree of corrosion in steel boilers can be significantly reduced by chemically treating the water.

Cast-iron boilers can be classified on the basis of their construction design as either round boilers with horizontal pancake sections or square (or rectangular) boilers with either horizontal or vertical sections. A coal-fired round vertical boiler commonly used in older heating installations is shown in Fig. 7.

A square shaped boiler with horizontal section design is shown in Fig. 8. The heating surface of each cast-iron section is exposed at right angles to the rising flue gases (Fig. 9). The water travels in a zig-zag path from section to section in a manner similar to the flow of water in a steel tube boiler (Fig. 10).

CAST IRON SECTIONS

UNIT STEEL JACKET

FIBERGLASS INSULATION

HORIZONTAL WATER PASSAGES

COMBINATION GAS VALVE

GAS BURNER RAISED DRILL PORTS

GAS BURNER MIXER TUBES

Courtesy *Hydrotherm, Inc.*

Fig. 8. Horizontal section design of a *Hydrotherm* gas-fired hydronic cast-iron boiler.

451

Oil-fired boiler.

Gas-fired boiler.

Courtesy Hydrotherm, Inc.

Fig. 9. Direction of heat travel.

Vertical cast-iron sections are available in a number of different designs, depending upon the manufacturer (Fig. 11). A typical arrangement of these sections is shown in Fig. 12.

Steel boilers may be classified with respect to the relative position of water and hot gases in the tubular heating surface. In fire-tube boilers, for example, the hot gases pass within the boiler tubes and the water necessary to produce the steam circulates around them. In water-tube boilers, the reverse is true.

Flexible steel tubes are used in *Bryan* steam and hot-water space heating boilers for the rapid circulation of the water around the heat rising from the fire (Figs. 13 and 14).

Boilers can also be classified according to the number of passes made by the hot gases (e.g. one pass, two passes, three passes, etc.) (Fig. 15.).

Oil-fired boiler.

Gas-fired boiler.

Courtesy Hydrotherm, Inc.

Fig. 10. Direction of water flow.

The *length* of travel of the hot gases is another method used for classifying boilers. The efficiency of a boiler heating surface depends, in part, upon the ratio of the cross sectional area of the passage to its length.

Among the various types of fuels used to fire boilers are oil, gas (natural and propane), coal and coke. Conversion kits for converting a boiler from one gas to another are available from some manufacturers. Changing from coal (or coke) to oil or gas can be accomplished by using conversion chambers and making certain of other modifications (see Chapter 16. BOILER AND FURNACE CONVERSIONS).

Electricity can also be used to fire boilers. One advantage in using electric-fired boilers is that the draft provisions required by boilers using combustible fuels is not necessary.

The classificatory criteria described above were selective and limited to the more common types in use. Considering the multiplicity of boiler types and designs available, it is extremely difficult to establish a classification system suitable for all of them.

453

Fig. 11. Vertical cast-iron section.

Gas-Fired Boilers

Gas-fired steam and hot-water space heating boilers generally consist of a number of closely placed cast iron sections or steel tubes with a series of gas burners (i.e. a gas burner assembly) placed beneath them. The flue gases pass upwards between the sections or tubes to the flue collector.

Draft losses are kept low in these boilers, because the pressure at which the gas is supplied is generally sufficient to draw in the amount of air necessary for combustion. The fact that there is no fuel bed resistance, as is the case with coal-fired boilers, also contributes to low draft loss.

Fig. 12. Cast-iron sections in a hydronic gas-fired boiler.

Fig. 13. Flexible steel water tube.

The draft in gas-fired boilers is generally nullified by the diverter; consequently, the resistance offered by the boiler passages is not an important variable. When there is a problem of excessive draft, it can be controlled by installing a sheet metal baffle in the

Courtesy *Bryan Steam Corp.*

Fig. 14. F-series gas-fired boiler with insulation and outer steel jacket removed.

flue connection at the boiler (see CONTROL OF EXCESSIVE DRAFT ON GAS-FIRED BOILERS in this chapter).

Most of the controls and accessories used to operate gas-fired boilers are described in considerable detail in Chapter 2 (GAS BURNERS), Chapter 5 (GAS AND OIL CONTROLS), and Chapter 6 of Volume 2 (OTHER AUTOMATIC CONTROLS). The exact placement of these controls and accessories on the steel boiler jacket may differ slightly from one manufacturer to another, but not significantly. The major difference will be in the types of controls and accessories used to govern the temperature, pressure, and

FORCED DRAFT FAN

LOW FURNACE

FOUR PASS GAS TRAVEL

PASS 4

PASS 3

PASS 1

PASS 2

Courtesy Cleaver-Brooks.

Fig. 15. A steam boiler with four pass travel.

flow rate of the heat conveying medium; and this is determined by whether it is a steam or hot-water space heating boiler (see STEAM BOILER FITTINGS AND ACCESSORIES and HOT WATER BOILER FITTINGS AND ACCESSORIES in this chapter). An exploded view of a *Hydrotherm* gas-fired boiler illustrating the location of various controls and accessories is shown in Fig. 16.

Oil-Fired Boilers

An oil-fired boiler contains a heat transfer surface consisting of either cast-iron sections or steel tubes, and a special combustion chamber shaped to meet the requirements of an oil burner.

The oil burner is a device designed to mix fuel oil with air under controlled conditions and to deliver the mixture to the combustion chamber for burning. The burner is mounted outside the chamber. The oil burners used in residential heating boilers are usually high-pressure atomizing type burners although other types are used on occasion.

It is possible to convert a coal-fired boiler to oil by redesigning the combustion chamber. However, boilers specifically designed to use oil as a fuel have proven to be more efficient and economical than coal-fired boilers converted to oil.

457

BOILER JACKET

VENT FITTING

TEMPERATURE-PRESSURE INDICATOR

HIGH LIMIT AQUASTAT

TIE ROD WITH NUT

CAST IRON SECTIONS

PUSH NIPPLE

PILOT BRACKET

MANIFOLD VALVE

PILOT LINE WITH FITTINGS

PILOT ASSEMBLY

MANIFOLD

BURNER ORIFICE

AIR SHUTTER WITH SCREW

ANCHOR BRACKET

MANIFOLD BRACKET

GAS BURNER COMPARTMENT

PRESSURE RELIEF VALVE

DRAFT HOOD

DOME BRACKET

DOME

GRID

WELL-CIRCULATOR CONTROL

UNIVERSAL CIRCULATOR FITTING

BASE DOOR

GAS BURNER

BRACKET (VENTURI)

GOOSENECK (VENTURI)

Courtesy *Hydrotherm, Inc.*

Fig. 16. Exploded view of a gas-fired hot-water space heating boiler.

Read Chapter 1 (OIL BURNERS), Chapter 5 (GAS AND OIL CONTROLS), and Chapter 6 of Volume 2 (OTHER AUTOMATIC CONTROLS) for a description of the controls and accessories used to operate oil-fired boilers. Also read the appropriate sections in this chapter (STEAM BOILER FITTINGS AND

ACCESSORIES and HOT WATER BOILER FITTINGS AND ACCESSORIES). An exploded view of a *Hydrotherm* oil-fired boiler illustrates the location of the various controls and accessories (Fig. 17).

BOILER JACKET

CROSS OVER ASSEMBLY

FLANGE GASKET

FLOW LIMITING VALVE

BAROMETRIC DRAFT CONTROLS

AUTOMATIC AIR VENT

TANKLESS HEATER COIL

COLLAR

DOME BRACKET

COIL GASKET

DOME

TEMPERATURE-PRESSURE INDICATOR

TRIDICATOR

CAST IRON SECTIONS

PRESSURE RELIEF VALVE

HIGH LIMIT AQUASTAT

TANK VENTURI

BURNER CONTROL

COMBUSTION CHANBER

OIL BURNER

DOOR ASSEMBLY

COMBUSTION CHAMBER BASE ASSEMBLY

Courtesy *Hydrotherm, Inc.*

Fig. 17. Exploded view of an oil-fired hot-water space heating boiler.

459

Coal-Fired Boilers

A typical coal-fired boiler contains a grate for the fuel bed located beneath metal heat transfer surfaces (e.g. cast-iron sections). The hot gases resulting from combustion rise through the passages transferring their heat to the water contained in them.

The controls and accessories used to operate a coal-fired boiler are described in Chapter 3, Volume 2 (COAL FIRING METHODS). The controls and accessories used to govern the temperature, pressure, and flow rate of the heat conveying medium are described in this chapter (see STEAM BOILER FITTINGS AND ACCESSORIES and HOT WATER BOILER FITTINGS AND ACCESSORIES).

A coal-fired boiler requires more draft than gas- or oil-fired boilers because of the resistance of the fuel bed. Special care must be given to the design of the chamber surrounding the grate to insure sufficient volume for a proper mixture of fuel and air. Provisions must also be made for introducing sufficient air through the grate (and fuel bed) for the combustion process.

Before the introduction of oil burners, coal was generally used for fuel and is still largely used especially for heating plants already built and in cases where the owner cannot afford the expense of converting to oil. Moreover, some types of cast-iron boilers designed to burn coal are not suited to burn oil, which adds to the expense of a conversion job. Finally, the increasing world-wide shortage of oil will find it more and more necessary to find applications for low polluting uses of coal as a heat source. An important use for coal will be steam and hot water heating systems equipped with devices to reduce the polluting effect of burning coal.

Coal burning boilers are usually arranged for hand firing, but may be fitted for automatic stoker operation. In many automatic stokers, fuel is carried from the hopper through a feed tube by means of a rotating worm. Intermittent action of the worm agitates the fuel bed, prevents arching of the coal in the retort and assures that the incoming air reaches every part of the fire at all times. An auxiliary air connection between the feed tube and wind box prevents gas accumulating in the tube and eliminates any tendency to "smoke back" through the hopper.

An underfeed stoker is one in which the fuel is fed upward from underneath. The action of a screw or worm carries the fuel back through a retort from which it passes upward, as the fuel is consumed, the ash being finally deposited on dead plates on either side of the retort from which it can be removed.

Only part of the fuel being burned is actually burned in the fuel bed. Under the influence of high temperature created in the fuel bed and lack of sufficient air, unburned gases are released above the retort and tuyeres and unless these gases are mixed with air and burned inside of the combustion chamber, they will leave the boiler in the unburned condition, carrying a large percentage of the heat originally contained in the coal.

Nearer the outside of the fuel bed, the fuel burns less violently and much more air is passed through the fuel bed than is necessary. It is this excess air that must be mixed with the unburned gases issuing from the central point of the fuel bed if high combustion efficiency is to be realized.

Coke can also be used to fire boilers, but it requires different handling. In order to be completely consumed, coke needs a greater volume of air per pound of fuel than coal, and therefore requires a stronger draft, which is increased by the fact that it can only burn economically in a thick bed.

Because less coke is burned per hour per square foot of grate than coal, a larger grate and a deep fire pot are required to accomodate the thick bed of coke.

The quick flaming combustion which characterizes coal is not produced by coke because the latter fuel contains very little hydrogen; however, a coke fire is more even and regular.

Electric-Fired Boilers

Compact electric-fired boilers enjoy widespread use in forced hot-water heating systems. A typical example of one of these boilers is shown in Fig. 18. Heat is generated by electric heating elements immersed in water contained in a waterproof cast-iron shell. Although small in size, these boilers are capable of generating as high as 90,000 Btu's per hour; enough heat for the average 8-room house.

The basic components of an electric-fired boiler are shown in Fig. 19. An automatic air vent (1) located above the cast-iron

461

Courtesy American Standard

Fig. 18. An electric-fired boiler for hot-water heating.

boiler shell (8) is used for bleeding off trapped air from the water system. Safety devices include a water pressure relief valve (2) for reducing system pressure; an adjustable limit control (3) which allows selection of maximum boiler water temperature for system design; and preset pressure controls (4) to guarantee safe operation within the prescribed pressure range.

The limit control is an immersion device which shuts off the boiler if the temperature exceeds a predetermined setting. The preset pressure controls consist of a high pressure switch and a preset safe-fill switch. The high pressure switch is designed to deenergize the boiler if the pressure reaches 28 psi. This switch is reinforced by the relief valve, which is preset to relieve system pressure at 30 psi. The preset safe-fill switch prevents the electric heating elements from being energized unless pressure in the system is 4 psi. This prevents any chance of the heating elements

AUTOMATIC AIR VENT

ADJUSTABLE LIMIT CONTROL

PRE-SET PRESSURE CONTROL

DRAIN VALVE

WATER PRESSURE RELIEF VALVE

TRANSFORMER

RELAY

ELECTRIC HEATING ELEMENT

DRAIN

TRANSFORMER

EXPANSION TANK

CIRCULATOR

Fig. 19. Basic components of an electric-fired boiler.

burning out should an attempt be made to operate the boiler while dry.

The circulator (5), cast-iron boiler shell (8), and expansion tank (9) are all enclosed within the outer steel jacket (10) of the boiler. The electric heating elements (6) are mounted inside the ends of the boiler casting. A drain valve (11) mounted below the boiler casting is designed for the connection of standard hose fittings.

All controls are prewired and mounted in place by the manufacturer. The only electrical connections that need to be made during the installation of the boiler are those to the main power supply and room thermostat.

The sequencing relay switch (7) provides for an incremental loading on the electric service line. This reduces line voltage fluctuation and prevents power surge during energizing.

Minimum installation clearances for the boiler described above are shown in Fig. 20. Two air openings of 108 sq. in. (6"X18") each are required for closet installation.

FRONT VIEW   SIDE VIEW

Courtesy American Standard

Fig. 20. Minimum installation clearances.

A typical piping arrangement for a series loop hot-water heating system using an electric-fired boiler of this capacity is shown in Fig. 21. An outline of the procedure used for filling this system is as follows:

1. Close all zone valves except zone to be purged.

2. Open boiler drain valve.

3. Open fill valve.

4. Close purge valve.

5. Vent air from boiler by manually opening relief valve.

6. Open second valve to be purged, and close first zone.

7. After second zone is purged, open third zone valve and close second zone.

8. Repeat the procedures described in Steps 6 and 7 for as many zones as in the system.

9. When all zones have been purged, close boiler drain cock.

10. When water appears at relief valve, release lever and allow to close.

11. Continue filling until pressure gauge reads approximately 12 psi. Close fill valve.

12. Open purge valves.

Courtesy American Standard

Fig. 21. Typical piping arrangement for a multizoned hot-water heating system.

Where multiple circuits are used without zone valves, balancing valves should be installed for balancing. These can be shut off to purge each circuit individually. For a single loop system, no additional vents, valves, drains or other accessories are required.

Smaller electric-fired boilers having outputs ranging from approximately 6000 to 20,000 Btu's are available for individual apartment heating or for residential zoned heating through the use of several units. The basic components of one of these smaller boilers is shown in Fig. 22. The piping arrangement for a series loop heating system using one of these boilers is illustrated in Fig. 23. The procedure for filling the system is similar to the one outlined above except for the elimination of Steps 6-9.

AUTOMATIC AIR VENT

ADJUSTABLE LIMIT CONTROL

PRE-SET PRESSURE CONTROL

DRAIN VALVE

WATER PRESSURE RELIEF VALVE

TRANSFORMER

EXPANSION TANK

RELAY

ELECTRIC HEATING ELEMENT

CIRCULATOR

Courtesy American Standard

Fig. 22. Basic components of an electric-fired boiler.

STEAM BOILER FITTINGS AND ACCESSORIES

The boilers used in steam heating systems are fitted with a variety of devices designed to insure the safe and proper operation of the boiler. These boiler fittings and accessories can be divided by function into two basic categories: (1) indicating or measuring devices, and (2) controlling devices. Sometimes both functions are combined in a single unit.

Indicating devices include water gauges, pressure gauges, and other similar types of equipment that provide information about the operating conditions in the boiler. They are designed to indicate temperatures, pressures, or a water level that fall outside the design limits of the boiler. Controlling devices include boiler equipment designed to cause changes in these conditions. For example, pressure relief valves are used to relieve excess pressure in the boiler.

Fig. 23. Typical piping arrangement for a hot-water heating system.

Typical fittings and accessories commonly found on steam boilers are illustrated in Figs. 24 and 25. A steam boiler is easily identified by an externally mounted low water cutoff and pump control.

STEAM BOILER PRESSURE RELIEF VALVES

A steam boiler is equipped with a pressure relief valve (or valves) which opens and releases excess steam at or below the maximum allowable working pressure of the boiler. These are pop safety relief valves designed to comply with the requirements of the ASME Boiler and Pressure Vessel Code.

On the low-pressure boilers used in residential steam heating systems, the pressure (or safety) relief valve is generally preset to open and release steam when a maximum pressure of 15 psi is reached in the boiler. The valve closes when the steam pressure once again falls below 15 psi.

WATER GAUGES

Water gauges are used on both steam and hot-water space heating boilers to indicate the water level. Some examples of the different types of water gauges used on boilers are shown in Fig.

FLOW NOZZLE

TOP VIEW

PRESSURE GAUGE

LOW WATER CUT OFF & PUMP CONTROL

PRESSURETROLS

WATER LEVEL

WATER GLASS

BLOW DOWN TAPPING

ACCESS PANEL

GAS BURNER CONNECTION

RIGHT SIDE VIEW

FRONT VIEW

Courtesy *Bryan Steam Corp.*

Fig. 24. Gas-fired steam boiler fittings.

26. Fig. 27 illustrates the typical location of a water gauge on a boiler.

· A water gauge is used to *visually* check the level of water in the boiler. If the water level in the boiler is high enough, the glass tube will be approximately half full. If no water is showing in the tube, the boiler must be turned off and refilled to the proper level. Do *not* add the water until the boiler has had time to cool off.

The construction of a typical water gauge is shown in Fig. 28. The glass tube of the water gauge must be long enough to cover the safe range of water level in the boiler. The ends of the water gauge are connected to the interior of the boiler by fittings located above the safe high water level and below the safe low water level.

STEAM GAUGE SIPHON CONTROLS

1"

WATER LINE
56" TO

350 MILLS

PUMP CONTROL AND
LOW WATER CUTOFF

Courtesy *The H.B. Smith Co., Inc.*

Fig. 25. Water column, piping, fittings and steam controls for gas-fired steam boiler.

WATER COLUMNS

A water column (Fig. 29) is a boiler fitting which combines tyro cocks, a water gauge, and an alarm whistle in a single unit. A float in the column activates an alarm whistle when the water drops below a safe level.

STEAM GAUGE

The difference between the pressure found on the inside of a boiler and the pressure on the outside of it is indicated by a steam gauge (Fig. 30). This is a gauge pressure reading and

Pressure up to 200PSI   Pressure up to 250PSI   Pressure up to 125PSI

Courtesy *Ernst Gage Co.*

Fig. 26. Various types of water gauges.

should not be confused with absolute pressure. A steam gauge is used to measure the steam pressure at the top of the boiler. Generally a reading of 12 psi (pounds per square inch) indicates a dangerous build up of pressure in low-pressure steam heating boilers. The boiler should be shut down before the pressure exceeds this level.

If the steam gauge is operating properly, the needle (or pointer) will move with each change of pressure inside the boiler. Shut off the steam and the gauge needle should drop to zero; turn on the steam and the needle should rise to the correct reading. It is very important that the steam gauge be regularly checked to insure that it is operating properly.

The steam gauges used on boilers operate on the bent-tube principle; that is to say, the tendency of a bent or curved tube to assume a straight position when pressure is applied. As shown in Fig. 31, one end of the curved tube is attached in a fixed position to a pigtail (connector tube) which is, in turn, attached to the

Fig. 27. Typical location of a water gauge.

boiler. The gauge needle is mounted on a rack-and-pinion gear attached to the *free* end of the curved tube. The pressure in the curved tube causes its free end to move slightly in its effort to assume a straight position. This slight movement is multiplied by the rack-and-pinion gear, causing the needle to rotate and indicate the steam pressure.

STEAM GAUGE PIGTAILS

A steam gauge pigtail (Fig. 32) is a length of tubing with one or more loops in it used to connect the steam gauge to the boiler. It functions as a protective device by preventing live boiler steam

471

BONNET RING
BONNET
PACKING NUT
PLUG
UPPER BODY
ROD PLATE
GLASS STUFFING BOX
GLASS PACKING
GLASS STUFFING-BOX WASHER
GLASS
ROD
GLAND
PACKING
STEM
HANDWHEEL LOCKNUT
HANDWHEEL
LOWER BODY
DRAIN-UNION TAIL
DRAIN-UNION RING

Courtesy Lunkenheimer Co.

Fig. 28. Constructional view of a typical water gauge.

Fig. 29. Water column.

Courtesy *Ernst Gage Co.*

8" STEAM GAUGE

NO. 157BSL
PUMP CONTROL &
LOW LEVEL C/O

WATER LINE CUTOFF

1-1/2"

Courtesy *The H.B. Smith Co., Inc.*

Fig. 30. Steam gauge.

from coming into contact with the working parts of the gauge. The steam is denied passage by condensation which forms in the loop (or loops) of the pigtail. It is recommended that the steam gauge pigtail be filled with water *before* it is attached to the boiler. It should be attached to the boiler in a position where heat and vibration will be at a minimum.

PRESSURE CONTROLLERS

When the room thermostat calls for heat in a steam heating system, the electrical current flows to the operating controls of

Fig. 31. Mechanism of a bent-tube steam gauge.

Fig. 32. Steam-gauge pigtail.

the automatic fuel burning equipment (oil burner, gas burner, or coal stoker) through a high-limit switch. In a steam boiler control system, the high-limit switch is activated by pressure, and is referred to as a *pressuretrol*.

Fig. 33 shows a pressuretrol operating controller which is designed to provide on-off and proportional control of steam boilers fired by proportional type burners. It can be field adjusted to operate either in unison (burner starts at high fire), or in sequence (burner starts at some firing rate other than high fire). A common bellows assembly located in the bellows housing actuates the stop-start snap switch (on-off operation) and the wiper of the 185 ohm potentiometer (proportional operation).

The pressure controller shown in Fig. 34 operates strictly as a high limit pressure safety control on steam heating boilers. It

MAIN SCALE PRESSURE ADJUSTMENT SCREW   DIFFERENTIAL ADJUSTMENT SCREW   CONDUIT KNOCKOUT

POTENTIOMETER COIL
SLIDING CONTACT
DIFFERENTIAL SCALE
DIFFERENTIAL INDICATOR
OPERATING LEVER
CONDUIT KNOCKOUT
BELLOWS HOUSING
U-S INDICATOR MARK
UNISON-SEQUENCE ADJUSTMENT DIAL

Courtesy Honeywell Tradeline Controls

Fig. 33. Pressuretrol operating controller.

MAIN SCALE SET POINT SCREW

INDICATOR

MAIN SCALE   DIFFERENTIAL SCREW

Courtesy Honeywell Tradeline Controls

Fig. 34. Pressuretrol controller.

breaks the electrical circuit on pressure rise. A variation of this controller (the *Honeywell* PA404B Pressuretrol Controller) is used for suspension type unit heaters.

Direct control for a proportional motor operating an automatic burner is obtained by using the pressure controller illustrated in Fig. 35. This unit contains two potentiometers operating in unison, which makes it possible to simultaneously control two motors. It is also provided with an adjustable throttling range.

PRESSURE ADJUSTMENT SCREW   THROTTLING RANGE ADJUSTMENT SCREW

CONDUIT KNOCKOUT

POTENTIOMETER COILS (2)

SLIDING CONTACTS (2)

THROTTLING RANGE INDICATOR

THROTTLING RANGE SCALE

OPERATING LEVER

CONDUIT KNOCKOUT

BELLOWS HOUSING

Courtesy Honeywell Tradeline Controls

Fig. 35. Proportional pressuretrol controller.

Fig. 36. Vacuumstat controller.

Courtesy Honeywell Tradeline Controls

Pressure controllers are also available for vapor or vacuum systems. A pressure controller with a bellows operated mercury switch is manufactured by *Honeywell* for use on vapor heating systems with pressures up to 4 psi. The pressure controller illustrated in Fig. 36 can be used as a boiler high-limit control with cut-in and cut-out settings in the vacuum range. In such installations, the heating system must include a vacuum pump and a siphon loop.

LOW-WATER CUTOFFS

Steam boilers must be provided with a water level control device that will shut off the automatic fuel burning equipment when the water level in the boiler drops to a level too low for safe operation. This water level control device is referred to as a *low-water cutoff*. The two types of low-water cutoffs used on steam boilers are the float type and the probe type.

A float type low-water cutoff device consists of a cutoff switch operating in conjunction with a float located in the boiler water or in a float chamber installed next to the boiler. The float is connected through a linkage to a switch which operates a feed water valve. As the water level falls, the float drops with it until it reaches a point at which the feed water switch is actuated. If the water level continues to fall, a second switch connected to the float by the linkage is actuated and the automatic fuel burning equipment is shut off.

A probe type low-water cutoff device depends on the flow of a low electrical current to control the operation of the automatic fuel burning equipment. The electrical current flows from the probe through the water to keep the relay energized. When the water level falls, there will be a point at which the probe loses direct contact with the water. As a result, the contact is broken and the flow of the electrical current is stopped. This, in turn, causes the relay to be energized, which results in shutting off the fuel burning equipment. A probe type low-water device *cannot* be used in the direct operation of feed water valves.

On some boilers, the low-water cutoff is combined with a water feeder to add water to the boiler when the water level falls below the safe operating limit. An example of one of these com-

GAUGE GLASS CONNECTION

COPPER
FLOAT

SWITCH
ASSEMBLY

STEAM SEAL
BELLOWS

WATER SEAL
BELLOWS

FEED VALVE

STRAINER

DRAIN VALVE
(GATE TYPE)

REMOVABLE SCREEN

Courtesy Watts Regulator Co.

Fig. 37. Basic components of a boiler water feeder and cutoff.

bined units is shown in Fig. 37. They are available with either single or dual switch assemblies. The single double-throw switch assembly provides a combination feeder and burner cutoff switch with an extra terminal for line voltage with single-pole double-throw service (Fig. 38). The dual switch assembly is used for line voltage burner service and for independent low (or high) voltage alarm, feed valve or pump starter service. When an emergency condition occurs, the switch interrupts the current to the burner and shuts it off. When the emergency has passed, the water feeder takes over and feeds make-up water when needed for normal operation. On units equipped with dual switch assemblies, the alarm, feed valve or pump starter switch actuates just before the burner switch cuts the firing. A typical installation of a *Watts* 60 LWD boiler water feeder and cutoff on a steam boiler is illustrated in Fig. 39.

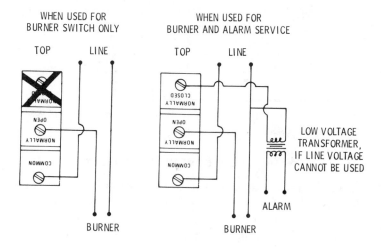

WHEN USED FOR
BURNER SWITCH ONLY

WHEN USED FOR
BURNER AND ALARM SERVICE

TOP   LINE   TOP   LINE

NORMAL

OPEN
NORMALLY

COMMON

CLOSED
NORMALLY

OPEN
NORMALLY

COMMON

LOW VOLTAGE
TRANSFORMER,
IF LINE VOLTAGE
CANNOT BE USED

ALARM

BURNER   BURNER

Fig. 38. A single switch assembly and wiring diagram.

A low-water cutoff *without* a boiler water feeder is usually adequate for providing automatic low water safety protection for most small boilers (Fig. 40). These units can be installed on any boiler having gauge glass connections.

Direct installation type low-water cutoffs are available for boilers with limited space in the boiler jacket (Fig. 41). Each unit contains a switch assembly and float, and is screwed into the boiler through a 2½ inch tapping.

479

SUPPLY

FEEDER CLOSING LEVEL

BY-PASS

Courtesy Watts Regulator Co.

Fig. 39. Typical installation of a boiler water feeder and cutoff.

INSTALLING A LOW WATER CUTOFF

The *Watts* No. 89 low-water cutoff illustrated in Fig. 40 is suitable for low pressure steam heating boilers with a maximum operating steam pressure of 15 lbs. Although installation is relatively simple, the instructions should be carefully followed. This is a boiler safety device, and it must operate properly or the boiler can be damaged.

Before installing the *Watts* low-water cutoff, the gauge glass and cocks must be removed from the boiler. When you have done this, proceed as follows:

1. Install the ½ inch tee (with long and short nipples) on the float chamber by inserting the short nipple in the float chamber tapping in the end opposite the switch. Screw the tee tight with the long nipple pointing in the correct direction, depending on the location of the boiler. (Fig. 42).

Fig. 40. A low water cutoff.

2. Insert the long nipple into the lower glass gauge tapping on the boiler.

3. Swing the entire float chamber until the nipple is made up tight. Line up the cutoff so that the top of the switch box is level (Fig. 43).

481

Fig. 41. Direct installation type low-water cutoff.

Fig. 42. Installing the short nipple in the float chamber topping.

Fig. 43. Installing long nipple into lower gauge glass topping in boiler.

4. Screw the nipple in the tee carrying the tubing connector into the upper gauge glass tapping and pull up tight. In-

FLARED CONNECTION

COMPRESSION COUPLING

Courtesy Watts Regulator Co.

Fig. 44. Installing compression coupling.

stall a compression coupling in the top float chamber tapping (Fig. 44).

5. Hold the tube bend in position and mark the tube on a level with the top of the hex on the compression coupling. Cut off the tube at this mark (Fig. 45).

6. Slide the ring and nut over the end of the connector tube.

7. Slide the end of the tube into the compression coupling in the float chamber, and tighten both couplings.

8. Install a drain valve in the bottom float chamber tapping (Fig. 46).

9. Reinstall the gauge cocks in the end of the tees and replace the gauge glass (Fig. 46).

The drain valve should be opened at least once every month (or oftener) during boiler operation to flush out sediment from the float chamber.

VACUUM RELIEF VALVES

A vacuum relief valve (Fig. 47) is an effective means of providing protection against the build up of excessive vacuum condi-

Courtesy Watts Regulator Co.

Fig. 45. Cutting tube on level with top of hex.

tions in steam heating and steam processing systems. They are also used in combination with temperature-pressure relief valves on water heaters.

Vacuum conditions can often occur after the supply line has been shut off. Steam condenses and a vacuum can be created which not only affects system operation but can also cause damage to equipment. The vacuum relief valve automatically admits air to the system in order to break up the vacuum. These valves operate under conditions up to a maximum temperature of 250°F. The valve disc should be constructed from a material capable of withstanding high temperatures, particularly when the valve is used in a steam processing system.

STEAM BOILER INJECTORS

A steam boiler injector is a device used on some boilers to create accelerated steam circulation in a steam heating system. The essential parts of a boiler injector are shown in Fig. 48.

This device operates on the induction principle. In operation, steam from the boiler, entering the steam nozzle, passes through it, through the space between the steam nozzle and combining

GAUGE-COCK

GAUGE GLASS

GAUGE COCK

DRAIN VALVE

Fig. 46. Installing drain valve.

tube, and then out through the overflow. This produces a vacuum which draws in the water through the water inlet. The incoming cold water condenses the steam in traversing the combining tube, and the water jet thus formed is driven at first out through the overflow. As the velocity of the water jet increases, sufficient momentum is obtained to overcome the boiler pressure, with the result that the water enters the delivery tube and passes the main check valve into the boiler.

The induction principle of the steam injector is the same as the operating principle of hydraulic boosters used in forced hot-water heating systems with the exception that fast flowing hot water is used instead of steam to obtain the inductive action to accomplish accelerated circulation.

485

STEAM

STEAM

PIGTAIL

PIGTAIL NOT NEEDED
FOR PRESSURE
UNDER 5 LBS.

KETTLE

TRAP

TRAP

TRAP

Courtesy Watts Regulator Co.

Fig. 47. Vacuum relief valve used in combination with a unit heater in a steam heating system and with a jacketed kettle in a steam processing system.

TRY COCKS

Try cocks are small valves installed on steam boilers at the safe high-water level and at the safe low-water level (Fig. 49). When the boiler water column is inoperable, try cocks function as a back up system to determine the water level. The water level is determined by opening *slightly* first one try cock and then the other. The water level is indicated by the color of the plume escaping from the try cock orifice. A steam plume is characteristically colorless, and will indicate that the water level is too low (i.e. below the level of the try cock orifice). Water, on the other hand, is characterized by a white plume. A false water level reading will be obtained if the try cock is opened too wide when the level of the water is only slightly below the level of the try cock orifice. The violent agitation of the water caused by a wide open try cock

Fig. 48. Cross-sectional view of a typical injector.

Courtesy *Ernest Gage Co.*

Fig. 49. Try cocks.

will result in some of the water escaping, giving the false impression that the level of the water in boiler is at or above the try cock fitting.

On large steam boilers, a third try cock is often installed between the other two. Sometimes a try cock on these larger boilers is at a level too high to reach. When this is the case, chain operated lever-type try cocks are installed.

BLOW-OFF VALVE

Sediments, contaminants and other impurities found in water will settle out over a period of time and accumulate at a low point in the bottom of the boiler. These can be removed by means of a blow-off valve placed in a line connected to the lowest part of the boiler. The blow-off valve is opened periodically and the accumulated sediments are drained off.

FUSIBLE PLUGS

A fusible plug is used on some boilers as a protection against dangerous low water conditions. Examples of fusible plugs are shown in Fig. 50. They are generally made of bronze and filled

(A) Inside type.

(B) Outside type.

Water in Boiler

Water in Boiler

Courtesy Lunkenheimer Co.

Fig. 50. Fusible plugs.

with pure tin. When the temperature (and pressure) in the boiler builds up to about 450°F (the approximate melting point of pure tin), the tin core melts and relieves the pressure within the boiler.

Fusible plugs must comply with ASME standards, and are available in several sizes for use on boilers having a steam pressure of less than 250 psi.

STEAM LOOP

A steam loop (Fig. 51) is a piping arrangement used in a steam heating system to return condensation to the boiler. The principal components of a steam loop piping arrangement are:

1. Riser,
2. Condenser,
3. Drop leg,
4. Check valve.

The *riser* carries a mixture of steam and water to a horizontal *condenser* where the steam portion of the mixture is condensed. The condensation of the steam causes a reduction of pressure, and it is this reduced pressure that provides the motive force for

Fig. 51. Essential parts of a steam loop.

the steam and water mixture in the riser. The goose neck in the riser prevents any backflow of the steam. The condensation accumulates in the *drop leg* to a height such that its weight will ballance the weight of the mixture in the riser. The *check valve* on the drop leg allows condensation to pass into the boiler but prevents steam in the boiler from backing into the pipe.

HOT WATER BOILER FITTINGS AND ACCESSORIES

The boilers used in hot-water heating systems are also fitted with a variety of devices designed to insure the safe and proper operation of the boiler. Some of these fittings and accessories are similar to those used on steam boilers (e.g. pressure relief valves and low-water cutoffs). They also share the same measuring and controlling functions described for steam boiler fittings and accessories. Figs. 52 and 53 illustrate their arrangement on a hot-water space heating boiler.

Courtesy Hydrotherm, Inc.

Fig. 52. Piping, fittings and controls for an oil-fired hot-water space heating boiler.

ASME RATED WATER PRESSURE RELIEF VALVES

Pressure relief valves are used on hot-water space heating boilers to relieve pressure created by two different conditions: (1) water thermal conditions and (2) steam pressure conditions.

As thermal conditions develop inside a hot-water space heating boiler, pressures may be built up to the setting of the relief

MANUAL VENT VALVE
1" COMPRESSION TANK CONNECTION
1" PLUG
7" THERMOMETER
CLEANOUT
1" TAP

PRESSURE-ALTITUDE GAUGE
LOW WATER CUTOFF
1"
COCK
1" TAP
JACKET
WATER LINE
1"

3"
10-7/8"
24-3/4"
15-1/2"

Courtesy *The H.B. Smith Co., Inc.*

Fig. 53. Piping, fittings and controls for a gas-fired hot-water space heating boiler.

valve. When the pressure reaches the safety limit setting, the valve functions as a *water* relief valve and discharges the small amount of water which is expanded in the system (Fig. 54).

If *both* water and steam are present in a hot-water space heating boiler, the firing controls are probably malfunctioning or have completely broken down. This may result in runaway firing of the boiler which could cause the boiler water to reach steam-forming

491

Courtesy A.W. Cash Valve Mfg. Corp.

Fig. 54. Typical ASME pressure relief valves.

temperatures, creating a steam pressure condition. Under these circumstances, the relief valve functions as a steam safety relief valve. The steam is discharged at a rate or faster than the boiler can generate it, thus restoring system pressure to a safe level. Although these valves are steam rated and have an emergency Btu steam discharge capacity if runaway firing conditions occur, they should *not* be used for continuous steam service.

In order to be completely effective, the safety relief valves used on hot-water space heating boilers must be designed to discharge the excessive water pressure created by thermal expansion and also the excessive steam pressure resulting from runaway emergency temperature conditions.

ALTITUDE GAUGES

An altitude gauge mounted on the boiler is used to indicate the water level in *open* hot-water heating systems. The level (altitude) of the water is indicated on the gauge by the relative positions of two hands or arrows. One hand (usually black) is stationary, and is permanently set when the boiler is filled. The movable hand (usually red or white) is initially set in the same position as the stationary hand after the boiler has been filled with water. Its position will change as the water level in the boiler changes. This movable hand indicates the true level of the water. Efficient operation is being provided as long as the movable hand is directly above the stationary one.

Fig. 55. Pressure gauge, thermometer and altimeter combined in one unit.

Fig. 55 shows a pressure gauge, thermometer and altimeter (altitude gauge) combined in one unit. These *Marsh* Ther-Alti-Meters are used on residential hot-water space heating boilers. They provide water pressure readings from 0 to 30 psi (red field to 50 psi), corresponding altitude from 0 to 70 ft., and water temperature from 60° to 320°F.

In *closed* hot-water heating systems, automatic valves are used to control boiler water level (see PRESSURE RELIEF VALVES and PRESSURE-REDUCING VALVES in this chapter).

AQUASTATS

An aquastat is a device commonly used on hot-water space boilers and some steam boilers to control temperature limit or to operate the circulator (hot-water heating system pump). It is similar in function to the pressure control on a steam boiler or the fan and limit control on a forced warm-air furnace.

493

Basically, an aquastat is an automatic switching device consisting of a metal or liquid filled heat sensitive element designed to detect temperature drop or rise of the boiler water. Aquastats can be strapped to the hot water supply riser or mounted so that the heat sensitive element is immersed in a boiler well.

The type of aquastat used in a heating installation generally depends upon whether it is designed to control temperature limit or to switch on the circulator. If the former is the case, the aquastat will close on temperature drop and open on temperature rise. The aquastat will close on temperature rise if it is used to operate the circulator.

An example of a strap-on type aquastat is the *ITT General Controls* L-53 Strap-On Hot Water Control illustrated in Fig. 56. This type of aquastat responds to water temperature changes as

BASE OF CONTROL MUST FIT
SNUGLY AGAINST PIPE

INSERT BOLT THROUGH
CLAMPS AND HOLES IN
STRAP AND PULL UP

PASS MOUNTING STRAP
THROUGH SLOT IN BASE
AND BEND TO FIT
AROUND PIPE

CUT OFF EXCESS STRAP

Courtesy *ITT General Controls*

Fig. 56. Snap-on hot water control.

Fig. 57. Direct action switch operation.

Courtesy *ITT General Controls*

Fig. 58. Reverse action switch operation.

Courtesy *ITT General Controls*

conducted through the supply riser pipe wall to the temperature responsive base of the device. The enclosed switching is supplied in direct action (Fig. 57), reverse action (Fig. 58), and double action (Fig. 59) models. Consequently, these strap-on type aquastats can be divided into three basic types depending upon their operating principles: (1) direct action types, (2) reverse action types, and (3) double action types.

A direct action (N.C.) type aquastat used as a high limit control must always be located on the supply riser where it will be subjected to the maximum boiler water temperature. Its location on the riser will therefore have to be as close to the boiler as possible, but ahead of any line shutoff valves.

495

Fig. 59. Double action switch operation.

Courtesy *ITT General Controls*

Direct action type aquastats will open the circuit on temperature rise. Setting the adjustable scale pointer to a position on the scale will result in breaking the circuit (Fig. 60). The "cut-out" setting when used as a high limit control should be as low as possible and still insure proper heating in cold weather. An initial "cut-out" setting of 170°F is recommended for a gravity hot-water heating system. An initial setting of 200°F is suggested for a forced hot-water (hydronic) heating system. The mechanical differential of the control illustrated in Fig. 60 is nonadjustable and approximately 15°F.

Fig. 60. Indicator can be set for a temperature higher than 200°F. by removing the screws (A) holding stop bracket to scale plate, and removing the bracket (B).

Courtesy *ITT General Controls*

A reverse action (N.O.) type aquastat should be mounted ahead of any valves or traps *on the return line* when it is used on a unit heater installation. It should be mounted on the largest riser from the boiler if it is used to prevent circulator operation when boiler water temperature is low. A reverse action type aquastat closes the circuit on temperature rise.

A double action (SPDT) type aquastat is used to start circulator operation with a single switch actuation and to function as an operating control to maintain boiler water temperature. This type of aquastat should be mounted on the largest riser ahead of any valve, but at a point where it will be subjected to maximum boiler water temperatures. The double action type aquastat illustrated in Fig. 59 opens *R* to *B* contacts and closes *R* to *W* contacts on temperature rise at scale setting.

Another example of a strap-on aquastat is the *Honeywell* LA409A aquastat shown in Fig. 61. This is a nonimmersion, surface mounted aquastat which functions as a high limit or safety control device in hot-water and steam heating systems, and which requires no tapping of the boiler or draining of the system. This aquastat is also available in a reverse action model to operate unit heater fans.

Fig. 61. An aquastat controller.

Courtesy Honeywell Tradeline Controls

A strap-on aquastat can be mounted in any position. When mounting these aquastats, make absolutely certain that the pipe surface is clean and free of rust and corrosion. All rough and high spots should be filed smooth. Nothing should be allowed to interfere with the operation of the temperature responsive base of the control.

The *Honeywell* L6081A aquastat shown in Fig. 62 is used to control boiler water temperature in gas- and oil-fired hydronic heating systems. It has an immersion type liquid filled heat sensing element which actuates two snap switches. One switch operates as a high limit control, the other as a low limit and/or circulator control.

LOW-WATER CUTOFFS

A low-water cutoff device such as the one shown in Fig. 63 can also be used in a hot-water heating system to provide pro-

WELL CLAMP SCREW

STRIP GAUGE

LOW LIMIT/CIRCULATOR DIFFERENTIAL ADJUSTMENT

HI LIMIT SETPOINT ADJUSTMENT

LOW LIMIT SETPOINT ADJUSTMENT

JUMPER SLOT

Courtesy Honeywell Tradeline Controls

Fig. 62. Multiple aquastat controller.

tection against a low water level condition in the boiler resulting from runaway firing caused by malfunctioning controls or a break in the return piping. The low water cutoff device should be installed in the piping so that the raised line cast on the float chamber body is on a level with the top of the boiler (Fig. 64). The drain valve located directly below the low-water cutoff should be opened periodically to flush mud and sediment from the float chamber. It is recommended that this be done at least once a month.

WATER PRESSURE REDUCING VALVES

Most forced hot-water heating systems are equipped with a pressure-reducing valve which is designed to automatically feed water into the boiler when the pressure in the system drops below the valve setting. When the pressure returns to the minimum pressure setting, the valve automatically closes. Thus, the function of a water pressure reducing valve is to keep the system automatically filled with water at the desired operating pressure (Fig. 65).

Fig. 63. Low-water cutoff used in a forced hot-water heating system.

Fig. 64. Installation of a low-water cutoff in a forced hot-water heating system.

These valves are also referred to as *pressure reducing boiler feed valves* or *feed water pressure regulators*. Regardless of what they are called, their function remains the same; that is, to deliver water to the boiler at the required pressure. To perform this function, they are *always* installed in the cold water supply line. If they are part of a combination pressure reducing and pressure relief

Fig. 65. Water pressure reducing valve.

valve (dual control), the pressure relief part of the combination valve is always located between the boiler and the pressure reducing part.

A manually operated feed valve was used on older hot-water heating systems to provide the same function as the automatic water pressure reducing valve. On these older systems, the manually operated feed valve should be *partly* opened and water added to the boiler when the movable (white or red) arrow on the altitude gauge drops below the setting of the stationary (black) arrow (see ALTITUDE GAUGES in this chapter).

Some pressure reducing valves are equipped with a built in check valve in the supply inlet to prevent a back flow of contaminated boiler water into the domestic water supply. This back flow drainage of boiler water generally occurs when the supply pressure falls below the feeder valve setting. Not only can this drainage cause possible damage to the boiler, it can also contaminate the domestic water supply. Some pressure reducing valves are also equipped with integral strainers to trap foreign matter that may clog the valve. The strainer screen should be removed for cleaning at the beginning of each heating season. This can be done by removing the bottom plug from the strainer and withdrawing the screen. Clean and carefully replace screen and plug.

HIGH TEMPERATURE
RESISTING
DIAPHRAGM FOR
HOT OR COLD WATER

RENEWABLE NICKEL
ALLOY SEAT

BRONZE BODY
CONSTRUCTION

DISC HOLDER

INTEGRAL STAINLESS
STEEL SCREEN

Courtesy Watts Regulator Co.

Fig. 66. Basic components of a water pressure reducing valve with integral
stainless steel strainer.

The principal components of a water pressure reducing valve
are shown in Fig. 66. This particular valve is constructed with an
integral strainer. It is also possible to use separate units in combination such as Fig. 67 illustrates. The two units are joined by
a short threaded connection. A typical installation in which a
water pressure reducing valve is used is shown in Fig. 68.

As was mentioned before, a water pressure reducing valve
should be installed in the water supply line to the boiler. It should
also be installed at a level *above* the boiler and in a horizontal
position. Before installing the valve, flush out the supply pipe to
clear it of chips, scale, dirt, and other foreign matter that could

501

DISC HOLDER

HIGH TEMPERATURE
RESISTING
DIAPHRAGM FOR
HOT OR COLD WATER

BRONZE BODY
CONSTRUCTION

RENEWABLE NICKEL
ALLOY SEAT

STAINLESS STEEL
PERFORATED
STRAINER SCREEN

Courtesy Watts Regulator Co.

Fig. 67. Basic components of a water pressure reducing valve with a separate strainer unit.

interfere with its operation. Install a shut-off valve *ahead* of the regulator and then connect the supply line to the inlet (usually marked IN on the valve casting).

To fill the system, open the shut-off valve located ahead of the pressure reducing valve. Water will flow into the system until it is full and under pressure. The shut-off valve must always be kept open when the system is in operation.

Water pressure reducing valves are usually set by the manufacturer to feed water to the boiler at approximately 15 lbs. pressure. This pressure setting is sufficient for residences and houses up to a three story building in size. For higher buildings in which

Courtesy *Watts Regulator Co.*

Fig. 68. Typical installation diagram showing the location of a water pressure reducing valve.

the pressure may not be sufficient to lift the water to highest radiation, it may be necessary to reset the water pressure reducing valve for higher pressure. To do this, calculate the number of feet from the regulator to the top of the highest radiation. Multiply this by .43 and add 3 lbs. This will give the pressure needed to raise the water to the highest radiator and keep it under pressure. Loosen the lock nut on the valve and turn the adjustment screw clockwise slowly until the gauge indicates the pressure calculated. When the desired pressure is obtained tighten the lock nut.

COMBINATION VALVES (DUAL CONTROLS)

Combination or *dual control valves* (Figs. 69 and 70) are designed for use in hot water space heating systems. These valves consist of a pressure reducing and regulating valve, and a positive relief valve contained in one body. Sometimes an integral by-pass valve is also included in the same unit.

The purpose of a combination valve is to provide pressure regulation and safety control, and to reduce boiler pressure and assure automatic filling when conditions warrant. The valve is installed on the supply line of a boiler with the relief valve section closest to the boiler.

503

The regulator side of the combination valve is designed to reduce the incoming water supply to the boiler to the required

Fig. 69. A combination relief/regulator valve. (Type CQ)

Fig. 70. A combination relief/regulator valve. (Type A-1)

14 lbs. operating pressure for a one, two, or three story house. The pressure in the system builds up due to thermal expansion when the boiler is fired. Under normal operating conditions, the expansion tank will absorb the additional pressure. However, if the expansion tank is water-logged or if the system has no expan-

RELIEF SECTION

1. Lever Pin
2. Lift Lever
3. Pressure Screw Cap
4. Pressure Screw
5. Lock Nut
6. Spring Chamber
7. Spring Button
8. Pressure Spring, stainless steel
9. Pull Rod Nut
10. Pull Rod
11. Lock Washer
12. Pressure Plate
13. Gasket, brass
14. Diaphragm
15. Composition Seat Shell
16. Composition Seat
17. Composition Seat Screw
18. Body Seat

24. Pressure Spring, black
25. Gasket, brass
26. Diaphragm Assembly
27. Cylinder
28. Body
29. Piston Assembly
30. Piston Spring
31. Strainer
32. Gasket, red fiber
33. Bottom Plug

REGULATOR SECTION

19. Closing Cap
20. Pressure Screw
21. Lock Nut
22. Spring Chamber
23. Spring Button

Courtesy A.W. Cash Valve Mfg. Corp.

Fig. 71. An exploded view of type CQ combination valve.

RELIEF SECTION

1. Lift Lever Pin
2. Lift Lever
3. Pressure Screw Cap
4. Pressure Screw
5. Lock Nut
6. Spring Chamber
7. Spring Button
8. Pressure Spring
9. Pull Rod Nut
10. Pull Rod
11. Pressure Plate
12. Diaphragm Gasket, brass
13. Diaphragms, bronze
14. Diaphragm Gasket, black
15. Seat Shell Gasket, gray
16. Seat Shell
17. Composition Seat Disc
18. Seat Disc Screw
19. Body Seat
20. Body

REGULATOR SECTION

21. Pressure Screw
22. Lock Nut
23. Spring Chamber
24. Spring Button

25. Pressure Spring
26. Diaphragm Gasket, brass
27. Diaphragm Assembly
28. Diaphragm Gasket, black
29. Cylinder
30. Piston Assembly
31. Piston Spring
23. Strainer Screen
33. Bottom Plug Gasket, red
34. Bottom Plug

Courtesy A.W. *Cash Valve Mfg. Corp.*

Fig. 72. An exploded view of type A-1 combination valve.

sion tank, the relief side of the valve will open at 23 psi and drop the pressure back to within safe limits. If the pressure should

507

drop below 14 psi, the regulator will open again and automatically refill the system. A built-in check valve prevents the back flow

Courtesy A.W. Cash Valve Mfg. Corp.

Fig. 73. A combination relief/regulator valve. (Type CBL)

of contaminated boiler water into the potable boiler water supply.

Exploded views of combination valves are shown in Figs. 71 and 72. The *Cash-Acme* Type CBL Valve (Fig. 73) differs from the other two by having a built in by-pass valve. In all other respects, it is similar to the others except that it should be installed as close as possible to the top of the boiler (using close nipples). The principal advantages of having the by-pass valve included are:

1. Allows rapid filling of the system.
2. Permits an easy high pressure test for leaks and system purging.
3. Passes first filling dirty water around the valve seat, assuring a clean, good seating surface.
4. Permits the use of a wide opening, small seated regulator which prevents wire drawing and rapid wearing of the seat.

The most common types of problems encountered when using combination valves are the following:

1. Relief valve drips or cannot be shut off tightly.
2. Boiler pressure rises above the reducing valve setting.
3. Water escaping "weep hole" on top of the regulator.
4. Water escaping feeder side of the combination valve.
5. Foreign matter on the seat of the relief or regulator section of the valve.

Some of the problems listed above (particularly the first two) may *not* be the fault of the valve itself. For example, a dripping relief valve can be caused by an expansion tank being completely filled up with water; an undersized expansion tank; or a leak in the coil of a tankless or indirect water heat installed in the boiler. These possibilities should be checked *before* attempting to service or repair the valve. The procedures are outlined in this chapter.

If, by process of elimination, the problem can be traced to the valve, try tapping the side of the valve with a wrench. Sometimes a piece of foreign matter will become lodged and cause the regulator piston to stick. A sharp tap with a wrench may dislodge it and allow the valve to function properly.

Foreign matter such as dirt, pipe scale, or chips will often cause a valve to malfunction by lodging on a seating surface, or nicking or chipping the surface. Valve manufacturers will often provide replacements, or instructions for field servicing and repairing. The latter should be attempted only by a skilled and experienced worker with the proper tools and gauges.

Water seeping from the regulator "weep hole" or from the feeder side of a combination valve is usually an indication that there is a rupture or leak in the diaphragm. Follow the procedures described in the preceding paragraph for repairing the valve. An occasional flushing of the relief side of a combination valve will reduce the possibility of the type of lime or scale build up that can cause the valve to fuse shut.

AIR SEPARATORS

An *air separator* is a device used to trap, separate, and eliminate air from the water in a hot-water space heating boiler. Air

SUPPLY

SCREW AIR SEPARATOR INTO BOILER NOZZLE

TRAPPED AIR ESCAPES THROUGH THIS 3/4" TAPPING INTO EXPANSION TANK OR FLOAT VENT

RETURN

Courtesy *Hydrotherm, Inc.*

Fig. 74. Air separator.

separators are *not* installed on boilers used only for heating the domestic water supply.

The air separator shown in Fig. 74 is screwed into the boiler supply nozzle. The separator traps the air in the top of the boiler section where the water is hottest and where it travels a long horizontal path at low velocity, permitting the air bubbles to separate. The trapped air escapes through a ¾ inch tapping into an air cushion expansion tank or through an automatic float vent on systems using diaphragm expansion tanks.

An air separator can only expel air which reaches the boiler. However, sometimes air pockets remain trapped in the piping or radiation, thereby impeding the water flow and reducing the heating performance of the boiler. Such air pockets can be eliminated by installing several different purge fittings in the heating system. A manual air vent is installed on the highest point of the system (automatic air vents are not generally required). See below.

Courtesy *Weil-McLain*

Fig. 75. Built-in air separator.

Some boilers are constructed with built-in air separators, which make the addition of an external unit unnecessary. Fig. 75 illustrates the working principles of one of these integral units. In this design, the air is diverted to an automatic air vent. A variation of this design provides for the diversion of air to an expansion tank. The manufacturer refers to this device as an *air eliminator*. Its function is similar to the air eliminator used on pipe lines, but it differs in design and construction.

PURGING AIR FROM THE SYSTEM

Built-in or externally installed air separators are designed to separate air from the water in an *operating* boiler, and vent it either into an expansion tank or to the atmosphere through an automatic air vent.

Air is also removed *from a heating system* by purging it from the piping connections to the boiler. This type of purging is standard when the system is being filled with water. Purging instructions are usually included in the owner's manual or the manufacturer's installation guide. Fig. 76 illustrates purging instructions for a *Hydrotherm* gas-fired boiler.

AIR SUPPLY AND VENTING

An adequate air supply must be provided for combustion to boilers that are fired by gas, oil, or coal, and the products of combustion must be vented to the outside atmosphere.

Combustion air (i.e. the air supply) is normally supplied through venting ducts or openings in the walls. The requirements for the air supply will depend upon the location and enclosure of the boiler.

Boilers installed in unconfined areas usually obtain adequate air for combustion by means of normal infiltration. However, normal air infiltration will prove to be inadequate for this purpose, if the construction of the structure is unusually tight. Under these circumstances, provision must be made for the entry of additional air from outside the building. Unobstructed openings with a total free area of not less than one square inch per 5000 Btu/hr of the total input of the boiler will be necessary to provide an adequate air supply for combustion purposes.

Some boilers are installed in boiler rooms or enclosures which are supplied with combustion air from *inside* the structure. The air supply must enter and leave the boiler room through two openings in an *interior* wall or door. One opening is located near the ceiling, the other opening near the floor. Each opening should have a free area of not less than one square inch per 1000 Btu/hr of the total input of the boiler.

If a boiler is located in a boiler room or similar type enclosure which receives its air supply from *outside* the building, the two air openings are located on an *exterior wall*. Each opening must

Fig. 76. Purging instructions for a HYDROTHERM HC series gas-fired boiler.

have a free area of not less than one square inch per 4000 Btu/hr of the total input of the boiler.

The products of combustion must be vented to the outside atmosphere. In order to accomplish this, the boiler must be connected to a suitable venting system which should include a flue pipe and a chimney or stack of adequate size and capacity.

The chimney height is usually governed by the height of the structure. As a general rule-of-thumb, the chimney should extend beyond the high pressure area caused by the passage of 10 feet of the flashing. If the chimney is too short, it will not extend beyond the high pressure area caused by the passage of winds over the structure. As a result, a down draft caused by the pressure difference in the high pressure area will suck air down into the chimney, which will block the escape of flue gases and interfere with the combustion process. Figs. 77 and 78 illustrate these principles. The *correct* chimney height is shown by the dotted lines.

Courtesy Hydrotherm, Inc.

Fig. 77. Correct chimney design for a residence.

Fig. 79 shows a type of chimney construction that is referred to as a *Type A vent*. This type of chimney can be either masonry or factory built construction, but it must be designed and constructed in accordance with the standards set forth in national codes. Type A vents are suitable for venting of all gas appliances. These are the *only* vents suitable for venting oil burning equipment. Gas-fired boilers which produce temperatures at the draft

Courtesy *Hydrotherm, Inc.*

Fig. 78. Correct chimney design for a building.

Fig. 79. Type A vent.

Courtesy *Hydrotherm, Inc.*

diverter not in excess of 550°F may also use a *Type B vent* (Fig. 80), which employs a double wall metal vent pipe. A *Type C vent* (Fig. 81) is used to vent gas appliances in attic installations.

The boiler should be located so that the length of flue pipe connecting the boiler (or draft diverter) to the chimney is as short as possible. Additional information about chimneys and flues is contained in Chapter 11 (GAS-FIRED FURNACES).

CAP

PLASTIC ROOF CEMENT

STORM COLLAR

ROOF FLASHING

1" CLEARANCE

FIRESTOP

FRAMING BETWEEN JOISTS

JOISTS

TEE

TEE CAP

1" CLEARANCE

Courtesy Hydrotherm, Inc.

Fig. 80. Type B vent.

INDUCED DRAFT FANS

Sometimes a chimney is too small to provide enough updraft to convey the products of combustion from the boiler to the out-

CLEARANCE - 2" MIN
BUT NOT LESS THAN $\frac{A}{3}$

A = DIA OF FLUE
B = A + 4 INCHES
C = A + 4 INCHES
D = NOT LESS THAN 24" OR NOT LESS
    THAN 6" ABOVE ROOF COPING

FLASHING & ROOF CEMENT

ROOF LINE

ROOF BOARDS

2" OPEN AIR SPACE

Courtesy Hydrotherm, Inc.

Fig. 81. Type C vent.

side atmosphere. When it is evident that natural venting will be inadequate, a mechanical draft inducer can be used to increase the capacity of the chimney. These devices are used with both gas- and oil-fired boilers.

The two principal types of induced draft fans are: (1) direct fan type draft inducer and (2) induced flow type draft inducer. When a direct fan type draft inducer is used, the gases are drawn through an inlet located on the side, top or bottom of the unit and discharged as shown in Fig. 82. The moving parts of an induced flow type draft inducer (Fig. 83) are located outside the path of the hot flue gases. Cool outside air is introduced through the blower section of the unit. This creates suction at the throat of the inducer body which in turn increases the flow rate of the flue gases.

The draft created by the induced draft fan should closely match the demand. If the size of the draft inducer is correctly estimated, any dilution and excessive cooling of the flue gases (a condition that may result in condensation) will be greatly reduced.

517

Many boiler manufacturers will recommend the size draft inducer to be used with specific boiler models, and will provide data for making the appropriate calculations. Manufacturers of draft inducers are similarly helpful.

Fig. 82. Direct fan type draft inducer.

Fig. 83. Indirect fan type draft inducer.

CONTROL OF EXCESSIVE DRAFT ON GAS-FIRED BOILERS

Sometimes a boiler installation will have excessive draft conditions as a result of oversized chimneys or other factors. Draft conditions in excess of the draft design limits of the boiler will reduce the combustion efficiency of the boiler, resulting in higher fuel costs and possible pilot problems.

A suggested method of controlling the draft on such installations is to install a sheet metal baffle, or restrictor, in the flue connection at the boiler (Fig. 84). This baffle may be inserted in a tee as shown and secured with stove bolts after the proper draft is

THIS ABOVE DIMENSIONS FOR BAFFLE
MAY VARY DUE TO DIFFERENT TYPES
OF FLUE CONNECTIONS AND TEE SIZES.

Courtesy *Bryan Steam Corp.*

Fig. 84. Control of excessive draft of gas-fired boilers.

SUGGESTED BAFFLE DIMENSIONS			
BAROMTRIC SIZE DIAMETER	CUT SHEET		
	A	B	R
10"	10"	14"	5"
12"	12"	16 3/4"	6"
14"	14"	19 1/2"	7"
16"	16"	22 1/2"	8"
18"	18"	25"	9"
20"	20"	28"	10"
24"	24"	33 1/2"	12"

obtained by measurements with a draft gauge. If the barometric control is installed in the side of the breeching by means of a draft control collar, the same method of baffling may be used with the baffle restricting the breeching between the boiler and the barometric control. In this method of baffling, the breeching between the barometric control and breeching should not be restricted.

The baffle installation is correct if the gate of the barometric control is approximately half open while the gas burner is operating, and the draft measurement is at or below the operating design of the boiler. The draft limit for a boiler can be obtained from the manufacturer's specifications. For example, *Bryan* atmospheric gas boiler burner units are designed to operate at very low draft, never to exceed .01″ W.C.

519

Always check with the local gas company regarding the use of baffles. Caution should be exercised to insure against over restricting the flue. Always be certain that there is a very slight draft after the baffle is secured in position.

TANKLESS WATER HEATERS

Some boilers are manufactured with the option of using a *tankless water heater*. This device consists of an immersion coil inserted in a steam or hot-water space heating boiler to provide domestic and commercial service hot water. It is called a "tankless" water heater, because no storage tank is used to store the heated water during periods of low demand.

Courtesy *Bryan Steam Corp.*

Fig. 85. Heat exchanger coil.

The immersion coil is made of small diameter copper tubes which are either straight with U-bends at the end (Fig. 85) or formed in the shape of a spiral (Fig. 86). Typical locations on steam and hot-water space heating boilers are illustrated in Fig. 87. As shown, the tankless heater is installed in the nipple port

Courtesy The H.B. Smith Co., Inc.

Fig. 86. Heat exchanger coil used in gas-fired boilers.

STEAM HEATER

WATER HEATER

Courtesy Weil-McLain

Fig. 87. Location of tankless water heater on oil-fired hot-water boiler.

on cast-iron hot water boilers. On steam boilers, it is installed in the left hand side of a special back section at a point well below the waterline. A uniform water temperature is maintained by a

521

thermostatic three-way mixing valve installed in the supply line leading from the immersion coil.

Because these immersion coils function as a heat exchanger type device and operate on an indirect heating principle, they are also variously referred to as an *indirect water heater, indirect heat exchanger,* or *heat exchanger coil.*

In other applications, immersion coils provide hot water radiation from a steam boiler, hot water for snow melting, heated water for pools, and industrial process water.

Additional information about water heaters, including suggestions for estimating the hot water allowance, can be found in Chapter 4, Volume 3 (WATER HEATERS).

LEAKING COILS

On occasions, the coil of a tankless or indirect water heater installed in a boiler will leak, allowing high pressure water into the boiler water and resulting in a rise in the boiler presure. This, in turn, can cause the relief valve to open and fail to close again tightly, or it can cause the boiler pressure to rise above the setting of the pressure reducing valve. These symptoms are often wrongly attributed to a valve malfunction when the actual cause may be a leaking coil.

You can determine whether or not the problem is a leaking coil by taking the following steps:

1. Shut off the feed valve and/or break the connection to the coil.
2. Check the pressure reading in the boiler.
3. Wait about 8 hours and check the boiler pressure reading again.

If the pressure reading has remained approximately the same over the 8 hour period, it is a strong indication that the coil is leaking. On the other hand, if the boiler pressure continues to rise, the problem may lie elsewhere in the system.

BLOWING DOWN A BOILER

Foam sometimes forms on the surface of the boiler water and is usually indicated by drops of water appearing with the steam.

This condition is caused by the presence of oil, dissolved salts, or similar type organic matter in the water. One method of eliminating this problem is by draining off part of the water in the boiler, and adding an equal amount of fresh, clean water. Another method consists of blowing the foam from the water with a specially connected pipe or hose. The boiler should have a blow down tapping for this purpose. The first method is the easiest. The second one (i.e. blowing down the boiler) requires considerable experience.

BOILER OPERATION AND MAINTENANCE

Always follow the manufacturer's recommendations for boiler operation and maintenance. If you do not have these instructions (e.g. an owner's manual), you should contact the local dealer for advice or write to the manufacturer.

The sections that follow contain recommendations for operating and maintaining boilers. Because many of these recommendations specifically pertain to either steam or hot-water space heating boilers, they were divided and listed accordingly. The final three sections (BOILER WATER, CLEANING BOILERS, and INSULATING BOILERS) contain general recommendations for all boilers.

STEAM BOILERS

Operating and maintaining a steam boiler differs in certain respects from the procedures followed for hot-water space heating boilers. Many of these differences are evident in the recommendations that follow.

1. Keep a steam boiler filled to the water level recommended by the manufacturer when not in use.
2. Check the water level in the boiler before starting it. The heating surface can be damaged if the water level is too low.
3. Keep the water level at the center of the water gauge glass during operation. If the water level is too low, use the manual feed valves to add more water. These valves are found in most systems.
4. Always add water to the boiler gradually. If at all possi-

ble, avoid adding water to a hot boiler. *Never* add water to an operating cast-iron boiler. Cold water fed rapidly into such boilers may come in contact with the hot surface of the cast-iron heat exchanger, causing it to crack. Shut off the boiler and wait until it has cooled down before adding water. After the boiler has had sufficient time to cool, slowly add water through the cold water feed line.

5. Check the low-water cutoff at regular intervals. Sediment or rust accumulating under the float of the low-water cutoff can cause it to malfunction. As a result, a drop in the boiler water level would not register properly.

6. Check all boiler accessories to make certain they are functioning properly. Movable parts should be inspected and oiled regularly.

7. Never allow a low-pressure steam boiler to exceed the upper safe pressure limits recommended by the manufacturer (usually three to four lbs. pressure).

HOT-WATER SPACE HEATING BOILERS

Recommendations for operating and maintaining hot-water space heating boilers are as follows:

1. Keep the boiler *and the pipes* in the heating system filled with water when not in use. Keeping the pipes filled with water reduces the possibility of rust and corrosion.

2. Check the water level in the boiler before starting it. The heating surface can be damaged if the water level is too low.

3. Always add water to a boiler gradually. Never add water to a hot boiler. Shut the boiler down and allow it to cool first.

4. Check all boiler accessories to make certain they are functioning properly. Movable parts should be inspected and oiled regularly. Such maintenance should also include the pump in a *forced* hot-water heating system.

5. When operating a hot water space heating boiler, make certain all flow valves are open.

6. Never allow a boiler to exceed the upper safe temperature limit recommended by the manufacturer (usually about 200°F).

BOILER WATER

Boiler water should be clean and kept clean for efficient operation. This is true for all boilers. *Never* add dirty or rusty water to a boiler. Even hard water may eventually interfere with the efficient operation of a boiler, and should therefore be chemically treated before being added.

The Steel Boiler Institute has adopted a chemical conditioning compound for treating water used in low-pressure steam and hot-water space heating boilers. Many manufacturers provide this chemical compound with their boilers. *Always* treat the water immediately after the boiler and the heating system have been cleaned.

CLEANING BOILERS

A boiler should be inspected *at least once* every year for the accumulation of soot and other deposits that could impair its operation. This inspection should take place *before* the start of the heating season.

The accumulation of soot will result in improper combustion. The soot can be removed with a chemical cleaner or a flue brush. Many boiler manufacturers provide access to the heating surfaces

Courtesy Weil-McLain

Fig. 88. Access to heating surfaces through top of boiler.

525

through a removable top jacket panel and cover plates (Fig. 88). The flue brush is used to push the soot down between the sections or fins to collect in the combustion chamber below. It can then be removed from the combustion chamber without any great difficulty.

Oil burners, gas burners, and coal stokers should also be inspected for dust accumulations and cleaned. Always follow the manufacturer's recommendations for cleaning the boiler and automatic firing equipment. If you do not have an owner's manual, contact a field representative or write to the manufacturer.

TROUBLESHOOTING BOILERS

Boilers are subject to numerous problems, many of which are due to improper type and design, or poor servicing. The following are some of the problems usually encountered:

1. *Boiler does not deliver enough heat.* This is a very common complaint, but the boiler may not be causing the problem. The problem may be caused by the automatic fuel burning equipment (see Chapters 16 and 3 of Volume 2), the radiators or convectors (see Chapter 2, Volume 3), or the automatic controls (see Chapters 5-6, Volume 2). Each of these chapters contain troubleshooting sections. Even a pipeline with improper pitch will cause the heat from the boiler to be blocked or trapped. If the boiler is found to be the cause of insufficient heat, the problem can be traced to any one (or more) of the following causes:

   a. Boiler is too small for the heating system.

   b. Improper arrangement of boiler sections in cast-iron boilers.

   c. Poor draft.

   d. Poor fuel.

   e. Heating surfaces are covered with soot.

2. *Boiler delivers no heat.* The automatic controls should be checked first. Sometimes a low-water cut-off on steam boilers will shut off the burner or stoker before enough steam has formed. Another possible cause is that too much water is being fed into a steam boiler. As a result,

not enough space is provided for the steam to form at the top of the boiler. Too little water in a hot-water space heating boiler can be caused by the limit control moving down to a lower setting.

3. *Too much time required to get up steam in a steam boiler.* This common problem can be traced to the following possible causes:

    a. Too little or badly arranged heating surface.

    b. Heating surfaces covered with soot.

    c. Heating passages too short.

    d. Poor fuel or fuel firing.

    e. Poor draft.

    f. Boiler too small.

    g. Boiler defective.

4. *Boiler is slow to respond to the operation of the dampers.* Slow response to damper operation can be caused by any of the following:

    a. Air leakage into the chimney or stack.

    b. Poor fuel or fuel firing.

    c. Boiler too small.

    d. Clinkers on grate or ash pit full of ashes (coal-fired boilers).

5. *Water line is unsteady.* The problem of an unsteady water line may simply be due to connecting the water column to an extremely active section of the boiler. Therefore, it is extremely unlikely that the actual water level in the boiler can be read from the water column. Other possible causes of this problem are:

    a. Dirt or grease in the water.

    b. Varying pressure difference on the system.

    c. Excessive boiler output.

6. *Water disappears from the gauge glass.* The complete loss of water from the gauge glass could be due to priming (i.e. water globules being carried over into the steam). Other causes include:

527

      a. Foaming.

      b. Pressure drop too great in return line.

      c. Improper water gauge connection.

      d. Valve closed in the return line.

7. *Water is carried over into the steam main.* This problem is usually caused by one of the following:

      a. Priming or foaming.

      b. Water line is too high.

      c. Outlet connections from boiler too small.

      d. Steam liberating surface too small.

      e. Boiler output excessive.

8. *Flues require cleaning too frequently.* A frequent build up of soot and dirt in flues can be caused by any of the following:

      a. Combustion rate too slow.

      b. Poor draft.

      c. Smokey combustion.

      d. Excess air in firebox.

9. *Low carbon dioxide.* This condition can generally be traced to one of the following causes:

      a. Air leakage between cast-iron sections.

      b. Improper conversion job.

      c. Problem with burners.

10. *Smoke from boiler fire door.* The following conditions may be the cause of this problem:

      a. Dirty or clogged flues.

      b. Incorrect setting of dampers.

      c. Poor or defective draft in the chimney.

      d. Incorrect reduction in the breeching size.

BOILER REPAIRS

Repairs to the boiler itself should be done by an experienced and skilled worker. They should never be attempted while the boiler is under pressure. Shut down the boiler and allow it to cool.

Most structural repairs should be regarded as only temporary until the section or part can be replaced with a new one. Minor cracks or leaks can be repaired as follows:

1. Minor cracks in the sections of a cast-iron boiler can be repaired by welding or brazing.

2. Boiler leaks can be *temporarily* sealed with a synthetic compound similar to the type used to seal automobile radiators. Permanent repairs should be made by an experienced worker.

BOILER INSULATION

All uninsulated boilers suffer a certain amount of heat loss. In some respects, this loss of heat can be beneficial, because it does warm the basement and the floor above. However, if the boiler is also used to provide domestic or commercial service hot water, this heat loss can be uneconomical. This is particularly true for the summer months.

Steel jacketed type boilers are generally insulated by the manufacturer. For example, the boiler illustrated in Fig. 3 is insulated with a 1 inch thick layer of fiberglass insulation attached to the inside surface of the steel jacket. However, other types of boilers, particularly the old cast-iron round boilers, do require an external layer of insulation to reduce the amount of heat loss. *Only* a noncombustible material such as asbestos cement, asbestos paper, or magnesia blocks should be used for this purpose.

Magnesia blocks or sheets can be cut to fit the contours of a boiler, and wired to the surface. The cracks and insulating material are then covered with asbestos cement. Corrugated asbestos paper can be used in place of magnesia block, but it should be applied in three layers to obtain the desired thickness (about 1¼ inch). Each layer should be covered with asbestos cement.

Boiler Installation

All new boilers are shipped with a complete set of installation instructions. Usually these instructions will also contain an inventory list of the contents in the crates and cartons. *Always* check the contents off against this list before you do anything else. Missing or damaged parts should be immediately reported to the manufacturer or his local representative.

A boiler should always be located so that the connecting flue pipe between the boiler (or draft diverter) and the chimney is as short as possible. Another important consideration are the recommended minimum clearances between the boiler and combustible materials. These two factors plus the design considerations of the system will dictate where the boiler is located.

The manufacturer's installation guide will also contain instructions on how to light and operate the boiler. The instructions will differ in accordance with the type of automatic fuel burning equipment used (e.g. oil burner, gas burner, etc.).

After completing the installation of the boiler, inspect all the controls to make certain they are operating properly. Start and stop the burner or stoker several times by moving the room thermostat setting.

An important point to remember is that *any and all* requirements of national standards and codes, as well as local authorities having jurisdiction, *always* take precedence over the installation instructions provided by the manufacturer. For example, the latest edition of the American National Standard for Installation of Gas Appliances and Gas Piping Z21.30 should be consulted when installing gas-fired boilers.

WATER CHILLERS

One method of adding central air conditioning to a structure heated by either hot water or steam is to install a water chiller. A *water chiller* is a device consisting of a compressor, a condensor, a thermal expansion valve, and a refrigerant evaporator coil enclosed in a steel shell. The water in the system is pumped through the shell, where it is cooled by the evaporator coil, and circulated through the radiators or convectors for cooling and dehumidification.

The boiler and water chiller are available packaged in a single unit, or they may be installed as separate units. In a hot-water heating system, the same piping can be used to carry both the hot water and chilled water to the radiators or convectors. The room radiators or convectors are individualy controlled to permit temperature adjustments. Steam piping should not be used to carry the chilled water.

Few residential water chillers are in use. The most popular method of applying central air conditioning to a steam or hot-water space heating system is to install a split system.

When a water chiller is installed as a separate unit, the chilled water should be piped in parallel with the boiler. Appropriate valves should be installed to prevent the chilled water from entering the boiler (Fig. 89).

Courtesy *Hydrotherm, Inc.*

Fig. 89. Piping connections for boiler and water chiller. Valve H opened on heating with valve C closed. Reverse procedure on cooling.

CHAPTER 16

Boiler and Furnace Conversions

Sometimes it becomes necessary to convert an existing boiler or furnace to another fuel. This is usually the case when the fuel for which it was originally designed has become too expensive relative to others or does not meet the desired level of efficiency. During the early 1950's when coal was becoming more and more expensive as a domestic heating fuel and before oil and gas burners were offered as integral parts of heating units by manufacturers, converting boilers and furnaces was much more widespread than it is today.

A number of manufacturers have made available gas-fired and oil-fired burners specifically designed for boiler and furnace conversions. It would be impossible within the space limitations of this chapter to cover all the different models of conversion burners offered by these manufacturers. Consequently, this chapter will emphasize all the *ramifications* involved in converting a boiler or furnace from one fuel to another rather than examining the specifications and operating characteristics of one or more conversion burners.

PREPARATION FOR CONVERSION

Before constructing a combustion chamber and installing a burner, the heating system should be carefully checked for defects and cleanliness. A boiler or furnace in need of repairs will not give satisfactory results after the burner is installed. Be certain that *all* flue passages are cleaned so that the maximum amount of heat generated is absorbed by the boiler. Soot or ash are good insulators and both are always undesirable.

All doors should fit tightly and all other openings or cracks should be tightly cemented shut. The stack from the furnace to the chimney should have tight joints. Dampers should not close off more than 80% of the cross sectional area of the stack. Stacks should be inspected for leaks and obstructions of any kind.

Remove the old fuel handling parts (e.g. the coal handling parts in a coal-fired furnace or boiler) and thoroughly clean the interior surfaces, removing all soot, scale, tars and dirt.

Any cracks in the heat exchanger should be repaired or replaced. On boilers, any water leaks should be sealed, broken gauges replaced, and loose boiler doors repaired or replaced. Air leaks along the floor should be grouted.

No positive catches should be used on firing doors. File off any catches so that the firing door will open easily to relieve pressure. A spring loaded door holder is recommended.

If a conversion burner with a direct spark ignition system is installed, no provision need be made for top venting or horizontal or downdraft furnace. Check the local codes for further information.

Fig. 1 illustrates some of the typical problems encountered when converting a coal-fired *boiler* to oil. It is highly doubtful that a boiler would have all of these problems at the same time, but for your convenience they are included in the illustration to serve as a reference.

BASIC COMBUSTION CHAMBER REQUIREMENTS

An essential requirement for combustion chambers is that the flame be in the presence of refractory materials so that it will not come into contact with any of the relatively cold heating surfaces of the boiler.

When the flame is burned in suspension in a combustion chamber of refractory material, the refractory walls will reflect the heat back into the flame and thereby increase its temperature. This increase in flame temperature due to reflected heat greatly

Fig. 1. Boiler problems to be remedied before converting and installing an oil burner.

increases the rapidity of combustion and thereby makes possible the burning of every particle of fuel with the zone of combustion.

The combustion chamber walls should be high enough so that the flame will not come in contact with the relatively cool walls of the boiler. At the burner end, the refractory wall need not be higher than the grate line of the boiler.

Finally, a well designed combustion chamber should provide rapid heating at the beginning of each firing cycle.

COMBUSTION CHAMBERS FOR CONVERSION GAS BURNERS

Conversion gas burners are designed for firing into a refractory lined combustion chamber constructed in the ashpit of a boiler or furnace. It is recommended that built-up combustion chambers

Table 1. Recommended minimum wall thicknesses of combustion chambers using conversion gas burners.

Input BTU/Hr.	Floor Area Sq. Inches	Preferred Width and Length	Recommended Minimum Wall Thickness	Recommended Minimum Floor Construction
100,000	180	12 x 15	2½ insulating firebrick plus back up of 1½ or more loose insulation	2½ insulating firebrick plus ½ asbestos or magnesia block
150,000	200	12 x 16		
200,000	220	13 x 17		
250,000	235	13 x 18		
300,000	260	13 x 20		
350,000	270	14 x 21		
400,000	330	15 x 22	4½ insulating firebrick	2½ insulating firebrick plus 2¼ common brick or hard firebrick, plus ½ asbestos or magnesia block
500,000	400	15 x 27		
600,000	460	15 x 31		
700,000	520	15 x 35		
800,000	600	18 x 33	4½ insulating firebrick plus back up of 1½ or more loose insulation	4½ insulating firebrick plus ½ asbestos or magnesia block or 2¼ common brick or hard firebrick
900,000	650	18 x 36		
1,000,000	700	22 x 32		
1,100,000	750	22 x 34		
1,200,000	800	22 x 36		
1,600,000	1152	26 x 44		
2,000,000	1440	29 x 50		
2,500,000	1760	32 x 55		
3,000,000	2100	35 x 60		
3,500,000	2470	38 x 65		
4,000,000	3120	40 x 68		

Courtesy *Magic Servant Products* Co.

used for this purpose be constructed of 2300°F. insulating firebrick, and cemented with insulating firebrick mortar.

Recommended *minimum* wall thicknesses and floor areas of these combustion chambers are based on the maximum rated Btu capacity (Table 1).

The height of combustion chamber walls will usually be determined by the grate line. Build the side and front walls about 2 inches above the grate line. The grate lugs and base of the water

535

legs of boilers should be covered by about 3 or 4 inches to avoid heating sections that may be filled with sediment. Carry the back wall one or two brick courses higher than the side or front walls and allow it to overhang in order to deflect hot gases from impingement on the heat exchanger surfaces. Use a *hard* firebrick for the overhang section because the high velocity gases moving against it tend to erode a softer brick.

If the combustion chamber is to be placed directly on the floor of the ashpit, a layer of asbestos millboard (½ inch minimum thickness) should underlay the combustion chamber. Ordinary brick may be used if the floor of the combustion chamber is raised off the floor of the ashpit. The remaining open space should be filled with vermiculite or some other suitable type of loose insulation.

COMBUSTION CHAMBERS FOR CONVERSION OIL BURNERS

A combustion chamber for a conversion oil burner should be constructed of lightweight 2000°F. insulating firebrick. With lightweight refractory materials, there is no long smoky delay waiting for the fire box to reach the temperature necessary for complete combustion.

Different types of combustion chamber construction are shown in Figs. 2 and 3. The shape of the combustion chamber will be determined by the shape of the boiler or furnace.

The combustion chamber area (space) is equal to the inside width of the boiler times the length times the distance from the the combustion chamber floor to the crown sheet. It is recommended that one cubic foot be provided for each boiler HP, or three cubic feet per G.P.H. oil-fired.

The floor area of the combustion chamber should be based on 100 square inches per gallon of oil (see Table 2 and Figs. 4 and 5). The minimum side wall height should conform to the recommended measurements listed in Table 2.

An overhang in the form of stepped corbels will deflect hot gases from impingement on the heat exchanger surfaces, increase combustion chamber temperatures, and thereby promote good combustion.

PROVIDE CORBEL OR BAFFLE FOR MAX. WIPING OF PRODUCTS AND COMPLETE COMBUSTION

BUILD COMBUSTION CHAMBER UP TO 2" ABOVE COLD AIR RETURN DUCT

INSTALL SLEEVE AND PACK INSULATING CEMENT AROUND BURNER TUBE

SEAL FRONT WITH INSULATING CEMENT (ADD 10% PORTLAND)

2 1/2" TO 4 1/2" THICK LIGHT WEIGHT REFRACTORY COMBUSTION CHAMBER

BACK FILL AND FLOOR OF INSULATING CEMENT

Courtesy Stewart-Warner.

Fig. 2. Recommendations for constructing a combustion chamber in a furnace.

WHEN FLUE OUTLET AND TARGET WALL ARE ON SAME END, USE CORBEL AND LOCATE COMBUSTION CHAMBER FOR MAX. CONTACT OF PRODUCTS WITH CROWN SHEET

FLUE OUTLET

CORBEL

TARGET WALL OF BURNER

PROTECT WATER LEG

FRONT FIRING

REAR FIRING

Courtesy Stewart-Warner.

Fig. 3. Combustion chambers for boilers.

High flame retention burners provide an intense flame that eliminates the need for refractory combustion chambers above 3.0 G.P.H., and has greatly reduced the need for such chambers in the 1.00 to 3.00 G.P.H. range.

Courtesy *Stewart-Warner.*

Fig. 4. Side view of chamber.

Courtesy *Stewart-Warner.*

Fig. 5. Top view of chamber.

CONSTRUCTION MATERIALS

The construction materials used in building combustion chambers for conversion burners should consist of the best grade of

538

insulating firebrick and a good, high temperature firebrick cement. The shape of the bricks is also important, with arch or circled brick of 4½ inch wall thickness being preferred. The cement filled joints between the bricks should be not more than ⅛ inch. Always use the cement furnished or recommended by the brick manufacturer for cementing the insulating firebrick. The cement should be thinned to the consistency of a very thick cream before it is used. The usual method for applying the cement is to dip each brick into the cement and set it in place as you lay each course.

Table 2. Combustion chamber floor area dimensions (100 sq. in. per G.P.H).

G.P.H.	Round I.D.	Rectangular W x L	Nozzle Height "N"	Min. Side Height "H"
.50	8"	7"×8"	5"	12"
.75	9½"	8"×9"	5"	12"
.85	10½"	9"×9½"	5"	12"
1.00	11½"	9"×11"	5½"	13"
1.10	12"	10"×11"	5½"	13"
1.25	12"	10½"×12"	6"	14"
1.35	13"	11"×12"	6"	14"
1.50	14"	12"×12½"	6½"	15"
1.75	15"	13"×13½"	6½"	15"
2.00	16"	14"×14"	6½"	16"
2.25	17"	14"×16"	7"	16"
2.50	18"	14"×18"	7"	17"

Courtesy *Stewart-Warner*

The insulating firebricks are usually backed up by a second insulating layer which will consist of one of the following materials:

1. Asbestos,
2. Common brick,
3. Magnesia block insulation,
4. Hard firebrick,
5. Expanded mica products (e.g. *vermiculite* or *zonolite*),
6. Dry sand,

7. Dry sand mixed with an expanded mica product.

Approved insulating firebricks recommended for use in constructing combustion chambers are:

1. Babcock & Wilcox No. K-23 and No. K-26,
2. A.P. Green No. G-23 and No. G-26,
3. Armstrong Cork No. A-23 and No. A-26,
4. Johns-Manville No. JM-23 and No. JM-26.

The spaces between the outer edges of the firebrick should be filled with high temperature cement and small pieces of firebrick to obtain firm construction and prevent infiltration of vapors through the wall.

BUILDING A COMBUSTION CHAMBER

Fig. 6 illustrates the construction of a typical custom built combustion chamber. The steps involved in constructing this chamber are as follows:

Fig. 6. Construction of a custom built chamber.

1. Place split brick *A* on the ash pit floor to form the floor of the combustion chamber.

2. Mix the refractory binder in a shallow pan to the consistency of a heavy cream.

3. Dip the bottom and sides of brick *B* in the binder and place it in a circular position on the combustion chamber floor. You should start at the back and work toward the front of the chamber.

4. Follow the same procedure with the rest of the bricks in the tier (i.e. those containing brick *B*).

5. When the tier is finished, tighten it in position with bands suitable for this purpose.

6. Fill behind the first tier with asbestos or rock wool.

7. Install top tier *D* in the same manner described in Steps 1-5, however use only one band to tighten it.

8. Cap the chamber with a mixture of 3 parts asbestos and 1 part fire clay and trowel smooth.

9. Place metal sleeve *F* in the opening and cement it tightly in place.

10. Insert the draft tube tin metal sleeve and pack with asbestos rope or rock wool.

Fig. 7 shows an old round cast iron boiler that has been converted to oil. Note the height of the combustion chamber wall. The procedure for setting up the oil burner to the combustion chamber is illustrated in Fig. 8.

VENTILATION REQUIREMENTS

If the existing boiler or furnace is located in an open area (basement or utility area) and the ventilation is relatively unrestricted, there should be a sufficient supply of air for combustion and draft hood dilution. On the other hand, if the heating unit is located in an enclosed furnace room or normal air infiltration is effectively reduced by storm windows or doors, then certain provisions must be made to correct this situation. Fig. 9 illustrates the type of modification that should be made in a furnace room to provide an adequate supply of air. As shown, two permanent

USE SPLIT BRICK IF NECESSARY
TO MAINTAIN CORRECT SIZE
COMBUSTION CHAMBER

SEAL UP ALL CRACKS OR
SPACES WITH ASBESTOS
OR FURNACE CEMENT

BRING BRICK
2" TO 4" ABOVE
WATER LEG

SAND OR CEMENT

IF NECESSARY USE SPLIT BRICK TO GET
BURNER AT PROPER DISTANCE FROM FLOOR.
IF BOILER IS ON WOOD FLOOR INSULATE
WELL WITH ASBESTOS AND USE FULL SIZE

Fig. 7. Combustion chamber for an older cast-iron boiler.

WHEN PLASTERING AND FINISHING OF
FRONT PUT NEWSPAPER OVER AIRTUBE
TO KEEP ALL SPOTLESS

PLASTER UP WELL WITH ASBESTOS OR
FURNACE CEMENT

BRICK UP ALMOST FLUSH WITH END OF
BURNER NOSE PIECE

Fig. 8. Procedure for setting up burner to combustion chamber.

grilles are installed in the walls of the furnace room, each of a size equal to one square inch of free area per 1000 Btu per hour of burner output. One grille should be located near the ceiling, and the other near the floor.

If the boiler or furnace is located in an area of particularly tight construction, the heating unit should be *directly* connected to an outdoors source of air. A permanently open grille sized for

VENTILATING AIR OPENING 1 SQ. IN.
FOR EACH 1000 BTU/HR. INPUT

AIR INLET OPENING 1 SQ. IN.
FOR EACH 1000 BTU/HR. INPUT

Fig. 9. Showing air openings necessary to supply air for combustion when furnace is installed in an enclosed room.

at least one square inch of free area per 5000 Btu per hour of burner output should also be provided. Connection to an outside source of air is also recommended if the building contains a large exhaust fan.

A conversion burner equipped with a spark ignition system is not suited for use in furnaces or boilers located in areas subjected to sustained reverse draft or where large ventilating fans operate in combination with insufficient make-up air. There is always the danger of drawing flue gases into the structure. See INSTALLING A CONVERSION GAS BURNER (PILOTLESS) in this chapter.

FLUE PIPE AND CHIMNEY

The *flue pipe* is a pipe connecting the smoke outlet of the furnace or boiler with the flue of a chimney (Figs. 10 and 11).

543

Courtesy Mid-Continent Metal Products Co.

Fig. 10. Chimney and flue connection with horizontal draft hood.

It should *not* extend beyond the inner liner of the chimney, and should *never* be connected with the flue of an open fireplace. Furthermore, flue connections from two or more sources should never enter the chimney at the same level from opposing sides.

Flue pipe should be constructed from corrosive resistant metal and be designed with as few sharp turns as practical. The straightest and shortest possible passage should be provided for the existing flue gases.

Pitch the flue pipe with a rise toward the chimney of at least ¼ inch per foot. At approximate intervals, fasten the flue pipe securely with sheet metal screws to prevent sagging.

Masonry is recommended for the construction of chimneys (pre-fab chimneys are also found suitable). Outside metal stacks are generally found unsuitable for oil-fired burners.

The ordinary chimney *must* be at least 3 feet higher than the roof or 2 feet higher than any ridge within 10 feet of the building in order to avoid down drafts.

Aside from adapting the furnace or boiler for a conversion burner, changes must also be made in the passages formed by the heating surface.

DO NOT EXTEND FLUE PIPE
BEYOND INSIDE OF CHIMNEY

MINIMUM RISE
1/4" TO FOOT

A. G. A. LISTED
VERTICAL
DRAFT HOOD

CORROSION
RESISTANT
LINING

CHINMEY

REDUCER

CLEANOUT

Courtesy Mid-Continent Metal Products Co.

Fig. 11. Chimney and flue connection with vertical draft hood.

Extra large flue passages are *not* suited to the high temperature flue gas encountered with oil. In boilers having these large flue passages, baffling must be resorted to in order to slow down the high velocities of the extra hot gases, otherwise unburned particles of oil may lodge on the heating surface, resulting in carbon. Because carbon is an excellent insulator, the efficiency of the heating surface is lowered whenever it collects. See BAFFLING in this chapter.

545

BAFFLING

Baffling is a type of obstruction designed to deflect and regulate the speed of flue gases. They are especially necessary for round boilers and furnaces constructed so that the flue passes are almost direct from the fire box. Furthermore, boilers designed for burning coal are usually provided with relatively large flue passages which are not normally suited to the higher flue gas velocities encountered in oil firing.

Baffling will, in most cases, help to partly overcome the inefficient operation resulting from the usual excessively high stack gas temperatures. In such instances it is advisable to experiment with various methods of baffling as shown in Fig. 12. *Note:* any baffling of the flue passes of a nature which prevents efficient operation of the burner should not be done.

Another form of baffling is a *corbel* or step out arrangement of the brick work of the rear wall, forming a target wall which the flame strikes and is curled back to prevent short circuiting.

GAS PIPING AND PIPING CONNECTIONS

Figs. 13 and 14 illustrate the piping connections necessary for two different models of gas-fired conversion burners. *Always* refer to local installation codes for guidance, because your piping *must* comply with these codes. Other helpful sources of information are the following booklets from the *American Gas Association:*

1. Bulletin Z21.30, *"Installation of Gas Appliances and Gas Piping."*
2. Bulletin Z21.8, *"Installation of Domestic Gas Conversion Burners."*

Also helpful is NFPA Bulletin No. 54, *"Gas Appliances and Gas Piping,"* which is available from the *National Fire Protection Association.*

No matter what type of gas-fired conversion burner you decide to use, it *must* be allowed to develop its rated capacity. This can be accomplished by making certain the burner is connected to a gas supply containing sufficient pressure.

In addition to providing for a gas supply with sufficient pressure, the following recommendations are also offered:

BOILER SECTIONS

FIREBRICK

UPTAKE

BOILER SECTIONS

BRICKS

STACK

SECTIONAL TOP VIEW

BRICK

FLUES

BRICK

RETARDER SPIRAL (SHEET METAL)

Fig. 12. Various methods of baffling for sectional cast-iron boilers.

1. Provide for a separate gas supply line of 1 inch pipe size direct from the meter to the burner (ample for runs up to 60 feet in length).

MAIN
SHUT - OFF
(BY INSTALLER)

VALVE DIAL
"A"

LATCH "B"
ENGAGE ON FLANGE PORTION
OF KNOB "A" TO HOLD PILOT
ON DURING LIGHTING

MOTOR OIL
TUBES

PILOT
ADJUSTMENT
"D"

DRIP LEG

MAIN AIR
ADJUSTMENT
(HIDDEN)

REST VERTICAL
PIPING ON FLOOR

BACKPLATE

MANIFOLD
PRESSURE TAP
(UNDER VALVE)

UNION

GAS PRESSURE
REGULATOR

LIGHTER BUTTON
AND IGNITOR
SLOT "C"

Courtesy Mid-Continent Metal Products Co.

Fig. 13. Pipe connections for a conversion gas burner. (with pilot light).

2. Install an intermediate regulator if the line pressure exceeds 13.5 inch W.C.

3. Connect the burner to the piping as shown in Figs. 13 and 14.

4. Install a manually operated main shut-off valve 4 to 5 feet above the floor on the vertical pipe.

5. Make certain the pipe is clean and free of any scale.

6. Use malleable iron fittings.

7. Remove all burrs and scales from the pipe and clean it before installing it.

8. Do not tap off from the bottom of horizontal runs when branching from a pipe line.

MAIN SHUT-OFF
(BY INSTALLER)

MOTOR OIL TUBES

MAIN AIR
ADJUSTMENT

VALVE
DIAL

DRIP LEG

REST VERTICAL
PIPING ON FLOOR

UNION

BACKPLATE

MANIFOLD PRESSURE TAP
(UNDER VALVE)

GAS PRESSURE
REGULATOR

Courtesy Mid-Continent Metal Products Co.

Fig. 14. Pipe connections for a conversion gas burner (without pilot light).

GAS INPUT SETTING

The gas-fired conversion burner illustrated in Fig. 13 is manufactured to operate on natural gas, but kits are available for converting to propane. A very important factor to be considered when installing these burners is a correct determination of the gas input setting, which should be based on the following two factors:

1. The heating rate of the structure.
2. The rated maximum input of the burner.

The *heating rate* of a structure is essentially the amount of warm air that the heating unit must deliver in order to adequately heat it. Methods for calculating the amount of heat required (i.e. the heating rate) were described in Chapter 4 (HEATING CALCULATIONS).

Another important factor to consider is the *rated maximum input* of the burner. The gas input *cannot exceed* this rating. If

549

the burner maximum input of the burner is below the required minimum of heat indicated by the heating rate, then the burner is inadequate for your needs.

Table 3. Spud sizing for natural gas.

NATURAL			
Capacity BTU/Hr.		Spud Size	
Max. 3.5″ W.C. Man. Pr.	Min.	Drill No.	Dia.
75,000	50,000	#20	.161
125,000	75,000	# 4	.209
200,000	125,000	17/64	.266

Courtesy Mid-Continent Metal Products Co.

The gas input setting for a natural gas conversion burner is regulated by the spud sizing and the pressure regulator. A burner is commonly shipped with three different spud sizes, one for each of three capacity range (high, intermediate, and low. Table 3). The capacity range is changed by using a different spud sizing. After the appropriate spud has been installed, the final gas input setting is made by adjusting the pressure regulator. Be sure to carefully read the burner manufacturer's installation instructions.

A propane gas conversion burner operates directly on 11″ w.c. gas supply pressure as determined by the propane tank regulator. The portion of the combination control on the standard natural gas burner is blanked off with a plate provided in the propane conversion kit. Two spuds are provided: a minimum sized main spud and a pilot spud. The main spud must be redrilled for the required capacity (see Table 4).

INSTALLING A CONVERSION GAS BURNER

The conversion gas burners illustrated in Figs. 13 and 14 are used to convert coal-fired or oil-fired central heating plants to gas. They are adaptable for use in forced warm-air furnaces, gravity (updraft) furnaces, or boilers (Figs. 15, 16, 17 and 18).

Table 4. Spud sizing for propane.

PROPANE		
Capacity BTU/Hr.	Spud Size	
11" W.C. Man. Pr.	Drill No.	Dia.
50,000	#46	.081
75,000	#40	.098
100,000	#32	.116
125,000	⅛"	.125
150,000	#29	.136
175,000	#26	.147
200,000	#19	.166

Courtesy Mid-Continent Metal Products Co.

In addition to the basic combustion chamber requirements (see the appropriate section in this chapter), the following recommendations *must* be followed when installing a conversion gas burner.

1. The diameter of the burner opening into the combustion chamber must *not* be oversize. If the opening is oversized, slip the stainless steel sleeve (included with the burner) over the nozzle and fill in the remaining space with refractory cement.

2. Position the end of the burner nozzle so that it is at least 1 inch short of the inside surface of the combustion chamber. *Never* allow the nozzle to extend into the combustion chamber.

3. If the burner nozzle is too short to reach the combustion chamber, it may be lengthened with an extension tube.

4. Check the burner ports and pilot before permanently setting the burner in place, and remove any foreign matter blocking the openings.

STARTING A CONVERSION GAS BURNER (WITH PILOT LIGHT)

This section covers the starting of conversion gas burners equipped with pilot lights. Before attempting to start this type of

SUPPLY

RETURN

IN WARM AIR FURNACES
PROTECT THE HEAT
EXCHANGER SURFACES FROM
DIRECT FLAME CONTACT

SEAL FLUE
PIPE

FREE SWINGING ACCESS
DOOR WITH PEEP SIGHT

COVER NON-HEAT
TRANSFER SURFACES

CONVERSION
GAS BURNER

MOUNTING FLANGE

SEAL STAINLESS
STEEL SLEEVE
WITH ASBESTOS
CEMENT

2300° INSULATING FIREBRICK

$\frac{1"}{2}$ MIN ASBESTOS OR MAGNESIA
OR COMMON BRICK

DO NOT LET BURNER NOZZLE
EXTEND INTO COMBUSTION CHAMBER

Courtesy Mid-Continent Metal Products Co.

Fig. 15. Updraft forced warm-air furnace equipped with conversion gas burners.

burner, the piping should be checked for leaks. The basic procedure for doing this is as follows:

1. Shut off all other gas appliances (i.e. gas water heaters, etc.) that tie into the same system.
2. Turn on the main shut-off valve (Fig. 13).
3. Turn valve dial "A" to OFF (Fig. 13).
4. Turn on gas pressure to the gas supply line.
5. Watch the meter test dial.

If the meter test dial shows no movement for *at least* five minutes, it is safe to say that there are no leaks in the piping. Movement of the meter test dial, however, means that you must locate the leak (or leaks) before starting the burner. The soap suds test is a simple and effective method for doing this.

The gas line must be purged before starting the burner. To do this, the following steps are recommended:

DRAFT HOOD

SUPPLY   RETURN

IN WARM AIR FURNACES
PROTECT THE HEAT
EXCHANGER SURFACES FROM
DIRECT FLAME CONTACT

COVER NON - HEAT
TRANSFER SURFACES

FREE SWINGING ACCESS
DOOR WITH PEEP SIGHT

CONVERSION
GAS BURNER

MOUNTING FLANGE

SEAL STAINLESS
STEEL SLEEVE
WITH ASBESTOS
CEMENT

SEAL FLUE
PIPE

2300° INSULATING FIREBRICK
1/2" MIN. ASBESTOS OR MAGNESIA
OR COMMON BRICK.

DO NOT LET BURNER NOZZLE
EXTEND INTO COMBUSTION CHAMBER.

Courtesy Mid-Continent Metal Products Co.

Fig. 16. Downdraft forced warm-air furnace equipped with a conversion gas burner.

1. Disconnect the pilot tubing.

2. Depress valve dial "A" as far as it will go, and turn it *counter clockwise* to pilot (Fig. 13). Do *not* release the dial at this point.

3. With valve dial "A" still fully depressed, lock the dial in position by engaging the lighting latch "B" (Fig. 13).

4. Allow the air to escape until the gas line is completely purged (which will be indicated by gas replacing the air issuing from the pilot connection). Caution: *Never* purge the gas line into the combustion chamber.

5. Turn off valve dial "A" (depress and turn clockwise).

6. Reconnect the pilot tubing.

After you have checked the piping for leaks and have completely purged the gas lines, you are now ready to start the burner. First add a few drops of oil (#20 SAE oil is recommended) to

AVOID DIRECT
FLAME CONTACT ON
HEAT EXCHANGER.

FORMER OIL BURNER
COMBUSTION CHAMBER
CAN BE USED IF
SUITABLE.

SEAL STAINLESS STEEL SLEEVE
WITH ASBESTOS CEMENT.

LEAK PROOF FIRING PORT
SEE SECTION IV.

DRAFT HOOD

FREE SWINGING
ACCESS DOOR
WITH PEEP SIGHT

CONVERSION
GAS BURNER

Courtesy *Mid-Continent Metal Products Co.*

Fig. 17. Downdraft suspended heater equipped with a conversion gas burner.

the burner motor bearings. Then, check the setting of valve dial "A". At this point, it should be turned to OFF (see Step 5 in the purging procedure described above).

You are now ready for the initial start up procedure which is accomplished as follows:

1. Set the main air shutter at approximately one-quarter open.

2. Turn on the main line switch.

3. Set the room thermostat *above* room temperature (this will cause the burner motor to start).

4. The burner motor should be allowed to run for about five minutes to insure complete purging of the combustion chamber.

FILL DOORS WITH
ASBESTORS CEMENT

REMOVE LATCH FROM
FIRING DOOR AND
INSTALL SPRING CLOSER

2300°F. INSULATING
FIREBRICK

CONVERSION
GAS BURNER

SEAL DOOR OPENING

SEAL STAINLESS
STEEL SLEEVE WITH
ASBESTORS CEMENT

DO NOT EXTEND
BURNER NOZZLE INTO
COMBUSTION CHAMBER

COVER WATER LEG BASE

LOOSE INSULATION.
COVER ALL NON HEAT
TRANSFER SURFACES

COMMON BRICK OR 1/2" MIN.
ASBESTORS OR MAGNESIA

Courtesy Mid-Continent Metal Products Co.

Fig. 18. Hot water or steam boiler equipped with a conversion gas burner.

5. Shut off the burner motor by adjusting the thermostat setting *below* room temperature.

6. Fully depress valve dial "A" and turn it *counter clockwise* to PILOT. Keep it depressed.

7. Lock valve dial "A" in its depressed position by engaging the lighting latch "B".

8. Light a match and hold the flame under the igniter valve and at ignitor slot "C" while at the same time depressing button "C" for at least 10 seconds.

9. Release the lighter button. The pilot should now ignite. If it does not, then repeat Step 8 above.

10. After the pilot has been burning for at least one minute, release valve dial "A" and turn it *counter clockwise* to the ON setting.

STARTING A CONVERSION GAS BURNER (PILOTLESS)

Most of the instructions given in the previous section for starting a conversion gas burner equipped with a pilot light will also apply here. The major difference is the use of direct spark ignition rather than a pilot flame.

The basic procedure for starting a pilotless conversion gas burner is as follows (Fig. 14):

1. Check the piping for leaks.
2. Bleed (purge) the gas line.
3. Set the air shutter about three quarters open.
4. Depress the valve dial and turn it to the ON position.
5. Turn on the main line switch.
6. Set the room thermostat *above* the room temperature. This should cause the burner motor to start. If it does, burner ignition should occur after 30 seconds.
7. If the burner fails to light, turn off the main line switch and then turn it on again.
8. Readjust the air shutter for a quiet, soft flame. The flame should be blue at the burner, changing to orange at the tips.
9. Start and stop the burner several times with the thermostat to check its operation.

SERVICING A CONVERSION GAS BURNER

Always be sure that the manual gas valve and burner switch are turned off before attempting to service a gas burner. *Never* attempt to remove any parts for service before taking this precaution.

In direct spark ignition burners, the nozzle and electrode assembly can usually be removed as a unit. For example, the nozzle and electrode assembly of the *Economite Model DS20A* conversion gas burner can be completely removed in the following manner:

1. Remove the burner backplate.
2. Disconnect the pipe union.
3. Remove the curved orifice pipe.
4. Withdraw the nozzle assembly.
5. Disconnect the electrode leads.
6. Remove the unit.

The nozzle and electrode assembly can be reinstalled by reversing Steps 1-6. When reassembling, make certain the orifice pipe enters the nozzle casting.

With the nozzle and electrode assembly completely removed, you are now in a position to inspect and clean it. Check the electrode for insulator cracks and evidence of serious burning.

Removal of the burner back plate also provides access to the motor and blower, motor relay, low voltage transformer, and terminal board. These can be removed as a unit for servicing, but you will have to disconnect the terminal board and unplug the electrical pass-through fitting to do this. The following service may be required for this unit:

1. Cleaning of the blower wheel.
2. Cleaning the motor air vents.

Does the motor run without further burner sequencing taking place? This may indicate trouble in the centrifugal interlock switch. Some burners are provided with removable motor end caps to provide ready access to the centrifugal switch. If this should be the case, remove the cap and clean the contacts by burnishing. The contacts must be open when the motor is not operating. If necessary, the burner motor can be removed and replaced as follows:

1. Remove the blower wheel.
2. Remove the retainer clips at the motor grommets.
3. Pull the motor out of the brackets.
4. Repair or substitute a new motor.
5. Place motor in brackets.
6. Make certain grommets fit well in the bracket and replace retainer clips.
7. Replace blower wheel.

The sealed motor relay and low voltage transformer normally do not require servicing. They should be replaced when defective.

Check gas lines and valves for leaks with the soap suds test. *Never* use a flame to locate a gas leak. If the leakage occurs around the valve, the valve seat may need cleaning. Disassemble the valve, clean the seat, and reassemble it. If the valve malfunction is due to causes other than the seat, replace the entire valve.

Replace the gas pressure regulator if it fails to maintain a constant pressure. On the conversion gas burner illustrated in Fig. 14, the regulator is part of a combination valve (in this case a *Robertshaw* 24-volt combination gas valve), but is designed so that it can be replaced without replacing the entire assembly.

Pressure adjustment required for setting intermediate capacities can be accomplished as follows:

1. Remove the adjustment screw cap.

2. Take a screwdriver and turn the adjustment screw *counterclockwise* to reduce pressure.

3. Measure the pressure through the manometer connection adjacent to the outlet tapping.

Conversion gas burners equipped with a pilot light require inspection and servicing of the following:

1. The pilot orifice size will depend upon the type of gas used. For the conversion gas burner illustrated in Fig. 13, the pilot orifice will have a diameter of .018 inch for natural gas and .012 inch for propane.

2. Set the end of the burner nozzle at least 1 inch short of the inside of the combustion chamber (see Figs. 15, 16, 17 and 18). It must *not* be set flush with the chamber or be allowed to extend into the chamber (the pilot is often snuffed out in such cases by recirculated flue products).

3. Check for a high or low gas pressure. If over 7″ w.c., reset the pilot adjustment (see Fig. 13), to reduce the size of the pilot flame and to increase the heat on the thermocouple.

The thermocouple and pilot safety section of the main valve should be tested to determine if they are operating efficiently. The testing procedure is as follows:

1. Turn the burner switch to OFF.

2. Allow the combustion chamber to cool and the pilot light to operate for *at least* five minutes.

3. Turn off the pilot and listen for a "click" from the main valve.

4. If the "click" of the main valve is heard *more* than 30 seconds after you have shut off the pilot, then both the

thermocouple and the pilot safety section of the main valve are operating efficiently.

5. If the "click" occurs *less* than 30 seconds after you have shut off the pilot, than either the thermocouple or pilot safety section of the main valve is faulty.

6. Test the thermocouple by disconnecting it from the main valve and checking it with a millivoltmeter. Under normal conditions (i.e. when heated by a standby pilot), it should develop at least 15 millivolts.

7. If the thermocouple develops at least 15 millivolts but the main valve "click" still occurs less than 30 seconds after you have shut off the pilot, the pilot safety section is probably malfunctioning.

8. The thermocouple may also be tested while still connected to the valve (a closed circuit connection) by using an adapter to connect the millivoltmeter. Under normal operating conditions, the thermocouple should develop at least 7 millivolts and the valve 4 millivolts.

The wiring diagram for a conversion gas burner equipped with a pilot light is shown in Fig. 19. The installation and service instructions from most manufacturers will include a wiring diagram for the burner. Study it carefully before servicing or making repairs.

OIL TANKS AND OIL PIPING

Local codes and the *National Board of Fire Underwriters* will provide information for the installation of oil tanks and piping.

In oil piping, copper tubing is commonly used for suction and return lines. A ⅜ inch OD minimum is recommended for lines under 30 feet in length. If the suction and return lines exceed 30 feet, then ½ OD copper tubing is recommended.

Other suggestions for the installation of the oil storage tank and piping include:

1. Use a good quality pipe joint compound and make your pipe thread connections as tight as possible.

2. Keep the lines between the oil tank and burner as straight as possible (avoid kinks and traps).

Courtesy *Mid-Continent Metal Products Co.*

Fig. 19. Wiring diagram for a conversion gas burner equipped with a pilot light.

3. Keep the connections in the return line to a minimum.

4. Check tubing flares and connections for air leaks (an indication there is a poor seal resulting from improper meeting).

The two basic types of piping arrangements are: (1) the single line system and (2) the two-line system. These two piping arrangements are primarily distinguished from one another by the location of the oil storage tank with respect to the oil-fired conversion burner.

A *single line piping* arrangement is used only where the oil storage tank is located *above* the level of the oil burner. No return line is necessary. The oil is taken from an outlet in the bottom of the storage tank and fed by gravity to the burner. The slope of the pipe should be gradually downward (approximately ½ inch per foot) to a point directly below the burner connection. This downward slope in the line is designed to prevent the formation of air pockets and bubbles. The installation of a shut-off valve in the line is recommended.

560

A *two-line piping* arrangement is required where the oil storage tank is buried at a level *below* that of the burner.

INSTALLING A CONVERSION OIL BURNER

Conversion oil burners should be installed in accordance with the regulations of the *National Fire Protection Association* (NFPA Bulletin No. 31) and local codes and regulations. Read these regulations and the manufacturer's installation instructions before making any attempt to install the burner.

If the furnace or boiler was originally designed to burn solid fuel and the ash pit is *not* used as a part of the combustion chamber, the ash door should be removed to prevent the accumulation of vapors in the ash pit. If removal of the ash door is not feasible, then some other means of bottom ventilation must be provided.

Before installing the burner, check the condition of the boiler or furnace. It must be in good condition or repair. The flue gas passages and combustion chamber must be tight against leaks. Reseal or reset the sections of a cast-iron boiler or furnace. Replace any damaged parts. The boiler or furnace should also be as clean as possible before installing the burner.

A combustion chamber must be provided that is in accordance with the specifications of the burner manufacturer. Read the appropriate sections in this chapter (e.g. COMBUSTION CHAMBER REQUIREMENTS, etc.) for more information.

If you are satisfied that the boiler or furnace with which you are working satisfactorily meets the conditions mentioned in the previous paragraphs, then you are ready to install the conversion oil burner. The basic steps recommended for installation are as follows:

1. Position the burner on its mounting flange or pedestal so that the burner air tube is flush with the inside surface of the combustion chamber front wall. Do *not* allow the burner air tube to extend into the combustion chamber.

2. Adjust the burner tube on pedestal mounted burners so that it pitches downward about 1½ inches. Any fuel in the burner tube will then drain into the combustion chamber.

3. On flange mounted burners, the distance from the flange to the end of the air cone will depend upon the require-

ments of each installation. Remove the air cone before installing the flange.

4. The distance between the center of the nozzle and the floor of the combustion chamber must be correct. If it is too close, it will result in flame impingement and carbonization. On the other hand, a nozzle placed too high will result in excessive flame noise and poor combustion.

5. The fuel pump should be adjusted for the oil pressure recommended by the manufacturer.

6. Adjust the air supply to the burner with the air inlet band. Rotate the air inlet band to the required position (the one which will deliver the smallest amount of air and still maintain clean combustion) and secure it in position.

7. Make certain the size and spray angle of the nozzle is correct for the installation (check the U.L. rating plate).

8. Install the nozzle and check the electrode tip position in accordance with the manufacturer's instructions.

9. Prime the fuel pump (see the manufacturer's instructions).

STARTING A CONVERSION OIL BURNER

Before attempting to start the oil burner, you should first carefully read the manufacturer's instructions. Unfortunately, these are not always available on older heating installations. If this is the case and you do not possess the necessary expertise to operate the burner without an instruction manual, I would strongly recommend calling a professional for a service call.

Always test the oil lines for possible leaks and make the necessary repairs before starting the oil burner. Consult the manufacturer's instructions for bleeding the fuel pump. The bleed valve on the fuel pump is designed to facilitate air purging, cleaning, and priming. See Chapter 1, Volume 2 (OIL BURNERS) for additional information about fuel pumps.

Check the electrical wiring diagram (if one exists) to make certain that the burner is correctly wired. Problems with an oil burner can often be traced to incorrect wiring. All wiring must be in accordance with the *National Electrical Code* as well as local codes, standards, and regulations.

It is important to make certain that the primary and limit controls are operating properly. Test the primary control by first removing the motor lead from the burner and then energizing the ignition circuit by throwing the switch. *If* the primary control is operating properly, the ignition will shut off *within two minutes* after the switch is thrown. On the other hand, if the ignition fails to shut off after two minutes, the primary control should be replaced.

Check the fuel level in the oil storage tank. There must be enough oil to operate the burner. Open both oil valves (at the oil storage tank and at the boiler or furnace).

You *must* have an adequate air supply for efficient combustion. The basic ventilation requirements (see VENTILATION RE-QUIREMENTS in this chapter) must be provided for when installing the oil burner. It is also important to determine if the secondary air setting is correct. If there is smoke or soot present when the oil burner is operating, the air setting must be corrected.

If all the suggestions in the previous paragraphs have been followed and the necessary adjustments or repairs made, the oil burner can now be started. The procedure for doing this is basically as follows:

1. Make certain that all controls have been set in normal starting position.
2. Make certain that the oil valve in the oil supply line is open.
3. Set the room thermostat 10° above the room temperature.
4. Check the reset button on the primary control to make certain it has been reset.
5. Open the air control band on the burner to about half open.
6. Throw the main electrical switch to ON. The burner should now start, ignite and burn. If it does not, recheck Steps 1 through 5. If you still experience difficulty in starting the burner, consult TROUBLESHOOTING OIL BURNERS in Chapter 1, Volume 2 (OIL BURNERS).
7. Allow the oil burner to operate for about fifteen minutes, and then shut it down and give it time to cool off. After it has cooled off, restart it to be sure it operates properly on a cool start.

While the oil burner is operating (Step 7), you should take the opportunity to make a number of tests. Check the fuel pump pressure with an oil pressure gauge. These should be factory set. Adjust the air control band on the burner until 0+ smoke is obtained with a smoke tester. Adjust the draft control (if necessary) to obtain at least .02" w.c. of draft over the fire.

SERVICING A CONVERSION OIL BURNER

Protecting the oil burner motor from unnecessary wear is extremely important. *At least* once a year (but not more than three times a year), preferably at the beginning of each heating season, place about ten drops of #20 SAE oil in each oil port of the burner motor. After you have oiled the burner motor, clean the fuel strainer and change the oil filter cartridge. Clogged fuel strainers are a common source of trouble. In rotary- and gun-type oil burners, these components are usually located either at the point at which the oil line connects to the tank or at the oil line connection to the burner. Follow the manufacturer's instructions for removing the fuel strainer, and clean it with a stiff wire brush or a blast of compressed air.

Be sure that you are using the correct weight of oil. If the oil is too heavy, the burner motor will start but it may fail to establish a flame. Continued use of the wrong oil may cause damage to burner components.

Soot should be periodically removed from the burner and combustion chamber. Commercial soot removers are available for this purpose. When burning out a layer of soot, take care not to ignite a soot fire in the chimney. This could be very dangerous because it may lead to the igniting of combustionable materials near the chimney. The hard layer of carbon that frequently forms on the bottom of the fire pot in burners should also be removed.

Other servicing recommendations for conversion oil burners are as follows:

1. Check the furnace or boiler for possible air leaks. Any that are found should be sealed (an asbestos cement is recommended) because they interfere with combustion efficiency.

2. Check the burner nozzle in gun-type burners. If it is dirty, remove and clean it according to the manufacturer's instructions.

3. Check the position of the electrodes in rotary- and gun-type oil burners. They should be positioned near but out of the direct spray of the oil.

Professional and Trade Associations

Many professional and trade associations have been organized to develop and provide materials and/or services in connection with heating, ventilating, and air conditioning. These materials and services include:

1. Formulating and establishing specifications and professional standards.
2. Certifying equipment and materials.
3. Conducting product research.
4. Promoting interest in the product.

Much useful information can be obtained by contacting these associations. With that in mind, the names and addresses of the principal organizations have been included in this appendix. They are listed in alphabetical order.

AIR-CONDITIONING AND REFRIGERATION INSTITUTE (ARI)

1815 North Fort Meyer Drive
Arlington, Va. 22209.
Manufacturers of air conditioning, warm air heating, and commercial and industrial refrigeration equipment.

AIR-CONDITIONING AND REFRIGERATION WHOLESALERS (ARW)
22371 Newman Avenue,
Dearborn, Michigan 48124.
Air conditioning and refrigeration equipment wholesalers.

AIR DIFFUSION COUNCIL (ADC)
435 North Michigan Avenue,
Chicago, Ill. 60611.
Manufacturers of registers, grilles, diffusers, and related equipment.

AIR MOVING AND CONDITIONING ASSOCIATION (AMCA)
30 West University Drive,
Arlington Heights, Ill. 60004.
Manufacturers of air moving and conditioning equipment.

AIR FILTER INSTITUTE (see AIR-CONDITIONING AND REFRIGERATION INSTITUTE).

AMERICAN GAS ASSOCIATION (AGA)
1515 Wilson Blvd.,
Arlington, Va. 22209.
Residential gas operating and performance standards from distributors and transporters of natural, manufactured and mixed gas.

AMERICAN NATIONAL STANDARDS INSTITUTE (ANSI)
1430 Broadway,
New York, N.Y. 10018.
Clearinghouse for nationally coordinated safety, engineering and industrial standards.

AMERICAN SOCIETY FOR TESTING AND MATERIALS (ASTM)
1916 Race Street,
Philadelphia, Pa. 19103.
Engineering standards for materials.

AMERICAN SOCIETY OF HEATING,
REFRIGERATION AND AIR-CONDITIONING
ENGINEERS (ASHRAE)
345 East 47 Street,
New York, N.Y. 10017.
Professional association.

AMERICAN SOCIETY OF MECHANICAL
ENGINEERS (ASME)
345 West 47th Street,
New York, N.Y. 10017.
Develops safety codes and standards for various types of equipment (boiler and pressure vessel codes, etc.)

BETTER HEATING-COOLING COUNCIL (see
HYDRONICS INSTITUTE)

ELECTRICAL ENERGY ASSOCIATION (EEA)
90 Park Avenue,
New York, N.Y. 10016.
Association of electric utility companys and related organizations.

FIREPLACE ASSOCIATION OF AMERICA (FAA)
435 N. Michigan Avenue,
Chicago, Ill. 60611.
Manufacturers of modified fireplaces, components, and related equipment.

HEATING AND PIPING CONTRACTORS NATIONAL
ASSOCIATION (see MECHANICAL CONTRAC-
TORS ASSOCIATION OF AMERICA)

HOME VENTILATING INSTITUTE (HVI)
230 North Michigan Avenue,
Chicago, Ill. 60601
Develops performance standards for residential ventilating equipment.

HYDRONICS INSTITUTE
35 Russo Place,
Berkley Heights, N.J. 07922.
Manufacturers, installers, etc. of hot water and steam heating and cooling equipment. Formed by a merger of the Better Heating-Cooling Council (1956-1970) and the Institute of Boiler and Radiator Manufacturers (1915-1970).

INSTITUTE OF BOILER AND RADIATOR MANUFACTURERS (see HYDRONICS INSTITUTE)

INTERNATIONAL SOLAR ENERGY SOCIETY
National Science Center,
P. O. Box 52,
Parkville, Victoria 3052,
Australia.
Educational and research organization.

MECHANICAL CONTRACTORS ASSOCIATION OF AMERICA (MCAA)
5530 Wisconsin Avenue, NW,
Suite 750,
Washington, D.C. 20015.
Contractors of piping and related equipment used in heating, cooling, refrigeration, ventilating, and air conditioning.

NATIONAL ASSOCIATION OF PLUMBING-HEATING-COOLING CONTRACTORS
1016 20th Street, NW,
Washngton, D.C. 20005.
Local plumbing, heating, and cooling contractors association.

NATIONAL BOARD OF FIRE UNDERWRITERS (NBFU)
85 John Street,
New York, N.Y. 10036.
Safety standards.

NATIONAL BUREAU OF STANDARDS
U.S. Department of Commerce,
Washington, D.C. 20025.
Government regulatory agency.

NATIONAL ENVIRONMENTAL SYSTEMS
CONTRACTORS ASSOCIATION (NESCA)
221 North LaSalle Street,
Chicago, Ill. 60601.
Heating, air conditioning and refrigeration system contractors.

NATIONAL FIRE PROTECTION ASSOCIATION
(NFPA)
470 Atlantic Avenue,
Boston, Mass. 02110.
Fire safety standards.

NATIONAL LP-GAS ASSOCIATION (NLPGA)
79 West Monroe Street,
Chicago, Ill. 60603.
Association of LP-gas producers, distributors, and equipment
manufacturers.

NATIONAL OIL FUEL INSTITUTE (NOFI)
60 East 42nd Street,
New York, N.Y. 10017.
Oil heating equipment manufacturers, distributors, etc.

NATIONAL WARM AIR HEATING AND AIR
CONDITIONING ASSOCIATION (NWAH & ACA)
(see NATIONAL ENVIRONMENTAL SYSTEMS
CONTRACTORS ASSOCIATION)

PLUMBING-HEATING-COOLING INFORMATION
BUREAU (PHCIB)
35 East Wacker Drive,
Chicago, Ill. 60601.
Association of plumbing, heating, and cooling manufacturers,
contractors, wholesalers, etc.

REFRIGERATION AND AIR CONDITIONING
CONTRACTORS ASSOCIATION (see NATIONAL
ENVIRONMENTAL SYSTEMS CONTRACTORS
ASSOCIATION)

REFRIGERATION SERVICE ENGINEERS SOCIETY
(RSES)
2720 Des Plaines Avenue,
Des Plaines, Ill. 60018.
Association of refrigeration, air conditioning, and heating equipment installers, servicemen, and salesmen.

SHEET METAL AND AIR CONDITIONING
CONTRACTORS' NATIONAL ASSOCIATION
(SMACNA)
1611 North Kent Street,
Arlington, Va. 22209.
Sheet metal contractors who install ventilating, warm air heating and air handling equipment and systems.

STEAM HEATING EQUIPMENT MANUFACTURERS
ASSOCIATION
P. O. Box 207,
Newtonville, Mass. 02160.

STEEL BOILER INSTITUTE (defunct)

UNDERWRITERS' LABORATORIES
207 East Ohio Street,
Chicago, Ill. 60601.
Testing laboratory. Promotes safety standards for equipment.

Table 1. Equivalent length of new straight pipe for valves and fittings for turbulent flow.

Fittings												Pipe Size												
		1/4	3/8	1/2	3/4	1	1 1/4	1 1/2	2	2 1/2	3	4	5	6	8	10	12	14	16	18	20	24
Regular 90° Ell	Screwed	Steel	2.3	3.1	3.6	4.4	5.2	6.6	7.4	8.5	9.3	11	13										
		C. I.										9.0	11										
	Flanged	Steel			.92	1.2	1.6	2.1	2.4	3.1	3.6	4.4	5.9	7.3	8.9	12	14	17	18	21	23	25	30
		C. I.										3.6	4.8		7.2	9.8	12	15	17	19	22	24	28
Long Radius 90° Ell	Screwed	Steel	1.5	2.0	2.2	2.3	2.7	3.2	3.4	3.6	3.6	4.0	4.6										
		C. I.										3.3	3.7										
	Flanged	Steel			1.1	1.3	1.6	2.0	2.3	2.7	2.9	3.4	4.2	5.0	5.7	7.0	8.0	9.0	9.4	10	11	12	14
		C. I.										2.8	3.4		4.7	5.7	6.8	7.8	8.6	9.6	11	11	13
Regular 45° Ell	Screwed	Steel	.34	.52	.71	.92	1.3	1.7	2.1	2.7	3.2	4.0	5.5										
		C. I.										3.3	4.5										
	Flanged	Steel			.45	.59	.81	1.1	1.3	1.7	2.0	2.6	3.5	4.5	5.6	7.7	9.0	11	13	15	16	18	22
		C. I.										2.1	2.9		4.5	6.3	8.1	9.7	12	13	15	17	20
Tee-Line Flow	Screwed	Steel	.79	1.2	1.7	2.4	3.2	4.6	5.6	7.7	9.3	12	17										
		C. I.										9.9	14										
	Flanged	Steel			.69	.82	1.0	1.3	1.5	1.8	1.9	2.2	2.8	3.3	3.8	4.7	5.2	6.0	6.4	7.2	7.6	8.2	9.6
		C. I.										1.9	2.2		3.1	3.9	4.6	5.2	5.9	6.5	7.2	7.7	8.8

Table 1. (Continued)

Fitting	Type	Material																					
Tee-Branch Flow	Screwed	Steel	2.4	3.5	4.2	5.3	6.6	8.7	9.9	12	13	17	21								
		C.I.								14	17										
	Flanged	Steel			2.0	2.6	3.3	4.4	5.2	6.6	7.5	9.4	12	15	18	24	30	34	37	43	47	52	62
		C.I.								7.7	10	15	20	25	30	35	39	44	49	57
180° Return Bend	Screwed	Steel	2.3	3.1	3.6	4.4	5.2	6.6	7.4	8.5	9.3	11	13								
		C.I.								9.0	11										
	Reg. Flanged	Steel		.92	1.2	1.6	2.1	2.4	3.1	3.6	4.4	5.9	7.3	8.9	12	14	17	18	21	23	25	30
		C.I.								3.6	4.8		7.2	9.8	12	15	17	19	22	24	28
	Long Rad. Flanged	Steel		1.1	1.3	1.6	2.0	2.3	2.7	2.9	3.4	4.2	5.0	5.7	7.0	8.0	9.0	9.4	10	11	12	14
		C.I.								2.8	3.4		4.7	5.7	6.8	7.8	8.6	9.6	11	11	13
Globe Valve	Screwed	Steel	21	22	22	24	29	37	42	54	62	79	110								
		C.I.								65	86										
	Flanged	Steel		38	40	45	54	59	70	77	94	120	150	190	260	310	390					
		C.I.								77	99		150	210	270	330						
Gate Valve	Screwed	Steel	.32	.45	.56	.67	.84	1.1	1.2	1.5	1.7	1.9	2.5								
		C.I.								1.6	2.0										
	Flanged	Steel						2.6	2.7	2.8	2.9	3.1	3.2	3.2	3.2	3.2	3.2	3.2	3.2	3.2	3.2	3.2
		C.I.								2.3	2.4		2.6	2.7	2.8	2.9	3.0	3.0	3.0	3.0	3.0
	Screwed	Steel	12.8	15	15	15	17	18	18	18	18	18	18								
		C.I.								15	15										

Table 1. (Continued)

Fitting	Material	1	2	3	4	5	6	7	8	9	10	11	12	13	14	15	16	17	18	19	20	21
Angle Valve — Flanged	Steel		15	15	17	18	18	21	22	22	28	38	50	63	90	120	140	160	190	210	240	300
	C. I.										23	31		52	74	98	120	150	170	200	230	280
Angle Valve — Screwed	Steel	7.2	7.3	8.0	8.8	11	13	15	19	22	27	38										
	C. I.										22	31										
Swing Check Valve — Flanged	Steel			3.8	5.3	7.2	10	12	17	21	27	38	50	63	90	120	140					
	C. I.										22	31		52	74	98	120					
Coupling Or Union — Screwed	Steel	.14	.18	.21	.24	.29	.36	.39	.45	.47	.53	.65										
	C. I.										.44	.62										
Bell Mouth Inlet	Steel	.04	.07	.10	.13	.18	.26	.31	.43	.52	.67	.95	1.3	1.6	2.3	2.9	3.5	4.0	4.7	5.3	6.1	7.6
	C. I.										.55	.77		1.3	1.9	2.4	3.0	3.6	4.3	5.0	5.7	7.0
Square Mouth Inlet	Steel	.44	.68	.96	1.3	1.8	2.6	3.1	4.3	5.2	6.7	9.5	13	16	23	29	35	40	47	53	61	76
	C. I.										5.5	7.7		13	19	24	30	36	43	50	57	70
Reentrant Pipe	Steel	.88	1.4	1.9	2.6	3.6	5.1	6.2	8.5	10	13	19	25	32	45	58	70	80	95	110	120	150
	C. I.										11	15		26	37	49	61	73	86	100	110	140
Y-Strainer		4.6	5.0	6.6	7.7	18	20	27	29	34	42	53	61									

Sudden Enlargement

$$h = \frac{(V_1 - V_2)^2}{2g} \text{ Feet of Liquid; IF } V_2 = 0 \quad h = \frac{V^2}{2g} \text{ Feet of Liquid}$$

Courtesy The Hydraulic Institute (reprinted from the Standards of The Hydraulic Institute, Eleventh Edition. Copyright 1965).

Table 2. Schedule 80 pipe dimensions.

Size Inches	Diameters External Inches	Diameters Internal Inches	Nominal Thickness Inches	Transverse Areas External Sq. Ins.	Transverse Areas Internal Sq. Ins.	Transverse Areas Metal Sq. Ins.	Length of Pipe Per Sq. Foot of External Surface Feet	Length of Pipe Per Sq. Foot of Internal Surface Feet	Cubic Feet Per Foot of Pipe	Weight Per Foot Pounds	Number Threads Per Inch of Screw
1/8	.405	.215	.095	.129	.036	.093	9.431	17.750	.00025	.314	27
1/4	.540	.302	.119	.229	.072	.157	7.073	12.650	.00050	.535	18
3/8	.675	.423	.126	.358	.141	.217	5.658	9.030	.00098	.738	18
1/2	.840	.546	.147	.554	.234	.320	4.547	7.000	.00163	1.00	14
3/4	1.050	.742	.154	.866	.433	.433	3.637	5.15	.00300	1.47	14
1	1.315	.957	.179	1.358	.719	.639	2.904	3.995	.00500	2.17	11 1/2
1 1/4	1.660	1.278	.191	2.164	1.283	.881	2.301	2.990	.00891	3.00	11 1/2
1 1/2	1.900	1.500	.200	2.835	1.767	1.068	2.010	2.542	.01227	3.65	11 1/2
2	2.375	1.939	.218	4.430	2.953	1.477	1.608	1.970	.02051	5.02	11 1/2
2 1/2	2.875	2.323	.276	6.492	4.238	2.254	1.328	1.645	.02943	7.66	8
3	3.500	2.900	.300	9.621	6.605	3.016	1.091	1.317	.04587	10.3	8
3 1/2	4.000	3.364	.318	12.56	8.888	3.678	.954	1.135	.06172	12.5	8
4	4.500	3.826	.337	15.90	11.497	4.407	.848	.995	.0798	14.9	8
5	5.563	4.813	.375	24.30	18.194	6.112	.686	.792	.1263	20.8	8

Table 2. Schedule 80 pipe dimensions. (Cont'd.)

6	6.625	5.761	.432	34.47	26.067	8.300	.576	.673	.1810	28.6	8
8	8.625	7.625	.500	58.42	46.663	12.76	.442	.501	.3171	43.4	8
10	10.750	9.564	.593	90.76	71.84	18.92	.355	.400	.4989	64.4	8
12	12.750	11.376	.687	127.64	101.64	26.00	.299	.336	.7058	88.6	
14	14.000	12.500	.750	153.94	122.72	31.22	.272	.306	.8522	107.0	
16	16.000	14.314	.843	201.05	160.92	40.13	.238	.263	1.112	137.0	
18	18.000	16.126	.937	254.85	204.24	50.61	.212	.237	1.418	171.0	
20	20.000	17.938	1.031	314.15	252.72	61.43	.191	.208	1.755	209.0	
24	24.000	21.564	1.218	452.40	365.22	87.18	.159	.177	2.536	297.0	

Table 3. Schedule 40 pipe dimensions

Size Inches	Diameters External Inches	Diameters Internal Inches	Nominal Thickness Inches	Transverse Areas External Sq. Ins.	Transverse Areas Internal Sq. Ins.	Transverse Areas Metal Sq. Ins.	Length of Pipe Per Sq. Foot of External Surface Feet	Length of Pipe Per Sq. Foot of Internal Surface Feet	Cubic Feet Per Foot of Pipe	Weight Per Foot Pounds	Number Threads Per Inch of Screw
1/8	.405	.269	.068	.129	.057	.072	9.431	14.199	.00039	.244	27
1/4	.540	.364	.088	.229	.104	.125	7.073	10.493	.00072	.424	18
3/8	.675	.493	.091	.358	.191	.167	5.658	7.747	.00133	.567	18
1/2	.840	.622	.109	.554	.304	.250	4.547	6.141	.00211	.850	14
3/4	1.050	.824	.113	.866	.533	.333	3.637	4.635	.00370	1.130	14
1	1.315	1.049	.133	1.358	.864	.494	2.904	3.641	.00600	1.678	11½
1¼	1.660	1.380	.140	2.164	1.495	.669	2.301	2.767	.01039	2.272	11½
1½	1.900	1.610	.145	2.835	2.036	.799	2.010	2.372	.01414	2.717	11½
2	2.375	2.067	.154	4.430	3.355	1.075	1.608	1.847	.02330	3.652	11½
2½	2.875	2.469	.203	6.492	4.788	1.704	1.328	1.547	.03325	5.793	8
3	3.500	3.068	.216	9.621	7.393	2.228	1.091	1.245	.05134	7.575	8
3½	4.000	3.548	.226	12.56	9.886	2.680	.954	1.076	.06866	9.109	8
4	4.500	4.026	.237	15.90	12.73	3.174	.848	.948	.08840	10.790	8
5	5.563	5.047	.258	24.30	20.00	4.300	.686	.756	.1389	14.61	8

Table 3. Schedule 40 pipe dimensions. (Cont'd.)

6	6.625	6.065	.280	34.47	28.90	5.581	.576	.629	.2006	18.97	8
8	8.625	7.981	.322	58.42	50.02	8.399	.442	.478	.3552	28.55	8
10	10.750	10.020	.365	90.76	78.85	11.90	.355	.381	.5476	40.48	8
12	12.750	11.938	.406	127.64	111.9	15.74	.299	.318	.7763	53.6	
14	14.000	13.125	.437	153.94	135.3	18.64	.272	.280	.9354	63.0	
16	16.000	15.000	.500	201.05	176.7	24.35	.238	.254	1.223	78.0	
18	18.000	16.874	.563	254.85	224.0	30.85	.212	.226	1.555	105.0	
20	20.000	18.814	.593	314.15	278.0	36.15	.191	.203	1.926	123.0	
24	24.000	22.626	.687	452.40	402.1	50.30	.159	.169	2.793	171.0	

Table 4. Properties of saturated steam.

Gauge Pressure PSIG	Temperature °F	Heat in Btu/lb. Sensible	Latent	Total	Specific Volume Cu. ft. per lb.	Gauge Pressure PSIG	Temperature °F	Heat in Btu/lb. Sensible	Latent	Total	Specific Volume Cu. ft. per lb.
25 —┐	134	102	1017	1119	142	150	366	339	857	1196	2.74
20 In Vac.	162	129	1001	1130	73.9	155	368	341	855	1196	2.68
15 ─┘	179	147	990	1137	51.3	160	371	344	853	1197	2.60
10	192	160	982	1142	39.4	165	373	346	851	1197	2.54
5	203	171	976	1147	31.8	170	375	348	849	1197	2.47
0	212	180	970	1150	26.8	175	377	351	847	1198	2.41
1	215	183	968	1151	25.2	180	380	353	845	1198	2.34
2	219	187	966	1153	23.5	185	382	355	843	1198	2.29
3	222	190	964	1154	22.3	190	384	358	841	1199	2.24
4	224	192	962	1154	21.4	195	386	360	839	1199	2.19
5	227	195	960	1155	20.1	200	388	362	837	1199	2.14
6	230	198	959	1157	19.4	205	390	364	836	1200	2.09
7	232	200	957	1157	18.7	210	392	366	834	1200	2.05
8	233	201	956	1157	18.4	215	394	368	832	1200	2.00
9	237	205	954	1159	17.1	220	396	370	830	1200	1.96
10	239	207	953	1160	16.5	225	397	372	828	1200	1.92
12	244	212	949	1161	15.3	230	399	374	827	1201	1.89
14	248	216	947	1163	14.3	235	401	376	825	1201	1.85
16	252	220	944	1164	13.4	240	403	378	823	1201	1.81
18	256	224	941	1165	12.6	245	404	380	822	1202	1.78

Table 4. Properties of saturated steam. (Cont'd.)

20	259	227	939	1166	11.9	250	406	382	820	1202	1.75
22	262	230	937	1167	11.3	255	408	383	819	1202	1.72
24	265	233	934	1167	10.8	260	409	385	817	1202	1.69
26	268	236	933	1169	10.3	265	411	387	815	1202	1.66
28	271	239	930	1169	9.85	270	413	389	814	1203	1.63
30	274	243	929	1172	9.46	275	414	391	812	1203	1.60
32	277	246	927	1173	9.10	280	416	392	811	1203	1.57
34	279	248	925	1173	8.75	285	417	394	809	1203	1.55
36	282	251	923	1174	8.42	290	418	395	808	1203	1.53
38	284	253	922	1175	8.08	295	420	397	806	1203	1.49
40	286	256	920	1176	7.82	300	421	398	805	1203	1.47
42	289	258	918	1176	7.57	305	423	400	803	1203	1.45
44	291	260	917	1177	7.31	310	425	402	802	1204	1.43
46	293	262	915	1177	7.14	315	426	404	800	1204	1.41
48	295	264	914	1178	6.94	320	427	405	799	1204	1.38
50	298	267	912	1179	6.68	325	429	407	797	1204	1.36
55	300	271	909	1180	6.27	330	430	408	796	1204	1.34
60	307	277	906	1183	5.84	335	432	410	794	1204	1.33
65	312	282	901	1183	5.49	340	433	411	793	1204	1.31
70	316	286	898	1184	5.18	345	434	413	791	1204	1.29

APPENDIX B

Table 4. Properties of saturated steam. (Cont'd.)

75	320	290	895	1185	4.91	350	435	414	790	1204	1.28
80	324	294	891	1185	4.67	355	437	416	789	1205	1.26
85	328	298	889	1187	4.44	360	438	417	788	1205	1.24
90	331	302	886	1188	4.24	365	440	419	786	1205	1.22
95	335	305	883	1188	4.05	370	441	420	785	1205	1.20
100	338	309	880	1189	3.89	375	442	421	784	1205	1.19
105	341	312	878	1190	3.74	380	443	422	783	1205	1.18
110	344	316	875	1191	3.59	385	445	424	781	1205	1.16
115	347	319	873	1192	3.46	390	446	425	780	1205	1.14
120	350	322	871	1193	3.34	395	447	427	778	1205	1.13
125	353	325	868	1193	3.23	400	448	428	777	1205	1.12
130	356	328	866	1194	3.12	450	460	439	766	1205	1.00
140	361	333	861	1194	2.92	500	470	453	751	1204	.89
145	363	336	859	1195	2.84	550	479	464	740	1204	.82
						600	489	475	728	1203	.74

Courtesy Sarco Company, Inc.

Table 5. Friction loss for water in feet per 100 feet of Schedule 40 teel pipe.

U.S. Gal./Min.	Vel. Ft./Sec.	hf Friction	U.S. Gal./Min.	Vel. Ft./Sec.	hf Friction
	⅜" PIPE			½" PIPE	
1.4	2.25	9.03	2	2.11	5.50
1.6	2.68	11.6	2.5	2.64	8.24
1.8	3.02	14.3	3	3.17	11.5
2.0	3.36	17.3	3.5	3.70	15.3
2.5	4.20	26.0	4.0	4.22	19.7
3.0	5.04	36.6	5	5.28	29.7
3.5	5.88	49.0	6	6.34	42.0
4.0	6.72	63.2	7	7.39	56.0
5.0	8.40	96.1	8	8.45	72.1
6	10.08	136	9	9.50	90.1
7	11.8	182	10	10.56	110.6
8	13.4	236	12	12.7	156
9	15.1	297	14	14.8	211
10	16.8	364	16	16.9	270
	¾" PIPE			1" PIPE	
4.0	2.41	4.85	6	2.23	3.16
5	3.01	7.27	8	2.97	5.20
6	3.61	10.2	10	3.71	7.90
7	4.21	13.6	12	4.45	11.1
8	4.81	17.3	14	5.20	14.7
9	5.42	21.6	16	5.94	19.0
10	6.02	26.5	18	6.68	23.7
12	7.22	37.5	20	7.42	28.9
14	8.42	50.0	22	8.17	34.8
16	9.63	64.8	24	8.91	41.0
18	10.8	80.9	26	9.65	47.8
20	12.0	99.0	28	10.39	55.1
22	13.2	120	30	11.1	62.9
24	14.4	141	35	13.0	84.4
26	15.6	165	40	14.8	109
28	16.8	189	45	16.7	137
			50	18.6	168
	1¼" PIPE			1½" PIPE	
12	2.57	2.85	16	2.52	2.26
14	3.00	3.77	18	2.84	2.79
16	3.43	4.83	20	3.15	3.38
18	3.86	6.00	22	3.47	4.05
20	4.29	7.30	24	3.78	4.76

U.S. Gal./Min.	Vel. Ft./Sec.	hf Friction	U.S. Gal./Min.	Vel. Ft./Sec.	hf Friction
22	4.72	8.72	26	4.10	5.54
24	5.15	10.27	28	4.41	6.34
26	5.58	11.94	30	4.73	7.20
28	6.01	13.7	35	5.51	9.63
30	6.44	15.6	40	6.30	12.41
35	7.51	21.9	45	7.04	15.49
40	8.58	27.1	50	7.88	18.9
45	9.65	33.8	55	8.67	22.7
50	10.7	41.4	60	9.46	26.7
55	11.8	49.7	65	10.24	31.2
60	12.9	58.6	70	11.03	36.0
65	13.9	68.6	75	11.8	41.2
70	15.0	79.2	80	12.6	46.6
75	16.1	90.6	85	13.4	52.4
			90	14.2	58.7
			95	15.0	65.0
			100	15.8	71.6

2″ PIPE			2½″ PIPE		
25	2.39	1.48	35	2.35	1.15
30	2.87	2.10	40	2.68	1.47
35	3.35	2.79	45	3.02	1.84
40	3.82	3.57	50	3.35	2.23
45	4.30	4.40	60	4.02	3.13
50	4.78	5.37	70	4.69	4.18
60	5.74	7.58	80	5.36	5.36
70	6.69	10.2	90	6.03	6.69
80	7.65	13.1	100	6.70	8.18
90	8.60	16.3	120	8.04	11.5
100	4.34	2.72	200	5.04	2.61
120	11.5	28.5	160	10.7	20.0
140	13.4	38.2	180	12.1	25.2
160	15.3	49.5	200	13.4	30.7
			220	14.7	37.1
			240	16.1	43.8

3″ PIPE			4″ PIPE		
50	2.17	.762	100	2.52	.718
60	2.60	1.06	120	3.02	1.01
70	3.04	1.40	140	3.53	1.35
80	3.47	1.81	160	4.03	1.71
90	3.91	2.26	180	4.54	2.14
100	3.34	2.75	200	5.04	2.61
120	5.21	3.88	220	5.54	3.13

U.S. Gal./Min.	Vel. Ft./Sec.	hf Friction	U.S. Gal./Min.	Vel. Ft./Sec.	hf Friction
140	6.08	5.19	240	6.05	3.70
160	6.94	6.68	260	6.55	4.30
180	7.81	8.38	280	7.06	4.95
200	8.68	10.2	300	7.56	5.63
220	9.55	12.3	350	8.82	7.54
240	10.4	14.5	400	10.10	9.75
260	11.3	16.9	450	11.4	12.3
280	12.2	19.5	500	12.6	14.4
300	13.0	22.1	550	13.9	18.1
350	15.2	30	600	15.1	21.4
	5″ PIPE			6″ PIPE	
160	2.57	.557	220	2.44	.411
180	2.89	.698	240	2.66	.482
200	3.21	.847	260	2.89	.560
220	3.53	1.01	300	3.33	.733
240	3.85	1.19	350	3.89	.980
260	4.17	1.38	400	4.44	1.25
300	4.81	1.82	450	5.00	1.56
350	5.61	2.43	500	5.55	1.91
400	6.41	3.13	600	6.66	2.69
450	7.22	3.92	700	7.77	3.60
500	8.02	4.79	800	8.88	4.64
600	9.62	6.77	900	9.99	5.81
700	11.2	9.13	1000	11.1	7.10
800	12.8	11.8	1100	12.2	8.52
900	14.4	14.8	1200	13.3	10.1
1000	16.0	18.2	1300	14.4	11.7
			1400	15.5	13.6

Courtesy Sarco Company, Inc.

Table 6. Flow of water through Schedule 40 steel pipe.

Pressure Drop per 1000 Feet of Schedue 40 Steel Pipe, in Pounds per Square Inch.

Discharge Gals. per Min.	1" Vel. Ft per Sec.	1" Pres. sure Drop	1¼" Vel. Ft per Sec.	1¼" Pres. sure Drop	1½" Vel. Ft per Sec.	1½" Pres. sure Drop	2" Vel. Ft per Sec.	2" Pres. sure Drop	2½" Vel. Ft per Sec.	2½" Pres. sure Drop	3" Vel. Ft per Sec.	3" Pres. sure Drop	3½" Vel. Ft per Sec.	3½" Pres. sure Drop	4" Vel. Ft per Sec.	4" Pres. sure Drop	5" Vel. Ft per Sec.	5" Pres. sure Drop	6" Vel. Ft per Sec.	6" Pres. sure Drop	8" Vel. Ft per Sec.	8" Pres. sure Drop
1	.37	0.49																				
2	.74	1.70	0.43	0.45																		
3	1.12	3.53	0.64	0.94	0.47	0.44																
4	1.49	5.94	0.86	1.55	0.63	0.74																
5	1.86	9.02	1.07	2.36	0.79	1.12																
6	2.24	12.25	1.28	3.30	0.95	1.53	.57	0.46														
8	2.98	21.1	1.72	5.52	1.20	2.63	.76	0.75														
10	3.72	30.8	2.14	8.34	1.57	3.86	.96	1.14	.67	0.48												
15	5.60	64.6	3.21	17.6	2.36	8.13	1.43	2.33	1.00	0.99												
20	7.44	110.5	4.29	29.1	3.15	13.5	1.91	3.86	1.34	1.64	.87	0.59										
25			5.36	43.7	3.94	20.2	2.39	5.81	1.68	2.48	1.08	0.67	.81	0.42								
30			6.43	62.9	4.72	29.1	2.87	8.04	2.01	3.43	1.30	1.21	.97	0.60								
35			7.51	82.5	5.51	38.2	3.35	10.95	2.35	4.49	1.52	1.58	1.14	0.79	.88	0.42						
40					6.30	47.8	3.82	13.7	2.68	5.88	1.74	2.06	1.30	1.00	1.01	0.53						
45					7.08	60.6	4.30	17.4	3.00	7.14	1.95	2.51	1.46	1.21	1.13	0.67						
50					7.87	74.7	4.78	20.6	3.35	8.82	2.17	3.10	1.62	1.44	1.26	0.80						
60							5.74	29.6	4.02	12.2	2.60	4.29	1.95	2.07	1.51	1.10						
70							6.69	38.6	4.69	15.3	3.04	5.84	2.27	2.71	1.76	1.50	1.12	0.48				
80							7.65	50.3	5.37	21.7	3.48	7.62	2.59	3.53	2.01	1.87	1.28	0.63				
90							8.60	63.6	6.04	26.1	3.91	9.22	2.92	4.46	2.26	2.37						
100							9.56	75.1	6.71	32.3	4.34	11.4	3.24	5.27	2.52	2.81	1.44	0.80	1.11	0.39		
125									8.38	48.2	5.45	17.1	4.05	7.86	3.15	4.38	1.60	0.95	1.39	0.56		
150									10.06	60.4	6.51	23.3	4.86	11.3	3.78	6.02	2.00	1.48	1.67	0.78		
175									11.73	90.0	7.59	32.0	5.67	14.7	4.41	8.20	2.41	2.04	1.94	1.06		
200											8.68	39.7	6.48	19.2	5.04	10.2	3.21	3.46	2.22	1.32		
225											9.77	50.2	7.29	23.1	5.67	12.9	3.61	4.37	2.50	1.66	1.44	0.44
250											10.85	61.9	8.10	28.5	6.30	15.9	4.01	5.14	2.78	2.05	1.60	0.55
275											11.94	75.0	8.91	34.4	6.93	18.3	4.41	6.22	3.06	2.36	1.76	0.63
300											13.02	84.7	9.72	40.9	7.56	21.8	4.81	7.41	3.33	2.80	1.92	0.75

APPENDIX B

Steam pipe capacity table — velocity and pressure drop per 100 ft for each pipe size. (Values read from the original rotated table; each pipe size lists velocity (Vel.) and pressure drop (Drop).)

Pipe sizes 5″ – 12″

Flow (lb/hr)	5″ Vel.	5″ Drop	6″ Vel.	6″ Drop	8″ Vel.	8″ Drop	10″ Vel.	10″ Drop	12″ Vel.	12″ Drop
325	5.21	8.25	3.61	3.29	2.08	0.88				
350	5.61	9.57	3.89	3.62	2.24	0.97				
375	6.01	11.0	4.16	4.16	2.40	1.11				
400	6.41	12.9	4.44	4.72	2.56	1.27				
425	6.82	14.1	4.72	5.34	2.72	1.43				
450	7.22	15.0	5.00	5.96	2.88	1.60				
475	7.62	16.7	5.27	6.66	3.04	1.69	1.93	0.30		
500	8.02	18.5	5.55	7.39	3.20	1.87	2.04	0.63		
550	8.82	22.4	6.11	8.94	3.53	2.26	2.24	0.70		
600	9.62	26.7	6.66	10.6	3.85	2.70	2.44	0.86		
650	10.42	31.3	7.21	11.8	4.17	3.16	2.65	1.01		
700	11.22	36.3	7.77	13.7	4.49	3.69	2.85	1.18	2.01	0.48
750	12.02	41.6	8.32	15.7	4.81	4.21	3.05	1.35	2.15	0.55
800	12.82	44.7	8.88	17.8	5.13	4.79	3.26	1.54	2.29	0.62
850	13.62	50.5	9.44	20.2	5.45	5.11	3.46	1.74	2.44	0.70
900	14.42	56.6	10.00	22.6	5.77	5.73	3.66	1.94	2.58	0.79
950	15.22	63.1	10.55	23.7	6.09	6.38	3.87	2.23	2.72	0.88
1,000	16.02	70.0	11.10	26.3	6.41	7.08	4.07	2.40	2.87	0.98
1,100	17.63	84.6	12.22	31.8	7.05	8.56	4.48	2.74	3.16	1.18
1,200			13.32	37.8	7.69	10.2	4.88	3.27	3.45	1.40
1,300			14.43	44.4	8.33	11.3	5.29	3.86	3.73	1.56
1,400			15.54	51.5	8.97	13.0	5.70	4.44	4.02	1.80
1,500			16.65	55.5	9.62	15.0	6.10	5.11	4.30	2.07
1,600			17.76	63.1	10.26	17.0	6.51	5.46	4.59	2.36
1,800			19.98	79.8	11.54	21.6	7.32	6.91	5.16	2.98
2,000			22.20	98.5	12.83	25.0	8.13	8.54	5.73	3.47
2,500					16.03	39.0	10.18	12.5	7.17	5.41
3,000					19.24	52.4	12.21	18.0	8.60	7.31
3,500					22.43	71.4	14.25	22.9	10.03	9.95
4,000					25.65	93.3	16.28	29.9	11.48	13.0
4,500							18.31	37.8	12.90	15.4
5,000							20.35	46.7	14.34	18.9
6,000							24.42	67.2	17.21	27.3
7,000							28.50	85.1	20.08	37.2
8,000									22.95	45.1
9,000									25.80	57.0
10,000									28.63	70.4
12,000									34.38	93.6

Pipe sizes 14″ – 24″

Flow (lb/hr)	14″ Vel.	14″ Drop	16″ Vel.	16″ Drop	18″ Vel.	18″ Drop	20″ Vel.	20″ Drop	24″ Vel.	24″ Drop
900	2.02	0.43								
950	2.14	0.48								
1,000	2.25	0.53								
1,100	2.38	0.59								
1,200	2.61	0.68	2.18	0.40						
1,300	2.85	0.81	2.36	0.47						
1,400	3.09	0.95	2.54	0.54						
1,500	3.32	1.10	2.73	0.62						
1,600	3.55	1.19	2.91	0.71						
1,800	3.80	1.35	3.27	0.85	2.58	0.48				
2,000	4.27	1.71	3.63	1.05	2.88	0.56				
2,500	4.74	2.11	4.54	1.63	3.59	0.88				
3,000	5.92	3.09	5.45	2.21	4.31	1.27	3.45	0.73		
3,500	7.12	4.45	6.35	3.00	5.03	1.52	4.03	0.94		
4,000	8.32	6.18	7.25	3.92	5.74	2.12	4.61	1.22	3.19	0.51
4,500	9.49	7.92	8.17	4.97	6.47	2.50	5.19	1.55	3.59	0.60
5,000	10.67	9.36	9.08	5.72	7.17	3.08	5.76	1.78	3.99	0.74
6,000	11.84	11.6	10.88	8.24	8.62	4.45	6.92	2.57	4.80	1.00
7,000	14.32	15.4	12.69	12.2	10.04	6.06	8.06		5.68	1.36
8,000	16.60	21.0	14.52	13.6	11.48	7.34	9.23	4.57	6.38	1.78
9,000	18.98	27.4	16.32	17.2	12.92	9.20	10.37	5.36	7.19	2.25
10,000	21.35	34.7	18.16	21.2	14.37	11.5	11.53	6.63	7.96	2.78
12,000	23.75	42.9	21.80	30.9	17.23	16.5	13.83	9.54	9.57	3.71
14,000	28.50	61.8	25.42	41.6	20.10	20.7	16.14	12.0	11.18	5.05
16,000	33.20	84.0	29.05	54.4	22.96	27.1	18.43	15.7	12.77	6.60

Courtesy Sarco Company, Inc.

Table 7. Warm-up load in pounds of steam per 100 feet of steam main (ambient temperature 70°F).

Steam Pressure (PSIG)	2″	2½″	3″	4″	5″	6″	8″	10″	12″	14″	16″	18″	20″	24″	0°F. Correction Factor†
0	6.2	9.7	12.8	18.2	24.6	31.9	48	68	90	107	140	176	207	208	1.50
5	6.9	11.0	14.4	20.4	27.7	35.9	48	77	101	120	157	198	233	324	1.44
10	7.5	11.8	15.5	22.0	29.9	38.8	58	83	109	130	169	213	251	350	1.41
20	8.4	13.4	17.5	24.9	33.8	43.9	66	93	124	146	191	241	284	396	1.37
40	3.9	15.8	20.6	29.3	39.7	51.6	78	110	145	172	225	284	334	465	1.32
60	11.0	17.5	22.9	32.6	44.2	57.3	86	122	162	192	250	316	372	518	1.29
80	12.0	19.0	24.9	35.3	47.9	62.1	93	132	175	208	271	342	403	561	1.27
100	12.8	20.3	26.6	37.8	51.2	66.5	100	142	188	222	290	366	431	600	1.26
125	13.7	21.7	28.4	40.4	54.8	71.1	107	152	200	238	310	391	461	642	1.25
150	14.5	23.0	30.0	42.8	58.0	75.2	113	160	212	251	328	414	487	679	1.24
175	15.3	24.2	31.7	45.1	61.2	79.4	119	169	224	265	347	437	514	716	1.23
200	16.0	25.3	33.1	47.1	63.8	82.8	125	177	234	277	362	456	537	748	1.22
250	17.2	27.3	35.8	50.8	68.9	89.4	134	191	252	299	390	492	579	807	1.21
300	25.0	38.3	51.3	74.8	104.0	142.7	217	322	443	531	682	854	1045	1182	1.20
400	27.8	42.6	57.1	83.2	115.7	158.7	241	358	493	590	759	971	1163	1650	1.18
500	30.2	46.3	62.1	90.5	125.7	172.6	262	389	535	642	825	1033	1263	1793	1.17
600	32.7	50.1	67.1	97.9	136.0	186.6	284	421	579	694	893	1118	1367	1939	1.16

MAIN SIZE

Loads based on Schedule 40 pipe for pressures up to and including 250 PSIG and on Schedule 80 pipe for pressures above 250 PSIG.
†For outdoor temperature of 0°F, multiply load value in table for each main size by correction factor corresponding to steam pressure.

Courtesy Sarco Company, Inc.

587

Table 8. Condensation load in pounds per hour per 100 feet of insulated steam main (ambient temperature 70°F; insulation 80% efficient))

Steam Pressure (PSIG)	MAIN SIZE														0°F. Correction Factor†
	2"	2½"	3"	4"	5"	6"	8"	10"	12"	14"	16"	18"	20"	24"	
10	6	7	9	11	13	16	20	24	29	32	36	39	44	53	1.58
30	8	9	11	14	17	20	26	32	38	42	48	51	57	68	1.50
60	10	12	14	18	24	27	33	41	49	54	62	67	74	89	1.45
100	12	15	18	22	28	33	41	51	61	67	77	83	93	111	1.41
125	13	16	20	24	30	36	45	56	66	73	84	90	101	121	1.39
175	16	19	23	26	33	38	53	66	78	86	98	107	119	142	1.38
250	18	22	27	34	42	50	62	77	92	101	116	126	140	168	1.36
300	20	25	30	37	46	54	68	85	101	111	126	138	154	184	1.35
400	23	28	34	43	53	63	80	99	118	130	148	162	180	216	1.33
500	27	33	39	49	61	73	91	114	135	148	170	185	206	246	1.32
600	30	37	44	55	68	82	103	128	152	167	191	208	232	277	1.31

Chart loads represent losses due to radiation and convection for saturated steam.

†For outdoor temperature of 0°F, multiply load value in table foreach main size by correction factor corresponding to steam pressure.

Courtesy Sarco Company, Inc.

Table 9. Flange standards. All dimensions are in inches.

125 lb. CAST IRON — ASA B16.1

Pipe Size	½	¾	1	1¼	1½	2	2½	3	3½	4	5	6	8	10	12
Diameter of Flange			4¼	4⅝	5	6	7	7½	8½	9	10	11	13½	16	19
Thickness of Flange (min)[1]			7⁄16	½	9⁄16	⅝	11⁄16	¾	13⁄16	15⁄16	15⁄16	1	1⅛	1 3⁄16	1¼
Diameter of Bolt Circle			3⅛	3½	3⅞	4¾	5½	6	7	7½	8½	9½	11¾	14¼	17
Number of Bolts			4	4	4	4	4	4	8	8	8	8	8	12	12
Diameter of Bolts			½	½	½	⅝	⅝	⅝	⅝	⅝	¾	¾	¾	⅞	⅞

[1] 125 lb. flanges have plain faces.

250 lb CAST IRON — ASA B16.2

Pipe Size	½	¾	1	1¼	1½	2	2½	3	3½	4	5	6	8	10	12
Diameter of Flange			4⅞	5¼	6⅛	6½	7½	8¼	9	10	11	12½	15	17½	20½
Thickness of Flange (min)[2]			11⁄16	¾	13⁄16	⅞	1	1⅛	1 3⁄16	1¼	1⅜	1 7⁄16	1⅝	1⅞	2
Diameter of Raised Face			2 11⁄16	3 3⁄16	3⅞	4 3⁄16	4 15⁄16	5⅝	6⅝	6⅞	7¼	7⅞	9 11⁄16	11 11⁄16	14⅝	16⅞
Diameter of Bolt Circle			3½	3⅞	4½	5	5⅞	6⅝	7¼	7⅞	9¼	10⅝	13	15¼	17¾
Number of Bolts			4	4	4	8	8	8	8	8	8	12	12	16	16
Diameter of Bolts			⅝	⅝	⅝	¾	¾	¾	¾	¾	¾	¾	⅞	1	1⅛

[2] 250 lb. flanges have a 1⁄16" raised face which is included in the flange thickness dimensions.

150 lb. BRONZE — ASA B16.24

Pipe Size	½	¾	1	1¼	1½	2	2½	3	3½	4	5	6	8	10	12
Diameter of Flange	3½	3⅞	4¼	4⅝	5	6	7	7½	8½	9	10	11	13½	16	19
Thickness of Flange (min)[3]	5⁄16	11⁄32	⅜	13⁄32	7⁄16	½	9⁄16	⅝	11⁄16	¾	13⁄16	15⁄16	1	1 1⁄16	1 3⁄16
Diameter of Bolt Circle	2⅜	2¾	3⅛	3½	3⅞	4¾	5½	6	7	7½	8½	9½	11¾	14¼	17

(150 lb. bronze, continued)

Number of Bolts	4	4	4	4	4	4	4	4	8	8	8	8	12	12
Diameter of Bolts	½	½	½	½	½	⅝	⅝	⅝	⅝	¾	¾	¾	⅞	⅞

[3] 150 lb. bronze flanges have plain faces with two concentric gasket-retaining grooves between the port and the bolt holes.

300 lb. BRONZE — ASA B16.24

Pipe Size	½	¾	1	1¼	1½	2	2½	3	3½	4	5	6	8	10	12
Diameter of Flange	3¾	4⅝	4⅞	5¼	6⅛	6½	7½	8¼	9	10	11	12½	15	16	19
Thickness of Flange (min)[4]	½	17/32	19/32	⅝	11/16	¾	13/16	29/32	31/32	1 1/16	1⅛	1 3/16	1⅜	1 13/16	2 1/16
Diameter of Bolt Circle	2⅝	3¼	3½	3⅞	4½	5	5⅞	6⅝	7¼	7⅞	9¼	10⅝	13	14¼	17
Number of Bolts	4	4	4	4	4	8	8	8	8	8	8	8	12	12	12
Diameter of Bolts	½	⅝	⅝	⅝	¾	¾	¾	¾	¾	¾	¾	¾	⅞	⅞	⅞

[4] 300 lb. bronze flanges have plain faces with two concentric gasket-retaining grooves between the port and the bolt holes.

150 lb. STEEL — ASA B16.5

Pipe Size	½	¾	1	1¼	1½	2	2½	3	3½	4	5	6	8	10	12
Thickness of Flange (min)[5]			7/16	½	9/16	⅝	11/16	¾	13/16	15/16	15/16	1	1⅛	1 3/16	1¼
Diameter of Raised Face			2	2½	2⅞	3⅝	4⅛	5	6	6⅜	7⅝	8½	10⅝	12¾	15
Diameter of Bolt Circle			3⅛	3½	3⅞	4¾	5½	6	7	7½	8½	9½	11¾	14¼	17
Number of Bolts			4	4	4	4	4	4	4	8	8	8	8	12	12
Diameter of Bolts			½	½	½	⅝	⅝	⅝	⅝	⅝	¾	¾	¾	⅞	⅞

[5] 150 lb. steel flanges have a 1/16" raised face which is included in the flange thickness dimensions.

300 lb. STEEL — ASA B16.5

Pipe Size	½	¾	1	1¼	1½	2	2½	3	3½	4	5	6	8	10	12
Diameter of Flange			4⅞	5¼	6⅛	6½	7½	8¼	9	10	11	12½	15	17½	20½

(Continuation of 300 lb. STEEL — ASA B16.5; pipe-size column headings appear on the preceding page)

Measurement													
Thickness of Flange (min)[6]	11/16	3/4	13/16	7/8	1	1 1/8	1 3/16	1 1/4	1 3/8	1 7/16	1 5/8	1 7/8	2
Diameter of Raised Face	2	2 1/2	2 7/8	3 5/8	4 1/8	5	5 1/2	6 3/16	7 5/16	8 1/2	10 5/8	12 3/4	15
Diameter of Bolt Circle	3 1/2	3 7/8	4 1/2	5	5 7/8	6 5/8	7 1/4	7 7/8	9 1/4	10 5/8	13	15 1/4	17 3/4
Number of bolts	4	4	4	8	8	8	8	8	8	12	12	16	16
Diameter of Bolts	5/8	5/8	3/4	5/8	3/4	3/4	3/4	3/4	3/4	3/4	7/8	1	1 1/8

[6] 300 lb. steel flanges have a 1/16" raised face which is included in the flange thickness dimensions.

400 lb. STEEL — ASA B16.5

Pipe Size	1/2	3/4	1	1 1/4	1 1/2	2	2 1/2	3	3 1/2	4	5	6	8	10	12
Diameter of Flange	3 3/4	4 5/8	4 7/8	5 1/4	6 1/8	6 1/2	7 1/2	8 1/4	9	10	11	12 1/2	15	17 1/2	20 1/2
Thickness of Flange (min)[7]	9/16	5/8	11/16	13/16	7/8	1	1 1/8	1 1/4	1 3/8	1 3/8	1 1/2	1 5/8	1 7/8	2 1/8	2 1/4
Diameter of Raised Face	1 3/8	1 11/16	2	2 1/2	2 7/8	3 5/8	4 1/8	5	5 1/2	6 3/16	7 5/16	8 1/2	10 5/8	12 3/4	15
Diameter of Bolt Circle	2 5/8	3 1/4	3 1/2	3 7/8	4 1/2	5	5 7/8	6 5/8	7 1/4	7 7/8	9 1/4	10 5/8	13	15 1/4	17 3/4
Number of Bolts	4	4	4	4	4	8	8	8	8	8	8	12	12	16	16
Diameter of Bolts	1/2	5/8	5/8	5/8	3/4	5/8	3/4	3/4	7/8	7/8	7/8	7/8	1	1 1/8	1 1/4

[7] 400 lb. steel flanges have a 1/4" raised face which is NOT included in the flange thickness dimensions.

600 lb. STEEL — ASA B16.5

Pipe Size	1/2	3/4	1	1 1/4	1 1/2	2	2 1/2	3	3 1/2	4	5	6	8	10	12
Diameter of Flange	3 3/4	4 5/8	4 7/8	5 1/4	6 1/8	6 1/2	7 1/2	8 1/4	9	10 3/4	13	14	16 1/2	20	22
Thickness of Flange (min)[8]	9/16	5/8	11/16	13/16	7/8	1	1 1/8	1 1/4	1 3/8	1 1/2	1 3/4	1 7/8	2 3/16	2 1/2	2 5/8
Diameter of Raised Face	1 3/8	1 11/16	2	2 1/2	2 7/8	3 5/8	4 1/8	5	5 1/2	6 3/16	7 5/16	8 1/2	10 5/8	12 3/4	15
Diameter of Bolt Circle	2 5/8	3 1/4	3 1/2	3 7/8	4 1/2	5	5 7/8	6 5/8	7 1/4	8 1/2	10 1/2	11 1/2	13 3/4	17	19 1/4
Number of Bolts	4	4	4	4	4	8	8	8	8	8	8	12	12	16	20
Diameter of Bolts	1/2	5/8	5/8	5/8	3/4	5/8	3/4	3/4	7/8	7/8	1	1	1 1/8	1 1/4	1 1/4

[8] 600 lb. steel flanges have a 1/4" raised face which is NOT included in the flange thickness dimensions.

Table 10. Pressure drop in Schedule 40 pipe.

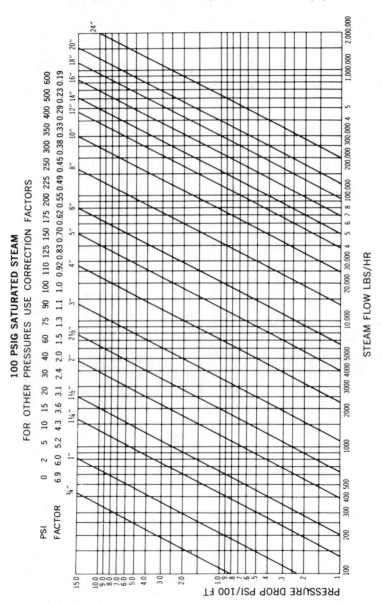

Courtesy Sarco Company, Inc.

Table 11. Steam velocity chart.

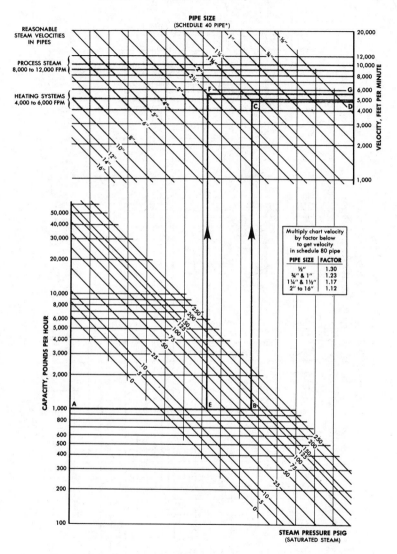

Table 12. Pressure drop in Schedule 80 pipe.

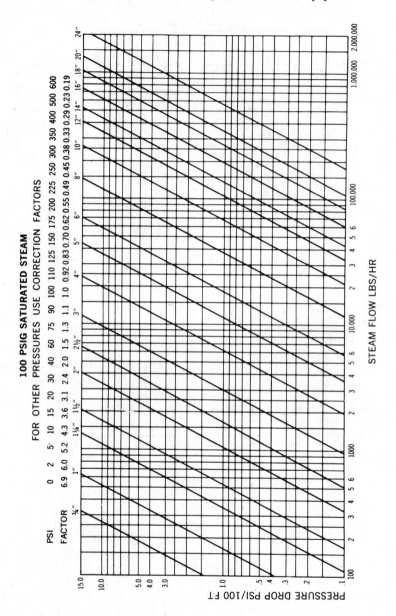

Courtesy *Sarco* Company, Inc.

Table 13. Chimney connector and vent connector clearance from combustible materials.

Description of Appliance	Minimum Clearance, Inches (See Note 1)
RESIDENTIAL TYPE APPLIANCES	
Single-Wall Metal Pipe Connectors	
Electric, Gas and Oil Incinerators	18
Oil and Solid-Fuel Appliances	18
Oil Appliances Listed as Suitable For Use With Type L Venting Systems, but only when connected to chimneys	9
Type L Venting System Piping Connectors	
Electric, Gas, and Oil Incinerators	9
Oil and Solid-Fuel Appliances	9
Oil Appliances Listed As Suitable For Use With Type L Venting Systems	(See Note 2)
COMMERCIAL-INDUSTRIAL TYPE APPLIANCES	
Low-Heat Appliances	
Single-Wall Metal Pipe Connectors	
Gas, Oil, and Solid-Fuel Boilers, Furnaces, and Water Heaters	18
Ranges, Restaurant Type	18
Oil Unit Heaters	18
Other Low-Heat Industrial Appliances	18
Medium-Heat Appliances	
Single-Wall Metal Pipe Connectors	
All Gas, Oil, and Solid-Fuel Appliances	36

NOTE 1: These clearances apply except if the listing of an appliance specifies different clearance, in which case the listed clearance takes precedence.

NOTE 2: If listed Type L venting system piping is used, the clearance may be in accordance with the venting system listing.

NOTE 3: If listed Type B or Type L venting system piping is used, the clearance may be in accordance with the venting system listing.

The clearances from connectors to combustible materials may be reduced if the combustible material is protected in accordance with Table 1C.

Courtesy National Oil Fuel Institute.

Table 14. Standard clearances for heat producing appliances in residential installations.

Residential Type Appliances For Installation in Rooms Which Are Large (See Note 1)		APPLIANCE				
		Above Top of Casing or Appliance	From Top and Sides of Warm-Air Bonnet or Plenum	From Front See Note 2	From Back	From Sides
Boilers and Water Heaters						
Steam Boilers — 15 psi; Water Boilers — 250°F; Water Heaters — 200°F; All Water Walled or Jacketed	Automatic Oil or Comb. Gas-Oil	6	—	24	6	6
Furnaces — Central						
Gravity, Upflow, Downflow, Horizontal and Duct Warm-Air — 250°F Max.	Automatic Oil or Comb. Gas-Oil	6^3	6^3	24	6	6
Furnaces — Floor						
For Mounting in Combustible Floors	Automatic Oil or Comb. Gas-Oil	36	—	12	12	12

Courtesy National Oil Fuel Institute.

1. Rooms which are large in comparison to the size of the appliance are those having a volume equal to at least 12 times the total volume of a furnace and at least 16 times the total volume of a boiler. If the actual ceiling height of a room is greater than 8 ft., the volume of a room shall be figured on the basis of a ceiling height of 8 ft.

2. The minimum dimension should be that necessary for servicing the appliance including acess for cleaning and normal care, tube removal, etc.

3. For a listed oil. combination gas-oil, gas, or electric furnace this dimension may be 2 in. if the furnace limit control cannot be set higher than 250°F or this dimension may be 1 in. if the limit control cannot be set set higher than 200°F.

Table 15. Standard clearances for heat producing appliances in commercial and industrial installations.

Commercial-Industrial Type Low Heat Appliances — Any and All Physical Sizes Except As Noted		APPLIANCE				
		Above Top of Casing or Appliance See Note 1	From Top and Sides of Warm-Air Bonnet or Plenum	From Front	From Back See Note 1	From Sides See Note 1
Boilers and Water Heaters						
100 cu. ft. or less Any psi. Steam	All Fuels	18	—	48	18	18
50 psi. or less, Any Size	All Fuels	18	—	48	18	18
Unit Heaters						
Floor Mounted or Suspended — Any Size	Steam or Hot Water	1	—	—	1	1
Suspended — 100 cu. ft .or less	Oil or Comb. Gas-Oil	6	—	24	18	18
Suspended — Over 100 cu. ft.	All Fuels	18	—	48	18	18
Floor Mounted Any Size	All Fuels	18	—	48	18	18

1. If the appliance is encased in brick, the 18 in. clearance above and at sides and rear may be reduced to not less than 12 in.

Courtesy National Oil Fuel Institute

Table 16. Clearance (in inches) with specified forms of protection.

Input Heat Units KWH. ELECT.	Eff. %	Usable Btuh.	GPH 100°F. Rise	Tank Size, Gallons	AVAILABLE HOT WATER STORAGE PLUS RECOVERY — 100°F. RISE —				Continuous Draw, GPH
					15 Min.	30 Min.	45 Min.	60 Min.	
1.5	92.5%	4,750	5.7*	20	21.4	22.8	24.3	25.7	5.7
2.5	92.5%	7,900	9.5*	20	32.4	34.8	37.1	39.5	9.5
4.5	92.5%	14,200	17.1*	30	44.3	48.6	52.9	57.1	17.1
4.5	92.5%	14,200	17.1*	50	54.3	58.6	62.9	67.1	17.1
6.0	92.5%	19,000	22.8*	66	71.6	77.2	82.8	88.8	22.8
7.0	92.5%	22,100	26.5	80	86.6	93.2	99.8	106.5	26.5

*Assumes Simultaneous Operation of 'Upper and Lower Elements

GAS BTUH.									
34,000	75%	25,500	30.6	30	37.7	45.3	53.0	55.6*	25.6*
42,000	75%	31,600	38.0	30	39.5	49.0	58.8	61.7*	31.7*
50,000	75%	37,400	45.0	40	51.3	62.6	73.9	77.6*	37.6*
60,000	75%	45,000	54.0	50	63.5	77.0	90.5	95.0*	45.0*

OIL, GPH.									
.50	75%	52,500	63.0	30	45.8	61.6	77.4	82.5*	52.5*
.75	75%	78,700	94.6	30	53.6	77.2	100.8	109.0*	79.0*
.85	75%	89,100	107.0	30	57.7	83.4	110.1	119.1*	89.0*
1.00	75%	105,000	126.0	50	81.5	113.0	144.5	155.0*	105.0*
1.20	75%	126,000	151.5	50	87.9	125.8	163.7	176.0*	126.0*
1.35	75%	145,000	174.0	50	93.5	137.0	180.5	195.0*	145.0*
1.50	75%	157,000	188.5	85	132.1	179.2	226.3	242.0*	157.0*
1.65	75%	174,000	204.5	85	136.1	187.2	238.4	259.0*	174.0*

* Based on 50 Minute Per Hour Operation

Courtesy *National Oil Fuel Institute*

Table 17. Performance of storage water heaters.

TYPE OF PROTECTION — Applied to the combustible material unless otherwise specified and covering all surfaces within the distance specified as the required clearance with no protection (see Fig. 1A). Thicknesses are minimum.	Where the required Clearance with no protection is:														
	36 Inches			**18 Inches**			**12 Inches**			**9 Inches**			**6 Inches**		
	Above	Sides & Rear	Chimney or Vent Connector	Above	Sides & Rear	Chimney or Vent Connector	Above	Sides & Rear	Chimney or Vent Connector	Above	Sides & Rear	Chimney or Vent Connector	Above	Sides & Rear	Chimney or Vent Connector
(a) ¼ in asbestos millboard spaced out 1 in.†	30	18	30	15	9	12	9	6	9	6	6	6	3	2	3
(b) 28 gage sheet metal on ¼ in. asbestos millboard	24	18	24	12	9	12	9	6	9	4	6	4	3	2	2
(c) 28 gage sheet metal spaced out 1 in.†	18	12	18	9	6	9	6	4	6	4	4	4	2	2	2
(d) 28 gage sheet metal on ⅛ in. asbestos millboard spaced out 1 in.†	18	12	18	9	6	9	6	4	6	4	4	4	2	2	2
(e) 1½ in. asbestos cement covering on heating appliance	18	12	36	9	6	18	6	4	6	2	4	9	2	1	6
(f) ¼ in. asbestos millboard on 1 in. mineral fiber bats reinforced with wire mesh or equivalent	18	12	18	6	6	6	4	4	6	2	4	4	2	2	2

Table 17. Performance of storage water heaters. (Cont'd.)

(g) 22 gage sheet metal on 1 in. mineral fiber bats reinforced with wire or equivalent	18	12	12	4	3	3	2	2	2	2	2
(h) ¼ in. asbestos millboard	36	36	18	18	18	12	12	12	9	4	4
(i) ¼ in. cellular asbestos	36	36	18	18	18	12	12	9	3	3	3

*Except for the protection described in (e), all clearances should be measured from the outer surface of the appliance to the combustible material disregarding any intervening protection applied to the combustible material.

†Spacers shall be of noncombustible material.

NOTE: Asbestos millboard referred to above is a different material from asbestos cement board. It is not intended that asbestos cement board be used in complying with these requirements when asbestos millboard is specified.

Courtesy *National Oil Fuel Institute*

Table 18. Conversion tables.

TO CONVERT	INTO	MULTIPLY BY
A		
Acres	sq. feet	43,560.0
Atmospheres	cms. of mercury	76.0
Atmospheres	ft. of water (at 4°C)	33.90
Atmospheres	in. of mercury (at 0°C)	29.92
Atmospheres	kgs./sq. cm.	1.0333
Atmospheres	pounds/sq. in.	14.70
B		
Barrels (U.S., liquid)	gallons	31.5
Barrels (oil)	gallons (oil)	42.0
Btu	foot-lbs.	778.3
Btu	gram-calories	252.0
Btu	horsepower-hrs.	3.931×10^{-4}
Btu	kilowatt-hrs.	2.928×10^{-4}
Btu/hr.	horsepower	3.931×10^{-4}
Btu/hr.	Watts	0.2931
C		
Calories, gram (mean)	B.T.U. (mean)	3.9685×10^{-3}
Centigrade	Fahrenheit	(C° + 40) 9/5-40
Centimeters	feet	3.281×10^{-2}
Centimeters	inches	0.3937
Centimeters	mils	393.7
Centimeters of mercury	atmospheres	0.01316
Centimeters of mercury	feet of water	0.4461
Centimeters of mercury	pounds/sq. in.	0.1934
Circumference	radians	6.283
Cubic centimeters	cu. feet	3.531×10^{-5}
Cubic centimeters	cu. inches	0.06102
Cubic centimeters	gallons (U.S. liq.)	2.642×10^{-4}
Cubic feet	cu. cms.	28,320.0
Cubic feet	cu. inches	1,728.0
Cubic feet	gallons (U.S. liq.)	7.481
Cubic feet	liters	28.32
Cubic feet	quarts (U.S. liq.)	29.92
Cubic feet/min.	gallons/sec.	0.1247
Cubic feet/min.	pounds of water/min.	62.43
Cubic inches	cu. cms.	16.39
Cubic inches	gallons	4.329×10^{-3}
Cubic inches	quarts (U.S. liq.)	0.01732
Cubic meters	cu. feet	35.31

Table 18. Conversion Tables (Continued)

TO CONVERT	INTO	MULTIPLY BY
Cubic meters	gallons (U.S. liq.)	264.2
Cubic yards	cu. feet	27.0
Cubic yards	cu. meters	0.7646
Cubic yards	gallons (U.S. liq.)	202.0

D

Degrees (angle)	radians	0.01745
Drams (apothecaries' or troy)	ounces (avoidupois)	0.13714
Drams (apothecaries' or troy)	ounces (troy)	0.125
Drams (U.S., fluid or apoth.)	cubic cm.	3.6967
Drams	grams	1.772
Drams	grains	27.3437
Drams	ounces	0.0625

F

Fahrenheit	centigrade	(F+40) 5/9-40
Feet	centimeters	30.48
Feet	kilometers	3.048×10^{-4}
Feet	meters	0.3048
Feet	miles (naut.)	1.645×10^{-4}
Feet	miles (stat.)	1.894×10^{-4}
Feet of water	atmospheres	0.02950
Feet of water	in. of mercury	0.8826
Feet of water	kgs./sq. cm.	0.03045
Feet of water	kgs./sq. meter	304.8
Feet of water	pounds/sq. ft.	62.43
Feet of water	pounds/sq. in.	0.4335
Foot-pounds	Btu	1.286×10^{-3}
Foot-pounds	gram-calories	0.3238
Foot-pounds	hp.-hrs.	5.050×10^{-7}
Foot-pounds	kilowatt-hrs.	3.766×10^{-7}
Foot-pounds/min.	Btu/min.	1.286×10^{-3}
Foot-pounds/min.	horsepower	3.030×10^{-5}
Foot-pounds/sec.	Btu-hr.	4.6263
Furlongs	miles (U.S.)	0.125
Furlongs	feet	660.0

G

Gallons	cu. cms.	3,785.0
Gallons	cu. feet	0.1337

Table 18. Conversion Tables (Continued)

TO CONVERT	INTO	MULTIPLY BY
Gallons	cu. inches	231.0
Gallons	cu. meters	3.785×10^{-3}
Gallons	cu. yards	4.951×10^{-3}
Gallons	liters	3.785
Gallons (liq. Br. Imp.)	gallons (U.S. liq.)	1.20095
Gallons (U.S.)	gallons (Imp.)	0.83267
Gallons of water	pounds of water	8.3453
Gallons/min.	cu. ft./sec.	2.228×10^{-3}
Gallons/min.	liters/sec.	0.06308
Gallons/min.	cu. ft./hr.	8.0208
Grains (troy)	grains (avdp.)	1.0
Grains (troy)	grams	0.06480
Grains (troy)	ounces (avdp.)	2.286×10^{-3}
Grains (troy)	pennyweight (troy)	0.04167
Grains/U.S. gal.	parts/million	17.118
Grains/U.S. gal.	pounds/million gal.	142.86
Grains/Imp. gal.	parts/mililon	14.286
Grams	grains	15.43
Grams	ounces (avdp.)	0.03527
Grams	ounces (troy)	0.03215
Grams	poundals	0.07093
Grams	pounds	2.205×10^{-3}
Grams/liter	parts/million	1,000.0
Gram-calories	Btu	3.9683×10^{-3}
Gram-calories	foot-pounds	3.0880
Gram-calories	kilowatt-hrs.	1.1630×10^{-6}
Gram-calories	watt-hrs.	1.1630×10^{-3}

H

Horsepower	Btu/min.	42.40
Horsepower	foot-lbs./min.	33,000.
Horsepower	foot-lbs./sec.	550.0
Horsepower (metric) (542.5 ft. lb./sec.)	horsepower (550 ft. lb./sec.)	0.9863
Horsepower (550 ft. lb./sec.)	horsepower (metric) (542.5 ft. lb./sec.)	1.014
Horsepower	kilowatts	0.7457
Horsepower	watts	745.7
Horsepower (boiler)	Btu/hr.	33.520
Horsepower (boiler)	kilowatts	9.803
Horsepower-hrs.	Btu	2,547
Horsepower-hrs.	foot-lbs.	1.98×10^6
Horsepower-hrs.	kilowatt-hrs.	0.7457

Table 18. Conversion Tables (Continued)

TO CONVERT	INTO	MULTIPLY BY
	I	
Inches	centimeters	2.540
Inches	meters	25.40
Inches	millimeters	2.540×10^{-2}
Inches	yards	22.778×10^{-2}
Inches of mercury	atmospheres	0.03342
Inches of mercury	feet of water	1.133
Inches of mercury	kgs./sq. cm.	0.03453
Inches of mercury	kgs./sq. meter	345.3
Inches of mercury	pounds/sq. ft.	70.73
Inches of mercury	pounds/sq. in.	0.4912
Inches of water (at 4°C)	atmospheres	2.458×10^{-3}
Inches of water (at 4°C)	inches of mercury	0.07355
Inches of water (at 4°C)	kgs./sq. cm.	2.538×10^{-3}
Inches of water (at 4°C)	ounces/sq. in.	0.5781
Inches of water (at 4°C)	pounds/sq. ft.	5.204
Inches of water (at 4°C)	pounds/sq. in.	0.03613
	J	
Joules	Btu	9.480×10^{-4}
	K	
Kilograms	grams	1,000.0
Kilograms	pounds	2.205
Kilograms/cu. meter	pounds/cu. ft.	0.06243
Kilograms/cu. meter	pounds/cu. in.	3.613×10^{-5}
Kilograms/sq. cm.	atmospheres	0.9678
Kilograms/sq. cm.	feet of water	32.84
Kilograms/sq. cm.	inches of mercury	28.96
Kilograms/sq. cm.	pounds/sq. ft.	2,048.
Kilograms/sq. cm.	pounds/sq. in.	14.22
Kilograms/sq. meter	atmospheres	9.678×10^{-5}
Kilograms/sq. meter	feet of water	3.281×10^{-3}
Kilograms/sq. meter	inches of mercury	2.896×10^{-3}
Kilograms/sq. meter	pounds/sq. ft.	0.2048
Kilograms/sq. meter	pounds/sq. in.	1.422×10^{-3}
Kilograms/sq. mm.	kgs./sq. meter	10^6
Kilogram-calories	Btu	3.968
Kilogram-calories	foot-pounds	3,088.
Kilogram-calories	hp-hrs.	1.560×10^{-3}
Kilogram-calories	kilowatt-hrs.	1.163×10^{-3}
Kilogram meters	Btu	9.294×10^{-3}
Kilometers	centimeters	10^5

Table 18. Conversion Tables (Continued)

TO CONVERT	INTO	MULTIPLY BY
Kilometers	feet	3,281.
Kilometers	miles	0.6214
Kilowatts	Btu/min.	56.87
Kilowatts	foot-lbs./min.	4.426×10^4
Kilowatts	foot-lbs./sec.	737.6
Kilowatts	horsepower	1,341
Kilowatts	watts	1,000.0
Kilowatt-hrs.	Btu	3,413.
Kilowatt-hrs.	foot-lbs.	2.655×10^6
Kilowatt-hrs.	horsepower-hrs.	1.341
Knots	statute miles/hrs.	1.151

L

Liters	cu. cm.	1,000.0
Liters	cu. feet	0.03531
Liters	cu. inches	61.02
Liters	gallons (U.S. liq.)	0.2642

M

Meters	centimeters	100.0
Meters	feet	3.281
Meters	inches	39.37
Meters	millimeters	1,000.0
Meters	yards	1.094
Microns	inches	39.37×10^{-6}
Microns	meters	1×10^{-6}
Miles (statute)	feet	5,280.
Miles (statute)	kilometers	1.609
Miles/hr.	cms./sec.	44.70
Miles/hr.	feet/min.	88.
Mils	inches	0.001
Mils	yards	2.778×10^{-5}

N

Nepers	decibels	8.686

O

Ohms	megohms	10^{-6}
Ohms	microhms	10^6
Ounces (avoirdupois)	drams	16.0
Ounces (avoirdupois)	grains	437.5
Ounces (avoirdupois)	grams	28.35
Ounces (avoirdupois)	pounds	0.0625

Table 18. Conversion Tables (Continued)

TO CONVERT	INTO	MULTIPLY BY
Ounces (avoirdupois)	ounces (troy)	0.9115
Ounces (troy)	grains	480.0
Ounces (troy)	grams	31.10
Ounces (troy)	ounces (avdp.)	1.09714
Ounces (troy)	pounds (troy)	0.08333

P

Parts/million	grains/U.S. gal.	0.0584
Parts/million	grains/Imp. gal.	0.07016
Parts/million	pounds/million gal.	8.33
Pounds (avoirdupois)	ounces (troy)	14.58
Pounds (avoirdupois)	drams	256.
Pounds (avoirdupois)	grains	7,000.
Pounds (avoirdupois)	grams	28.35
Pounds (avoirdupois)	kilograms	0.02835
Pounds (avoirdupois)	ounces	16.0
Pounds (avoirdupois)	tons (short)	0.0005
Pounds (troy)	ounces (avdp.)	13.1657
Pounds of water	cu. feet	0.01602
Pounds of water	cu. inches	27.68
Pounds of water	gallons	0.1198
Pounds of water/min.	cu. ft./sec.	2.670×10^{-4}
Pounds/cu. ft.	kgs./cu. meter	0.01602
Pounds/cu. ft.	grams/cu. cm.	16.02
Pounds/cu. ft.	pounds/cu. in.	5.787×10^{-4}
Pounds/cu. in.	pounds/cu. ft.	1,728.
Pounds/sq. ft.	atmospheres	4.725×10^{-4}
Pounds/sq. ft.	feet of water	0.01602
Pounds/sq. ft.	inches of mercury	0.01414
Pounds/sq. in.	atmospheres	0.06804
Pounds/sq. in.	feet of water	2.307
Pounds/sq. in.	inches of mercury	2.036
Pounds/sq. in.	kgs./sq. meter	703.1
Pounds/sq. in.	pounds/sq. ft.	144.0

R

Radians	degrees	57.30
Revolutions/min.	degrees/sec.	6.0
Revolutions/min.	radians/sec.	0.1047
Revolutions/min.	revs./sec.	0.01667

S

Square centimeters	sq. feet	1.076×10^{-3}

Table 18. Conversion Tables (Continued)

TO CONVERT	INTO	MULTIPLY BY
Square centimeters	sq. inches	0.1550
Square centimeters	sq. meters	0.0001
Square centimeters	sq. millimeters	100.0
Square feet	acres	2.296×10^{-5}
Square feet	sq. cms.	929.0
Square feet	sq. inches	144.0
Square feet	sq. miles	3.587×10^{-8}
Square inches	sq. cms.	6.452
Square inches	sq. feet	6.944×10^{-3}
Square inches	sq. yards	7.716×10^{-4}
Square meters	sq. feet	10.76
Square meters	sq. inches	1,550.
Square meters	sq. millimeters	10^6
Square meters	sq. yards	1.196
Square millimeters	sq. inches	1.550×10^{-3}
Square yards	sq. feet	9.0
Square yards	sq. inches	1,296.
Square yards	sq. meters	0.8361

T

TO CONVERT	INTO	MULTIPLY BY
Temperature ($^\circ$C) + 273	absolute temperature ($^\circ$C)	1.0
Temperature ($^\circ$C) + 17.78	temperature ($^\circ$F)	1.8
Temperature ($^\circ$F) + 460	absolute temperature ($^\circ$F)	1.0
Temperature ($^\circ$F)−32	temperature ($^\circ$C)	5/9
Tons (long)	kilograms	1,016.
Tons (long)	pounds	2,240.
Tons (long)	tons (short)	1.120
Tons (metric)	kilograms	1,000.
Tons (metric)	pounds	2,205.
Tons (sshort)	kilograms	907.2
Tons (short)	pounds	2,000.
Tons (short)	tons (long)	0.89287
Tons of water/24 hrs.	pounds of water/hr.	83.333
Tons of water/24 hrs.	gallons/min.	0.16643
Tons of water/24 hrs.	cu. ft./hr.	1.3349

W

TO CONVERT	INTO	MULTIPLY BY
Watts	Btu/hr.	3.4129
Watts	Btu/min.	0.05688

Table 18. Conversion Tables (Continued)

TO CONVERT	INTO	MULTIPLY BY
Watts	horsepower	1.341×10^{-3}
Watts	horsepower (metric)	1.360×10^{-3}
Watts	kilowatts	0.001
Watts (Abs.)	B.T.U. (mean)/min.	0.056884
Watt-hours	Btu	3.413
Watt-hours	horsepower-hrs.	1.341×10^{-3}

Y

Yards	centimeters	91.44
Yards	kilometers	9.144×10^{-4}
Yards	meters	0.9144

Index

The Audel® Mail Order Bookstore

Here's an opportunity to order the valuable books you may have missed before and to build your own personal, comprehensive library of Audel books. You can choose from an extensive selection of technical guides and reference books. They will provide access to the same sources the experts use, put all the answers at your fingertips, and give you the know-how to complete even the most complicated building or repairing job, in the same professional way.

Each volume:
- **Fully illustrated**
- **Packed with up-to-date facts and figures**
- **Completely indexed for easy reference**

APPLIANCES

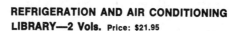

HOME APPLIANCE SERVICING, 3rd Edition
A practical book for electric & gas servicemen, mechanics & dealers. Covers the principles, servicing, and repairing of home appliances. 592 pages; 5¼ x 8¼; hardbound. **Price: $12.95**

REFRIGERATION AND AIR CONDITIONING
LIBRARY—2 Vols. Price: $21.95

REFRIGERATION: HOME AND COMMERCIAL
Covers the whole realm of refrigeration equipment from fractional-horsepower water coolers, through domestic refrigerators to multi-ton commercial installations. 656 pages; 5½ x 8¼; hardbound. **Price: $12.95**

AIR CONDITIONING: HOME AND COMMERCIAL
A concise collection of basic information, tables, and charts for those interested in understanding, troubleshooting, and repairing home air conditioners and commercial installations. 464 pages; 5½ x 8¼; hardbound. **Price: $10.95**

OIL BURNERS, 3rd Edition
Provides complete information on all types of oil burners and associated equipment. Discusses burners—blowers—ignition transformers—electrodes—nozzles—fuel pumps—filters—Controls. Installation and maintenance are stressed. 320 pages; 5½ x 8¼; hardbound. **Price: $9.95**

Use the order coupon on the back page of this book.

All prices are subject to change without notice.

AUTOMOTIVE

AUTOMOBILE REPAIR GUIDE, 4th Edition

A practical reference for auto mechanics, servicemen, trainees, and owners Explains theory, construction, and servicing of modern domestic motorcars. 800 pages; 5½ x 8¼; hardbound. **Price: 14.95**

CAN-DO TUNE-UP™ SERIES

Each book in this series comes with an audio tape cassette. Together they provide an organized set of instructions that will show you and talk you through the maintenance and tune-up procedures designed for your particular car. All books are softcover.

AMERICAN MOTORS CORPORATION CARS

(The 1964 thru 1974 cars covered include: Matador. Rambler. Gremlin. and AMC Jeep (Willys).) 112 pages; 5½ x 8½; softcover. **Price: $7.95**

CHRYSLER CORPORATION CARS

(The 1964 thru 1974 cars covered include: Chrysler, Dodge, and Plymouth.) 112 pages; 5½ x 8½; softcover. **Price: $7.95**

FORD MOTOR COMPANY CARS

(The 1954 thru 1974 cars covered include: Ford, Lincoln, and Mercury.) 112 pages; 5½ x 8½; softcover. **Price: $7.95**

GENERAL MOTORS CORPORATION CARS

(The 1964 thru 1974 cars covered include: Buick, Cadillac, Chevrolet, Oldsmobile, and Pontiac.) 112 pages; 5½ x 8½; softcover. **Price: $7.95**

PINTO AND VEGA CARS

1971 thru 1974. 112 pages. 5½ x 8½; softcover. **Price: $7.95**

TOYOTA AND DATSUN CARS

1964 thru 1974. 112 pages; 5½ x 8½; softcover. **Price: $7.95**

VOLKSWAGEN CARS

(The 1964 thru 1974 cars covered include: Beetle. Super Beetle. and Karmann Ghia.) 96 pages; 5½ x 8½; softcover. **Price: $7.95**

AUTOMOTIVE AIR CONDITIONING

You can easily perform most all service procedures you've been paying for in the past. This book covers the systems built by the major manufacturers, even after-market installations. Contents: introduction—refrigerant—tools—air conditioning circuit—general service procedures—electrical systems—the cooling system—system diagnosis—electrical diagnosis—troubleshooting. 232 pages; 5½ x 8½; softcover. **Price: $7.95**

Use the order coupon on the back page of this book.

All prices are subject to change without notice.

DIESEL ENGINE MANUAL, 3rd Edition

A practical guide covering the theory, operation, and maintenance of modern diesel engines. Explains diesel principles—valves—timing—fuel pumps—pistons and rings—cylinders—lubrication—cooling system—fuel oil and more. 480 pages; 5½ x 8¼; hardbound. **Price: $10.95**

GAS ENGINE MANUAL, 2nd Edition

A completely practical book covering the construction, operation, and repair of all types of modern gas engines. 400 pages; 5½ x 8¼; hardbound. **Price: $9.95**

BUILDING AND MAINTENANCE

ANSWERS ON BLUEPRINT READING, 3rd Edition

Covers all types of blueprint reading for mechanics and builders. This book reveals the secret language of blueprints, step-by-step in easy stages. 312 pages; 5½ x 8¼; hardbound. **Price: $9.95**

BUILDING MAINTENANCE, 2nd Edition

Covers all the practical aspects of building maintenance. Painting and decorating; plumbing and pipe fitting; carpentry; heating maintenance; custodial practices and more. (A book for building owners, managers, and maintenance personnel.) 384 pages; 5½ x 8¼; hardbound. **Price: $9.95**

COMPLETE BUILDING CONSTRUCTION

At last—a *one-volume* instruction manual to show you how to construct a frame or brick building from the footings to the ridge. Build your own garage, tool shed, other outbuilding—even your own house or place of business. Building construction tells you how to lay out the building and excavation lines on the lot; how to make concrete forms and pour the footings and foundation; how to make concrete slabs, walks, and driveways; how to lay concrete block, brick and tile; how to build your own fireplace and chimney: It's one of the newest Audel books, clearly written by experts in each field and ready to help you every step of the way. 800 pages; 5½ x 8¼; hardbound. **Price: 19.95**

GARDENING & LANDSCAPING

A comprehensive guide for homeowners and for industrial, municipal, and estate groundskeepers. Gives information on proper care of annual and perennial flowers; various house plants; greenhouse design and construction; insect and rodent controls; and more. 384 pages; 5½ x 8¼; hardbound. **Price: $9.95**

CARPENTERS & BUILDERS LIBRARY, 4th Edition (4 Vols.)

A practical, illustrated trade assistant on modern construction for carpenters, builders, and all woodworkers. Explains in practical, concise language and illustrations all the principles, advances, and shortcuts based on modern practice. How to calculate various jobs. **Price: $35.95**

> Vol. 1—Tools, steel square, saw filing, joinery cabinets. 384 pages; 5½ x 8¼; hardbound. **Price: $10.95**
>
> Vol. 2—Mathematics, plans, specifications, estimates 304 pages; 5½ x 8¼; hardbound. **Price: $10.95**
>
> Vol. 3—House and roof framing, laying out foundations. 304 pages; 5½ x 8¼; hardbound. **Price: $10.95**
>
> Vol. 4—Doors, windows, stairs, millwork, painting. 368 pages; 5½ x 8¼; hardbound. **Price: $10.95**

Use the order coupon on the back page of this book.

All prices are subject to change without notice.

CARPENTRY AND BUILDING

Answers to the problems encountered In today's building trades. The actual questions asked of an architect by carpenters and builders are answered in this book. 448 pages; 5½ x 8¼; hardbound. **Price: $10.95**

WOOD STOVE HANDBOOK

The wood stove handbook shows how wood burned In a modern wood stove offers an immediate, practical, low-cost method of full-time or part-time home heating. The book points out that wood is plentiful, low in cost (sometimes free), and nonpolluting, especially when burned in one of the newer and more efficient stoves. In this book, you will learn about the nature of heat and its control, what happens inside and outside a stove, how to have a safe and efficient chimney, and how to install a modern wood burning stove. You will also learn about the different types of firewood and how to get it, cut it, split it, and store it. 128 pages; 8½ x 11; softcover. **Price: $7.95**

HEATING, VENTILATING, AND AIR CONDITIONING LIBRARY (3 Vols.)

This three-volume set covers all types of furnaces, ductwork, air conditioners, heat pumps, radiant heaters, and water heaters, Including swimming-pool heating systems. **Price: $32.95**

Volume 1

Partial Contents: Heating Fundamentals . . . Insulation Principles . . . Heating Fuels . . . Electric Heating System . . . Furnace Fundamentals . . . Gas-Fired Furnaces . . . Oil-Fired Furnaces . . . Coal-Fired Furnaces . . . Electric Furnaces. **Price: $11.95**

Volume 2

Partial Contents: Oil Burners . . . Gas Burners . . . Thermostats and Humidistats . . . Gas and Oil Controls . . . Pipes, Pipe Fitting, and Piping Details . . . Valves and Valve Installations 560 pages; 5½ x 8¼; hardbound. **Price: $11.95**

Volume 3

Partial Contents: Radiant Heating . . . Radiators, Convectors, and Unit Heaters . . . Stoves, Fireplaces, and Chimneys . . . Water Heaters and Other Appliances . . . Central Air Conditioning Systems . . . Humidifiers and Dehumidifiers. 544 pages; 5½ x 8¼; hardbound. **Price: $11.95**

HOME MAINTENANCE AND REPAIR: Walls, Ceilings, and Floors

Easy-to-follow instructions for sprucing up and repairing the walls, ceiling, and floors of your home. Covers nail pops, plaster repair, painting, paneling, ceiling and bathroom tile, and sound control, 80 pages; 8½ x 11; softcover. **Price: $6.95**

HOME PLUMBING HANDBOOK , 2nd Edition

A complete guide to home plumbing repair and installation. 200 pages; 8½ x 11; softcover. **Price: $7.95**

MASONS AND BUILDERS LIBRARY—2 Vols.

A practical, illustrated trade assistant on modern construction for bricklayers, stonemasons, cement workers, plasterers, and tile setters. Explains all the principles, advances, and shortcuts based on modern practice—including how to figure and calculate various jobs. **Price: $17.95**

Vol. 1—Concrete, Block, Tile, Terrazzo. 368 pages; 5½ x 8¼; hardbound. **Price: $9.95**

Vol. 2—Bricklaying, Plastering, Rock Masonry, Clay Tile. 384 pages; 5½ x 8¼; hardbound. **Price: 9.95**

Use the order coupon on the back page of this book.

All prices are subject to change without notice.

PLUMBERS AND PIPE FITTERS LIBRARY—3 Vols.

A practical, illustrated trade assistant and reference for master plumbers, journeymen and apprentice pipe fitters, gas fitters and helpers, builders, contractors, and engineers. Explains in simple language, illustrations, diagrams, charts, graphs, and pictures, the principles of modern plumbing and pipe-fitting practices. **Price: $26.95**

Vol. 1—Materials, tools, roughing-in. 320 pages; 5½ x 8¼; hardbound. **Price: $9.95**

Vol. 2—Welding, heating, air-conditioning. 384 pages; 5½ x 8¼; hardbound. **Price: $9.95**

Vol. 3—Water supply, drainage, calculations. 272 pages; 5½ x 8¼; hardbound. **Price: $9.95**

PLUMBERS HANDBOOK

A pocket manual providing reference material for plumbers and/or pipe fitters. General information sections contain data on cast-iron fittings, copper drainage fittings, plastic pipe, and repair of fixtures. 288 pages; 4 x 6; softcover. **Price: $9.95**

QUESTIONS AND ANSWERS FOR PLUMBERS EXAMINATIONS, 2nd Edition

Answers plumbers' questions about types of fixtures to use, size of pipe to install, design of systems, size and location of septic tank systems, and procedures used in installing material. 256 pages; 5½ x 8¼; softcover. **Price: $8.95**

TREE CARE MANUAL

The conscientious gardener's guide to healthy, beautiful trees. Covers planting, grafting, fertilizing, pruning, and spraying. Tells how to cope with insects, plant diseases, and environmental damage. 224 pages; 8½ x 11; softcover. **Price: $8.95**

UPHOLSTERING

Upholstering is explained for the average householder and apprentice upholsterer. From repairing and reglueing of the bare frame, to the final sewing or tacking, for antiques and most modern pieces, this book covers it all. 400 pages; 5½ x 8¼; hardbound. **Price: $9.95**

WOOD FURNITURE: Finishing, Refinishing, Repairing

Presents the fundamentals of furniture repair for both veneer and solid wood. Gives complete instructions on refinishing procedures, which includes stripping the old finish, sanding, selecting the finish and using wood fillers. 352 pages; 5½ x 8¼; hardbound. **Price: $9.95**

ELECTRICITY/ELECTRONICS

ELECTRICAL LIBRARY

If you are a student of electricity or a practicing electrician, here is a very important and helpful library you should consider owning. You can learn the basics of electricity, study electric motors and wiring diagrams, learn how to interpret the NEC, and prepare for the electrician's examination by using these books.

Electric Motors, 3rd Edition. 528 pages; 5½ x 8¼; hardbound. **Price: $10.95**

Guide to the 1981 National Electrical Code. 608 pages; 5½ x 8¼; hardbound. **Price: $13.95**

House Wiring, 5th Edition. 256 pages; 5½ x 8¼; hardbound. **Price: $9.95**

Practical Electricity, 3rd Edition. 496 pages; 5½ x 8¼; hardbound. **Price: $10.95**

Questions and Answers for Electricians Examinations, 7th Edition. 288 pages; 5½ x 8¼; hardbound. **Price: $9.95**

ELECTRICAL COURSE FOR APPRENTICES AND JOURNEYMEN

A study course for apprentice or journeymen electricians. Covers electrical theory and its applications. 448 pages; 5½ x 8¼; hardbound. **Price: $10.95**

Use the order coupon on the back page of this book.

All prices are subject to change without notice.

RADIOMANS GUIDE, 4th Edition

Contains the latest information on radio and electronics from the basics through transistors. 480 pages; 5½ x 8¼; hardbound. **Price: $11.95**

TELEVISION SERVICE MANUAL, 4th Edition

Provides the practical information necessary for accurate diagnosis and repair of both black-and-white and color television receivers. 512 pages; 5½ x 8¼; hardbound. **Price: $11.95**

ENGINEERS/MECHANICS/ MACHINISTS

MACHINISTS LIBRARY, 3rd Edition

Covers modern machine-shop practice. Tells how to set up and operate lathes, screw and milling machines, shapers, drill presses, and all other machine tools. A complete reference library. **Price: $29.95**

Vol. 1—Basic Machine Shop. 352 pages; 5½ x 8¼; hardbound. **Price: $10.95**

Vol. 2—Machine Shop. 480 pages; 5½ x 8¼; hardbound. **Price: $10.95**

Vol. 3—Toolmakers Handy Book. 400 pages; 5½ x 8¼; hardbound. **Price: $10.95**

MECHANICAL TRADES POCKET MANUAL

Provides practical reference material for mechanical tradesmen. This handbook covers methods, tools, equipment, procedures, and much more. 256 pages; 4 x 6; softcover. **Price: $8.95**

MILLWRIGHTS AND MECHANICS GUIDE, 2nd Edition

Practical information on plant installation, operation, and maintenance for millwrights, mechanics, maintenance men, erectors, riggers, foremen, inspectors, and superintendents. 960 pages; 5½ x 8¼; hardbound. **Price: $16.95**

POWER PLANT ENGINEERS GUIDE, 2nd Edition

The complete steam or diesel power-plant engineer's library. 816 pages; 5½ x 8¼; hardbound. **Price: $15.95**

QUESTIONS AND ANSWERS FOR ENGINEERS AND FIREMANS EXAMINATIONS, 3RD EDITION

Presents both legitimate and "catch" questions with answers that may appear on examinations for engineers and firemans licenses for stationary, marine, and combustion engines. 496 pages; 5½ x 8¼; hardbound. **Price: $10.95**

WELDERS GUIDE, 2nd Edition

This new edition is a practical and concise manual on the theory, practical operation, and maintenance of all welding machines. Fully covers both electric and oxy-gas welding. 928 pages; 5½ x 8¼; hardbound. **Price: $14.95**

WELDER/FITTERS GUIDE

Provides basic training and instruction for those wishing to become welder/fitters. Step-by-step learning sequences are presented from learning about basic tools and aids used in weldment assembly, through simple work practices, to actual fabrication of weldments. 160 pages· 8½ x 11; softcover. **Price: $7.95**

Use the order coupon on the back page of this book.

All prices are subject to change without notice.

FLUID POWER

PNEUMATICS AND HYDRAULICS, 3rd Edition

Fully discusses installation, operation, and maintenance of both HYDRAULIC AND PNEUMATIC (air) devices. 496 pages; 5½ x 8¼; hardbound. **Price: $10.95**

PUMPS, 3rd Edition

A detailed book on all types of pumps from the old-fashioned kitchen variety to the most modern types. Covers construction, application, installation, and troubleshooting. 480 pages; 5½ x 8¼; hardbound. **Price: $10.95**

HYDRAULICS FOR OFF-THE-ROAD EQUIPMENT

Everything you need to know from basic hydraulics to troubleshooting hydraulic systems on off-the-road equipment. Heavy-equipment operators, farmers, fork-lift owners and operators, mechanics—all need this practical, fully illustrated manual. 272 pages; 5½ x 8¼; hardbound. **Price: $8.95**

HOBBY

COMPLETE COURSE IN STAINED GLASS

Written by an outstanding artist in the field of stained glass, this book is dedicated to all who love the beauty of the art. Ten complete lessons describe the required materials, how to obtain them, and explicit directions for making several stained glass projects. 80 pages; 8½ x 11; softbound. **Price: $6.95**

BUILD YOUR OWN AUDEL DO-IT-YOURSELF LIBRARY AT HOME!

Use the handy order coupon today to gain the valuable information you need in all the areas that once required a repairman. Save money and have fun while you learn to service your own air conditioner, automobile, and plumbing. Do your own professional carpentry, masonry, and wood furniture refinishing and repair. Build your own security systems. Find out how to repair your TV or Hi-Fi. Learn landscaping, upholstery, electronics and much, much more.

All prices are subject to change without notice.

HERE'S HOW TO ORDER

1. Enter the correct title(s) and author(s) of the book(s) you want in the space(s) provided.

2. Print your name, address, city, state and zip code clearly.

3. Detach the order coupon below and mail today to:

   Theodore Audel & Company
   4300 West 62nd Street
   Indianapolis, Indiana 46206
   ATTENTION: ORDER DEPT.

   All prices are subject to change without notice.

ORDER COUPON

Please rush the following book(s).

Title_____

Author_____

Title_____

Author_____

NAME_____

ADDRESS_____

CITY_____ STATE_____ ZIP_____

☐ Payment enclosed _____
   (No shipping and   Total
   handling charge)

☐ Bill me (shipping and handling charge will be added)
Add local sales tax where applicable.

Litho in U.S.A.

HERE'S HOW TO ORDER

Select the Audel book(s) you want, fill in the order card below, detach and mail today. Send no money now. You'll have 15 days to examine the books in the comfort of your own home. If not completely satisfied, simply return your order and owe nothing.

If you decide to keep the books, we will bill you for the total amount, plus a small charge for shipping and handling.

1. Enter the correct title(s) and author(s) of the book(s) you want in the space(s) provided.

2. Print your name, address, city, state and zip code clearly.

3. Detach the order card below and mail today. No postage is required.

Detach postage-free order card on perforated line

FREE TRIAL ORDER CARD

☐ Please rush the following book(s) for my free trial. I understand if I'm not completely satisfied, I may return my order within 15 days and owe nothing. Otherwise, you will bill me for the total amount plus a small postage & handling charge.

Title_____

Author_____

Title_____

Author_____

NAME_____

ADDRESS_____

CITY_____ STATE_____ ZIP_____

☐ Save postage & handling costs. Full payment enclosed (plus sales tax, if any).

Cash must accompany orders under $5.00.
Money-back guarantee still applies.

DETACH POSTAGE-PAID REPLY CARD BELOW AND MAIL TODAY!

Just select your books, enter the titles and authors on the order card, fill out your name and address, and mail. There's no need to send money.

15-Day Free Trial On All Books . . .